Management of
CONTAMINATED
SITE PROBLEMS

Management of
CONTAMINATED
SITE PROBLEMS

D. Kofi Asante-Duah, Ph.D., C.E.
Tetra Tech, Inc.
Environmental Consulting Engineers & Scientists
Santa Barbara, California

LEWIS PUBLISHERS

Boca Raton   New York   London   Tokyo

Library of Congress Cataloging-in-Publication Data

Asante-Duah, D. Kofi.
    Management of contaminated site problems / D. Kofi Asante-Duah.
        p.   cm.
    Includes bibliographical references and index.
    ISBN 1-56670-079-5 (alk. paper)
    1. Hazardous waste sites--Management.   I. Title.
TD1052.A83   1995
628.5--dc20   95-11258
                                                                                CIP

© 1996 by CRC Press, Inc.
Lewis Publishers is an imprint of CRC Press

No claim to original U.S. Government works
International Standard Book Number 1-56670-079-5
Library of Congress Card Number 95-11258
Printed in the United States of America 1 2 3 4 5 6 7 8 9 0
Printed on acid-free paper

PREFACE

Several environmental media, contaminated to various degrees by numerous chemicals or pollutants exist in several locations globally. These impacted media must be tackled in an effective manner. The nature and extent of these environmental contamination problems is so diverse as to make it almost impractical to cover every aspect of it in one volume; however, the concepts presented in this book will generally be applicable to the broad spectrum of environmental contamination problems often encountered in practice. The focus of this volume is on the effective management of contaminated site problems, achievable through the design of effectual corrective action programs; this is necessary to abate the potential risks associated with such sites. The key to achieving this lies in the use of an informed strategy to develop and implement case-specific corrective action programs.

Contaminated sites and facilities may pose significant risks to the general public because of the potential health and environmental effects; to property owners and financiers due to possible financial losses and reduced property values; and to other potentially responsible parties because of possible financial liabilities that could result from their effects. The effective management of contaminated sites is, therefore, an important environmental issue. This book provides the relevant protocols to follow in making key decisions in corrective action assessment programs for potentially contaminated site problems. It provides a wealth of information that can be applied to the appraisal, assessment, and implementation of contaminated site restoration programs.

This book represents a collection and synthesis of the relevant elements of the corrective action assessment process pertaining to contaminated site problems. It specifically addresses issues relevant to the investigation and the management of potentially contaminated site problems, spanning problem diagnosis and site characterization to the development and implementation of site restoration programs. The book offers an elaboration of the procedures utilized in the planning, development, and evaluation of corrective action programs for potentially contaminated sites and facilities. It also includes specific example analyses and workplans that can be adapted for the evaluation and implementation of corrective measures for such sites and facilities. The book is organized into eight chapters, together with a bibliographical listing and eleven appendices.

Chapter 1, *Introduction,* includes background information on the sources, nature, and implications of contaminated site problems; it also presents a summary review of selected environmental legislations relevant to the management of contaminated site problems. Chapter 2, *Diagnostic Assessment of Contaminated Site Problems,* discusses the investigatory and assessment processes involved in mapping out the types and extents of contamination at potentially contaminated sites. Chapter 3, *Site Characterization Activities at Contaminated Sites,* elaborates on the types of activities usually undertaken in the conduct of environmental site assessments at potentially contaminated sites; this includes a discussion of the steps taken to develop comprehensive workplans in the corrective measure investigations at such sites. Chapter 4,

Corrective Action Assessment of Contaminated Site Problems, presents a number of logistical tools and concepts that are typically used to support the development and evaluation of corrective action plans and decisions. Chapter 5, *Risk Assessment Applications to Contaminated Site Problems,* consists of methods of approach used for completing risk assessments as an integral part of the investigation and management of contaminated site problems. Chapter 6, *Risk-Based Cleanup Criteria for Contaminated Site Problems,* elaborates the development of site-specific cleanup criteria and remediation goals for contaminated site problems. Chapter 7, *A Synopsis of Contaminated Site Restoration Methods and Technologies,* offers a general overview of the commonly used technologies and processes that find extensive application in contaminated site cleanup programs. Chapter 8, *Management Strategies for Contaminated Site Problems,* recapitulates several pertinent requirements for the effective management of environmental contamination problems; offers some prescriptions and decision protocols to utilize in the management of potentially contaminated site problems; and discusses the development, screening and selection of remedial alternatives used to address contaminated site problems.

A number of illustrative example problems, consisting of selected evaluation workplans and a variety of analyses relevant to the design of effectual corrective action programs for potentially contaminated site problems, are interspersed throughout the book. Also included are a set of appendices that contain a selected listing of abbreviations and acronyms; a glossary of selected terms and definitions; important environmental fate and transport properties of chemicals appearing at contaminated sites; health and safety requirements for the investigation of contaminated sites; and a listing of selected environmental models and databases potentially applicable to the management of environmental contamination problems (including an overview of the International Register of Potentially Toxic Chemicals [IRPTC] and the Integrated Risk Information System [IRIS]). The appendices also include equations for estimating potential receptor exposures to chemicals present at contaminated sites; carcinogen identification and classification systems; methods for deriving toxicity parameters used in human health risk assessments; equations for calculating human health risks associated with contaminated site problems; equations for calculating contaminated site cleanup levels for site restoration programs; and some selected units and measurements (of potential interest to the environmental professional, analyst, or decision maker).

This book should serve as a useful reference for many professionals encountering environmental contamination problems, in general, and contaminated site problems, in particular. It can also serve as a resource and text for students taking coursework programs in hazardous waste management and other studies relating to environmental contamination problems. In fact, the basic technical principles and concepts involved in the management of contaminated site problems will generally not differ in any significant way from one geographical region to another; the basic differences in strategies tend to lie in the regional policies and legislation. Consequently, this book will serve as a useful reference material for the global community dealing with environmental contamination problems. The protocols presented will aid environmental professionals to more efficiently manage contaminated sites and associated problems.

D. Kofi Asante-Duah, Ph.D., C.E.
Newport Beach, California
March, 1995

THE AUTHOR

D. Kofi Asante-Duah, Ph.D., C.E., is a senior scientist/engineer with Tetra Tech, Inc., Santa Barbara, California and a member of the Institute for Risk Research, University of Waterloo, Ontario, Canada.

Dr. Asante-Duah has served internationally in various capacities, including senior project engineer, project manager, principal investigator, and consultant, for several projects. He has several years of diversified experience with consulting firms and research institutions. His fields of expertise span several topics in the areas of environmental health studies; hazardous waste risk assessment and risk management; design of corrective action programs, including the development of cleanup criteria for site remediation; probabilistic risk assessment and dam safety evaluation; stochastic simulation model applications for solving water management problems; decision analysis approaches to environmental assessment; statistical evaluation for environmental impact assessment; and environmental policy analyses. He has worked on projects relating to leachate migration from landfills and other contaminated sites into aquifers; modeling surface water contamination from hazardous waste sites; review and evaluation of remedial action plans for inactive waste sites, culminating in recommendations for site remediation; and use of parametric statistical analyses to determine impacts of industrial discharges into streams. Dr. Asante-Duah's projects have also included studies of the effects of household hazardous wastes on human health and the environment, based on evaluation of alternative disposal practices; comparative evaluation of alternative remedial options for industrial facilities; policy analyses of the tradeoffs involved in the transboundary movements of hazardous wastes; risk management and prevention programs for facilities handling acutely hazardous materials; and human health and environmental risk assessments for industrial facilities and contaminated sites.

Dr. Asante-Duah has previously been on visiting appointments to the University of Pittsburgh, Civil Engineering Department, as visiting scholar/scientist in 1985/86, and the University of Waterloo, Civil Engineering Department, as visiting research assistant professor in 1990/91. He has previously been a research assistant at the Utah Water Research Laboratory, Utah State University (1986-88). Dr. Asante-Duah is a past UNESCO fellow (1984) and also a World Bank's McNamara fellow (1990-91). He has affiliations with several international professional associations.

Dr. Asante-Duah is the author of a book on hazardous waste risk assessment. He is also the author or co-author of several technical papers and presentations relating to water resources evaluation, risk assessment, and hazardous waste management.

ACKNOWLEDGMENTS

Sincere thanks are due to the Duah family of Abaam, Kade, and Nkwantanang, and several friends and colleagues who provided much-needed moral and enthusiastic support throughout preparation of the manuscript for this book. Review comments and suggestions on an earlier draft of the manuscript were provided by Dr. L. Douglas James, National Science Foundation, Washington, D.C. and Dr. Lou Levy, Consultant, Tetra Tech, Inc., Los Angeles, California. This book also benefited from review comments of several anonymous individuals, as well as from discussions with a number of professional colleagues. Any shortcomings that remain are, however, the sole responsibility of the author.

To all my families at Abaam, Kade, and Nkwantanang
To my mother, Alice Adwoa Twumwaa
To my father, George Kwabena Duah
To all the Duah brothers and sisters
To all the good friends

CONTENTS

1   INTRODUCTION

Industry has become an essential part of modern society, and waste production is an inevitable characteristic of the industrial activities. A responsible system for dealing with the wastes generated by industry is essential to sustain the modern way of life. In particular, the effective management of hazardous wastes, and the associated treatment, storage, and disposal facilities (TSDFs) is of major concern to both the industry generating wastes and to governments (representing the interests of individual citizens). This is because waste materials can cause adverse impacts on the environment and to public health. It is evident, however, that the proper management of hazardous wastes poses several challenges.

Waste management facilities apparently are responsible for most environmental contamination and contaminated site problems often encountered. The contributing waste disposal activities may relate to industrial wastewater impoundments, land disposal sites for solid wastes, land spreading of sludges, chemical spills, leaks from underground storage facilities, septic tanks and cesspools, disposal of mine wastes, or indeed a variety of TSDFs. This situation has, in a way, contributed to the widespread occurrence of contaminated sites and related environmental contamination problems globally.

The focus of this volume is on the effective management of contaminated site problems through the design of effectual corrective action programs. Many sites, contaminated to various degrees by several chemicals or pollutants, exist in a number of geographical locations and must be addressed. The management of these contaminated sites is of great concern in view of the risks associated with such sites. For instance, apart from their immediate and direct health and environmental threats, contaminated sites can contribute to the long-term contamination of the ambient air, soils, surface waters, groundwater resources, and the food chain. Such detrimental effects are inevitable if the impacted sites are not managed effectively, or if remedial actions are not taken in an effective manner.

In general, contaminated sites and facilities may pose significant risks to the public because of the potential health and environmental effects, to property owners and financiers due to possible financial losses and reduced property values, and to other potentially responsible parties (PRPs) because of possible financial liabilities that could result from their effects. The effective management of contaminated sites should, therefore, be considered as an important environmental issue. The key to achieving this lies in the use of an informed strategy to develop and implement case-specific corrective action programs.

To design an effectual corrective action program, contaminated sites have to be extensively studied to determine the areal extent of contamination, the quantities of contaminants that human and ecological receptors could potentially be exposed to, the human health and ecological risks associated with the site, and the types of corrective or remedial actions necessary to abate risks from such sites. Oftentimes, soils become the principal focus of attention in the investigation of contaminated sites. This is because soils at such sites not only serve as a medium of exposure to potential receptors, but also as a long-term reservoir for contaminants that may be released to other media.

Figure 1.1 is a conceptual representation of potential consequences related to the migration and chemical exposures from contaminated sites. This is a picture well digested by several individuals, even by lay persons. Armed with this information and knowledge, there is increasing public concern about the several problems and potentially dangerous situations associated with contaminated sites. Such concerns, together with the legal provisions of various legislative instruments and regulatory programs, have compelled both industry and governmental authorities to carefully formulate effective management plans for potentially contaminated sites. These plans include techniques and strategies needed to provide good preliminary assessments, site characterizations, impact assessments, and the development of cost-effective corrective action programs.

1.1   WHAT ARE THE SOURCES
OF ENVIRONMENTAL CONTAMINATION?

Environmental contamination will generally be due to the presence or release of hazardous materials or wastes at a site or TSDF. Hazardous waste/material is that product which has the potential to cause detrimental effects on human health and/or the environment. Such materials may be toxic organic or inorganic chemicals, bioaccumulative materials, nondegradable and persistent chemicals, and/or radioactive substances. Among the most persistent of the organic-based contaminants are the organo-halogens; organo-halogens are rarely found in nature and therefore relatively few biological systems can break them down, in contrast to compounds such as petroleum hydrocarbons which are comparatively more easily degraded (Fredrickson et al., 1993). In fact, the nature of waste constituents can significantly impact human health and the environment. The identification of the types of potentially hazardous contaminants present in the environment is therefore important in the investigation of potential risks associated with environmental contaminants.

Wastes are generated from several operations associated with industrial (e.g., manufacturing and mining), agricultural, military, commercial (e.g., auto repair shops, utility companies, gas stations, dry-cleaning facilities, transportation, and other service industries) and domestic activities. In fact, many of the environmental contamination problems encountered in a number of places are the result of waste generation associated with various forms of industrial activities. In particular, the chemicals and allied products are generally seen as the major sources of industrial hazardous waste generation. Table 1.1 is a listing of the major industrial sectors that are potential sources of waste generation. These industries generate several waste types, such as:

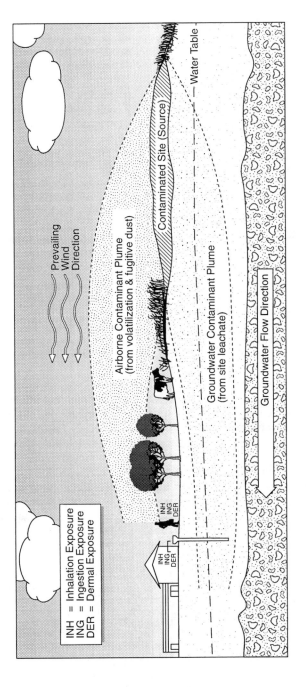

Figure 1.1 Conceptual illustration of the potential consequences associated with a contaminated site.

**Table 1.1 Typical Industries Potentially Contributing
to Environmental Contamination Problems**

Aerospace	Organic Chemicals
Automobile	Paints and Coatings
Batteries (storage and primary)	Perfumes and Cosmetics
Beverages	Pesticides and Herbicides
Computer Manufacture	Petroleum Refining
Electronic Components Manufacturing	Pharmaceuticals
Electroplating and Metal Finishing	Photographic Equipment and Supplies
Explosives	Printing
Food Processing and Dairy Products	Pulp and Paper Mills
Ink Formulation	Rubber Products, Plastic Materials, and
Inorganic Chemicals	Synthetics
Inorganic Pigments	Ship-building
Iron and Steel	Soap and Detergent Manufacturing
Leather Tanning and Finishing	Textile Mills
Metal Smelting and Refining	Wood Preservation and Processing

Source: Asante-Duah, 1993.

- Organic waste sludges and still bottoms containing chlorinated solvents, metals, oils, etc.
- Oil and grease contaminated with polychlorinated biphenyls (PCBs), polyaromatic hydrocarbons (PAHs), metals, etc.
- Heavy metal solutions of arsenic, cadmium, chromium, lead, mercury, etc.
- Pesticide and herbicide wastes
- Anion complexes containing cadmium, copper, nickel, zinc, etc.
- Paint and organic residuals
- Several miscellaneous chemicals and byproducts that need special handling/management

Table 1.2 summarizes the typical hazardous waste streams potentially generated by the major industrial operations. These waste streams have the potential to enter the environment by several mechanisms.

1.1.1 The Making of Contaminated Sites

Contaminated sites may arise in a number of ways, many of which result from manufacturing and other industrial activities/operations. Contaminant releases into the environment can occur via several processes and migration/transport pathways, including:

- Air (due to volatilization and fugitive dust emissions)
- Surface water (from surface runoff or overland flow and groundwater seepage)
- Groundwater (through infiltration, leaching, and leachate migration)
- Soils (due to erosion, including fugitive dust generation/deposition, and tracking)
- Sediments (from surface runoff/overland flow, seepage, and leaching)
- Biota (due to biological uptake and bioaccumulation)

Contaminated site problems typically are the result of soil contamination due to placement of wastes on or in the ground, as a result of spills, lagoon failures, or contaminated runoff, and/or from leachate generation.

Table 1.2 Typical Potentially Hazardous Waste Streams from Various Industrial Sectors

Sector/Source	Typical Hazardous Waste Stream
Agricultural and Food Production	Acids and Alkalis; Fertilizers (e.g., nitrates); Herbicides (e.g., dioxins); Insecticides; Unused pesticides (e.g., aldicarb, aldrin, DDT, dieldrin, parathion, toxaphene)
Airports	Hydraulic fluids; Oils
Auto/Vehicle Servicing	Acids and Alkalis; Heavy metals; Lead-acid batteries (e.g., cadmium, lead, nickel); Solvents; Waste oils
Chemical/Pharmaceutical Industry	Acids and Alkalis; Biocide wastes; Cyanide wastes; Heavy metals (e.g., arsenic, mercury); Infectious and Laboratory wastes; Organic residues; PCBs; Solvents
Domestic	Acids and Alkalis; Dry-cell batteries (e.g., cadmium, mercury, zinc); Heavy metals; Insecticides; Solvents (e.g., ethanol, kerosene)
Dry Cleaning/Laundries	Detergents (e.g., boron, phosphates); Dry cleaning filtration residues; Halogenated solvents
Educational/Research Institutions	Acids and Alkalis; Ignitable wastes; Reactives (e.g., chromic acid, cyanides; hypochlorites, organic peroxides, perchlorates, sulfides); Solvents
Electrical Transformers	Polychlorinated biphenyls (PCBs)
Equipment Repair	Acids and Alkalis; Ignitable wastes; Solvents
Leather Tanning	Inorganics (e.g., chromium, lead); Solvents
Machinery Manufacturing	Acids and Alkalis; Cyanide wastes; Heavy metals (e.g., cadmium, lead); Oils; Solvents
Medical/Health Services	Laboratory wastes; Pathogenic/Infectious wastes; Radionuclides; Solvents
Metal Treating/Manufacture	Acids and Alkalis; Cyanide wastes; Heavy metals (e.g., antimony, arsenic, cadmium, cobalt); Ignitable wastes; Reactives; Solvents (e.g., toluene, xylenes)
Military Training Grounds	Heavy metals
Mineral Processing/Extraction	High-volume/Low-hazard wastes (e.g., mine tailings); Red muds
Motor Freight/Railroad Terminals	Acids and Alkalis; Heavy metals; Ignitable wastes (e.g., acetone; benzene; methanol); Lead-acid batteries; Solvents
Paint Manufacture	Heavy metals (e.g., antimony, cadmium, chromium); PCBs; Solvents; Toxic pigments (e.g., chromium oxide)
Paper Manufacture/Printing	Acids and Alkalis; Dyes; Heavy metals (e.g., chromium, lead); Inks; Paints and Resins; Solvents
Petrochemical Industry/Fueling Stations	Benzo-*a*-pyrene (BaP); Hydrocarbons; Oily wastes; Lead; Phenols; Spent catalysts
Photofinishing/Photographic Industry	Acids; Silver; Solvents
Plastic Materials and Synthetics	Heavy metals (e.g., antimony, cadmium, copper, mercury); Organic solvents
Shipyards and Repair Shops	Heavy metals (e.g., arsenic, mercury, tin); Solvents
Textile Processing	Dyestuff; Heavy metals and compounds (e.g., antimony, arsenic, cadmium, chromium, mercury, lead, nickel); Halogenated solvents; Mineral acids; PCBs
Timber/Wood Preserving Industry	Heavy metals (e.g., arsenic); Non-halogenated solvents; Oily wastes; Preserving agents (e.g., creosote, chromated copper arsenate, pentachlorophenol)

Contaminants released to the environment are controlled by a complex set of processes including various forms of transport (e.g., intermedia transfers), transformation (e.g., biodegradation), and biological uptake (e.g., bioaccumulation or bioconcentration). Consequently, contaminated soils can potentially impact several other environmental matrices. For instance, atmospheric contamination may result from emissions of contaminated fugitive dusts and volatilization of chemicals present in soils; surface water contamination may result from contaminated runoff and overland flow of chemicals (from leaks, spills, etc.) and chemicals adsorbed to mobile sediments; ground-water contamination may result from the leaching of toxic chemicals from contaminated soils or the downward migration of chemicals from lagoons and ponds. Indeed, several different physical and chemical processes can affect contaminant migration from a contaminated site, as well as the intermedia transfer of contaminants at the site.

1.2 MINIMIZING ENVIRONMENTAL CONTAMINATION PROBLEMS

In the past, hazardous waste management practices were tantamount and synonymous to the simplistic rule of "out of sight, out of mind". This resulted in the creation of several contaminated sites that need to be addressed today. With the improved knowledge linking environmental contaminants to several human health problems and also to some ecological disasters, the "new society" has come to realize the urgent need to clean up for the "past sins". Proactive actions are being taken to minimize the continued creation of more contaminated site problems. Indeed, in recent years, and in several countries, social awareness of environmental problems and the need to reduce sources of contamination, as well as the urge to clean up contaminated sites, has been increasing. Nonetheless, some industries continue to generate and release large quantities of contaminants into the environment.

In fact, large quantities of wastes have always been generated by many industries, to the point that there is a near-crisis situation with managing such wastes. Consequently, waste/materials recycling and re-use, waste exchange, and waste minimization are becoming more prominent in the general waste management practices of several countries. For instance, the recovery of waste oils, solvents, and waste heat from incinerators has become a common practice in several countries. Also, operations exist in a number of countries to recycle heavy metals from various sources (such as silver from photofinishing operations, lead from lead-acid batteries, mercury from batteries and broken thermometers, heavy metals from metal finishing wastes, etc.). Indeed, waste/materials recycling has become an integral part of many modern industrial processes. Furthermore, waste exchange schemes exist in some countries to promote the use of one company's byproduct or waste as another's raw material.

Figure 1.2 illustrates the basic components of a typical waste management program for an industrial facility. The general trend of choice is for on-site waste management, with more emphasis for the future directed at waste minimization. Typical on-site waste minimization elements that can be applied to waste management programs will consist of:

- Waste recyling (i.e., the recovery of materials used or produced by a process for separate use or direct re-use in-house)
- Material recovery (i.e., the processing of waste streams to recover materials which can be used as feedstock for conversion to another product)

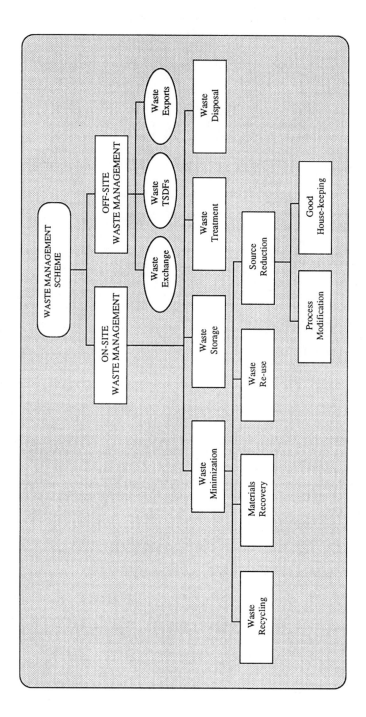

Figure 1.2 Essential elements of a typical waste management program.

- Waste re-use (i.e., the direct re-use of the wastes as is, without further processing)
- Reduction at source (that may comprise of process modifications and/or the introduction of good housekeeping methods)

All these aspects of the waste management program will likely dominate the scene in industrial waste management planning of the future. In particular, waste reduction at source will gain more prominence in the long term. Reduction at source will be accomplished through process modifications and through efficient or good housekeeping. The process modification aspect of the waste reduction at source will generally consist of the following typical elements:

- Modification of the process pathway
- Improved control measures
- Equipment modification
- Changes in operational setting/environment
- Increased automation
- Product substitution

Good and efficient housekeeping can be attained by implementing the following typical tasks:

- Material handling improvements
- Spill and leak monitoring, collection, and prevention
- Preventative maintenance
- Inventory control
- Waste stream segregation

Even when an extensive effort is made at minimizing wastes, there still will be some residual wastes remaining when all is said and done. Such wastes can be permitted for long-term storage on-site; or for the permanent disposal at on-site landfills, surface impoundments, etc., or for an end-of-pipe treatment on-site in order to reduce waste quantity or contaminant concentrations prior to off-site disposal.

In conclusion, it is apparent that the possibility of creating contaminated site problems will remain an inevitable concomitant of industrial activities in modern societies. However, their effects and extent can be minimized by taking the proactive measures identified here.

1.3 THE NATURE OF CONTAMINATED SITES

Site contamination occurs when chemicals are detected where such constituents are not expected and/or not desired. Contaminated sites may pose different levels of risk, depending on the degree of contamination present at the site. The degree of hazard posed by the contaminants involved will generally be dependent on several factors (Box 1.1). In fact, it is important to recognize the fact that there may be varying degrees of hazards associated with different contaminated site problems, and that there are good economic advantages for ranking potentially contaminated sites according to the level of hazard they present. Such a classification will facilitate the development

BOX 1.1

Factors Affecting the Degree of Hazard Posed by Contaminants Present in the Environment

- Physical form and composition of contaminants
- Quantities of contaminants
- Reactivity (fire, explosion)
- Biological and ecological effects (human toxicity, ecotoxicity)
- Mobility (i.e., transport in various environmental media, including leachability)
- Persistence (fate in environment, detoxification potential, etc.)
- Indirect health effects (from pathogens, vectors, etc.)
- Local site conditions (e.g., temperature, soil type, groundwater table conditions, humidity, light)

and implementation of efficient site management programs. Thus, in order to develop adequate site management strategies, potentially contaminated sites should be categorized in an appropriate manner.

A typical site categorization scheme for potentially contaminated sites will comprise of putting the "candidate" sites into groups or clusters, based on the potential health and environmental risks associated with these sites (e.g., high-, intermediate-, and low-risk sites, conceptually represented by Figure 1.3). In general, the high-risk sites will prompt the most concern, requiring immediate and urgent corrective measures that may include time-critical removal actions. A site is designated as high risk when site contamination represents a real or imminent threat to human health and/or to the environment. In this case, an immediate action will generally be required to reduce the threat.

1.4 HEALTH, ENVIRONMENTAL, AND SOCIOECONOMIC IMPLICATIONS OF CONTAMINATED SITE PROBLEMS

Unfortunate lessons from the past, such as the Love Canal incident in New York, clearly demonstrate the dangers arising from the presence of contaminated sites within or near residential communities. Love Canal was a disposal site for chemical wastes for about 25 to 30 years (e.g., Gibbs, 1982; Levine, 1982). Subsequent use of the site culminated in residents of a township in the area suffering from various health impairments; it is believed that several children in the neighborhood apparently were born with serious birth defects. Analogous incidents are recorded in other U.S. locations, in Europe, in Japan, and in other locations in Asia (Asante-Duah, 1993). The presence of potentially contaminated sites can therefore create potentially hazardous situations and pose significant risks of concern to society.

The mere existence of contaminated sites can result in contaminant releases and possible receptor exposures, resulting in both short- and long-term effects on a variety of populations potentially at risk. Exposure to chemical constituents present at

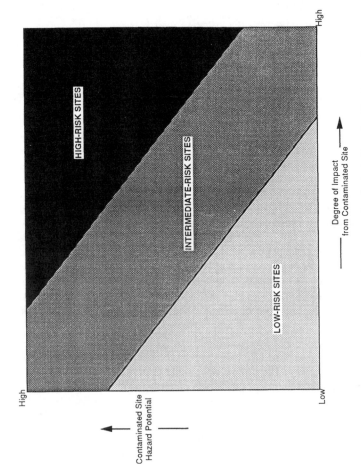

Figure 1.3 A conceptual representation for a contaminated sites classification system.

BOX 1.2
General Health Effects Associated with
Hazardous Materials in the Environment

- Cancer
- Genetic defects, including mutagenesis (i.e., causing heritable genetic damage)
- Reproductive abnormalities, including teratogenesis (i.e., causing damage to developing fetus)
- Alterations of immunobiological homeostasis
- Central nervous system (CNS) disorders
- Congenital anomalies

---

contaminated sites can indeed produce adverse effects in both human and nonhuman ecological receptors. In general, any chemical present at a contaminated site can cause severe health impairment or even death if taken by organisms (including humans) in sufficiently large amounts. On the other hand, there are those chemicals of primary concern which, even in small doses, can cause adverse health impacts.

Box 1.2 identifies a number of health effects of primary concern that may result from populations exposed to hazardous chemicals present at potentially contaminated sites. The potential for adverse health effects on populations contacting hazardous chemicals present at such a site can involve any organ system, depending on the specific chemicals contacted, the extent of exposure (i.e., dose or intake), the characteristics of the exposed individual (e.g., age, gender, body weight, psychological status, genetic make-up, immunological status, susceptibility to toxins, hypersensitivities), the metabolism of the chemicals involved, weather conditions (e.g., temperature, humidity, barometric pressure, season), and the presence or absence of confounding variables such as other diseases (Grisham, 1986).

Table 1.3 lists typical symptoms, health effects, and other biological responses that could be produced from specific toxic chemicals commonly encountered at contaminated sites. Invariably, exposures to chemicals escaping into the environment can result in a reduction of life expectancy and possibly a period of reduced quality of life (caused by anxiety from exposures, diseases, etc.). The existence of unregulated contaminated sites in society can therefore be perceived as a potential source of several health, environmental, and socioeconomic problems. In fact, this can actually reduce the future development potentials of a community. Conversely, it has been noted that sensible policies and actions that protect the environment can at the same time contribute to long-term economic progress (Schramm and Warford, 1989). Economic development and environmental protection should therefore complement each other, since improving one has the potential to enhance the other.

In general, potential health, ecological, and socioeconomic problems are averted by carefully implementing substantive corrective action and risk management programs for contaminated site problems. Several analysts have indeed concluded that environmental problems generally are inseparable from socioeconomic development problems, and that long-term economic growth depends on protecting the environment (The World Bank, 1989).

Table 1.3 Some Typical Toxic Manifestations and Effects Caused by Selected Environmental Contaminants

Chemical	Typical Health Effects & Toxic Manifestations
Arsenic and Compounds	Acute hepatocellular injury, Anemia, Angiosarcoma, Cirrhosis, Developmental disabilities, Embryotoxicity, Heart disease, Hyperpigmentation, Peripheral neuropathies
Asbestos	Asbestosis (scarring of lung tissue)/Fibrosis (lung and respiratory tract), Emphysema, Irritations, Pneumonia/Pneumoconiosis
Benzene	Aplastic anemia, CNS depression, Embryotoxicity, Leukemia and lymphoma, Skin irritant
Beryllium	Granuloma (lungs and respiratory tract)
Cadmium	Developmental disabilities, Kidney damage, Neoplasia (lung and respiratory tract), Neonatal death/Fetal death, Pulmonary edema; Bioaccumulates in aquatic organisms
Chromium and Compounds	Asthma, Cholestasis (of liver), Neoplasia (lung and respiratory tract), Skin irritant
Copper	Gastrointestinal irritant, Liver damage; Toxic to fish
Cyanide	Asthma, Asphyxiation, Hypersensitivity, Pneumonitis, Skin irritant; Toxicity to fish
Dichlorodiphenyl Trichloroethane (DDT)	Ataxic gait, Convulsions, Human infertility/Reproductive effects, Kidney damage, Neurotoxin, Peripheral neuropathies, Tremors; Bioaccumulation in aquatic organisms
Dieldrin	Convulsions, Kidney damage, Tremors; Bioaccumulates in aquatic organisms
Dioxins and Furans (PCDDs/PCDFs)	Hepatitis, Neoplasia, Spontaneous abortion/Fetal death; Bioaccumulative
Formaldehyde	Allergic reactions; Gastrointestinal upsets; Tissue irritation
Lead and Compounds	Anemia, Bone marrow depression, CNS symptoms, Convulsions, Embryotoxicity, Neoplasia, Neuropathies, Kidney damage, Seizures; Biomagnifies in food chain
Lindane	Convulsions, Coma and death, Disorientation, Headache, Nausea and vomiting, Neurotoxin, Paresthesias
Lithium	Gastroenteritis, Hyperpyrexia, Nephrogenic diabetes, Parkinson's disease
Manganese	Bronchitis, Cirrhosis (liver), Influenza (metal-fume fever), Pneumonia
Mercury and Compounds	Ataxic gait, Contact allergen, CNS symptoms, Developmental disabilities, Neurasthenia, Kidney and liver damage, Minamata disease; Biomagnification of methyl-mercury
Methylene Chloride	Anesthesia, Respiratory distress, Death
Naphthalene	Anemia
Nickel and Compounds	Asthma, CNS effects, Gastrointestinal effects, Headache, Neoplasia (lung and respiratory tract)
Nitrate	Methemoglobinemia (in infants)
Organochlorine Pesticides	Hepatic necrosis, Hypertrophy of endoplasmic reticulum, Mild fatty metamorphosis
Pentachlorophenol (PCP)	Malignant hyperthermia
Phenol	Asthma, Skin irritant
Polychlorinated Biphenyls (PCBs)	Embryotoxicity/Infertility/Fetal death, Dermatoses, Hepatic necrosis, Hepatitis, Immune suppression; Toxicity to aquatic organisms
Silver	Blindness, Skin lesions, Pneumoconiosis
Toluene	Acute renal failure, Ataxic gait, CNS depression, Memory impairment
Trichloroethylene (TCE)	CNS depression, Deafness, Liver damage, Paralysis, Respiratory and cardiac arrest, Visual effects
Vinyl Chloride	Leukemia and lymphoma, Neoplasia, Spontaneous abortion/Fetal death, Tumors, Death
Xylene	CNS depression, Memory impairment
Zinc	Corneal ulceration, Esophagus damage, Pulmonary edema

BOX 1.3
Questions Relevant to Defining Corrective Action Needs

- What is the nature of the contamination?
- What are the sources of the contamination?
- What is the extent of the contamination?
- What are the populations potentially at risk?
- What are the likely and significant exposure pathways and scenarios connecting contaminant source(s) to potential receptors?
- What is the likelihood for health and environmental impacts from the contamination?
- What interim measures, if any, are required as part of a risk management and/ or risk prevention program?
- What type of corrective action(s) may be appropriate to remedy the situation?
- What level of residual contamination will be acceptable for site restoration, if the site needs to be cleaned up?

1.5 ESTABLISHING THE NEED FOR CORRECTIVE ACTIONS

Corrective actions for contaminated sites are generally developed and implemented with the principal objective to protect public health and the environment. Typically, answers will have to be generated for several pertinent questions (Box 1.3) when one is confronted with a potentially contaminated site problem. The answers to these questions will help define the corrective action needs for the site. Ultimately, a thorough site investigation that establishes the nature and extent of contamination may be necessary, in order to arrive at an appropriate and realistic corrective action decision.

In a number of situations, it becomes necessary to implement interim corrective measures prior to the development and full implementation of a comprehensive site restoration program. Such preliminary corrective actions may consist of a variety of site control activities, such as the installation of security fences (to restrict access to the site), the construction of physical barriers (to restrict site access and also to minimize potential runoff from the site), the application of dust suppressants (to reduce airborne migration of contaminated soils to offsite locations), the removal of "hot-spots" and drums (where fire, explosion, or acute toxic exposure could possibly occur), etc.

1.6 LEGISLATIVE-REGULATORY CONTROLS AFFECTING ENVIRONMENTAL CONTAMINATION PROBLEMS

Several items of legislation have been formulated and implemented in most industrialized countries to deal with the regulation of toxic substances present in our modern societies. Variations in national legislations and controls do affect the options available for the management of contaminated site problems in different regions of the world. Such variances also affect the cleanup standards and costs necessary to achieve the appropriate program goals. In fact, the establishment of national and/or regional regulatory control programs within the appropriate legislative-regulatory and enforcement frameworks is a major step in developing effective waste and contaminated site management programs. In general, there tends to be improved management practices in countries or regions where regulatory programs are well established, in comparison to those without appropriate regulatory and enforcement programs.

Some of the relevant statutes and regulations which, directly or indirectly, affect contaminated site management decisions, policies, and programs in the U.S. include the *Clean Air Act* (CAA), the *Safe Drinking Water Act* (SDWA), the *Clean Water Act* (CWA), the *Federal Water Pollution Control Act* (FWPCA), the *Resource Conservation and Recovery Act* (RCRA), the *Comprehensive Environmental Response, Compensation, and Liability Act* of 1980 (CERCLA or "Superfund"), the *Superfund Amendments and Reauthorization Act* of 1986 (SARA), the *Toxic Substances Control Act* (TSCA) of 1976, the *Federal Insecticide, Fungicide and Rodenticide Act* (FIFRA), and the *Endangered Species Act* (ESA). These statutes and regulations are briefly annotated below, with further details to be found in the literature (e.g., USEPA, 1974; 1985; 1987; 1988; 1989).

Clean Air Act (CAA) — The objective of the CAA of 1970 is to protect and enhance air quality resources in order to promote and maintain public health and welfare, and the productive capacity of the population. Under Section 109, the CAA requires that National Ambient Air Quality Standards (NAAQS) be set and ultimately met for any air pollutant which, if present in the air, may reasonably be anticipated to endanger public health or welfare and whose presence in the air results from numerous or diverse mobile and/or stationary sources. Under Section 111 of the CAA, EPA sets New Source Performance Standards (NSPS) for new or modified stationary source categories whose emissions cause or significantly contribute to air pollution which may endanger public health or welfare. Section 112 of the CAA also requires the establishment of National Emission Standards for Hazardous Air Pollutants (NESHAP), where a hazardous air pollutant is defined as a pollutant not covered by a NAAQS and exposure to which may reasonably be anticipated to result in an increase in mortality or an increase in serious irreversible, or incapacitating reversible, illness. This covers all pollutants that may cause significant risks.

Safe Drinking Water Act (SDWA) — The SDWA was enacted in 1974 in order to assure that all people served by public water systems would be provided with a supply of high-quality water. The SDWA amendments of 1986 established new procedures and deadlines for setting national primary drinking water standards, and established a national monitoring program for unregulated contaminants, among others. Overall, the statute covers public water systems, drinking water regulations, and the protection of underground sources of drinking water. The SDWA offers regulations with regard to drinking water standards. It requires the Environmental Protection Agency (EPA) to specify drinking water contaminants that may have adverse human health effects and that are known or anticipated to occur in water.

Clean Water Act (CWA) — The CWA was enacted in 1977, and an amendment to this introduced the *Water Quality Act* (WQA) of 1987. The objective of the CWA is to restore and maintain the chemical, physical, and biological integrity of the nation's waters. This objective is achieved through the control of discharges of pollutants to navigable waters. This control is implemented through the application of federal, state, and local discharge standards. The statute covers the limits on waste discharge to navigable waters, standards for discharge of toxic pollutants, and the prohibition on discharge of oil or hazardous substances into navigable waters. A closely related legislation, the *Federal Water Pollution Control Act* (FWPCA), deals solely with the regulation of effluent and water quality standards.

Resource Conservation and Recovery Act (RCRA) — RCRA was enacted in 1976 (as an amendment to the *Solid Waste Disposal Act* of 1965, later amended in 1970 by the *Resource Recovery Act*) to regulate the management of hazardous waste, to ensure the safe disposal of wastes, and to provide for resource recovery from the environment by controlling hazardous wastes "from cradle to grave". This is a federal law designed to prevent new uncontrolled hazardous waste sites. It establishes a regulatory system to track hazardous substances from the time of generation to disposal. The law requires safe and secure procedures to be used in treating, transporting, sorting, and disposing of hazardous substances. It covers treatment methods, techniques, or processes; storage and holding periods; disposal methods; and federal authorization to seek an injunction or bring suit against owners/operators of facilities that endanger public health or the environment. Basically, RCRA regulates hazardous waste generation, storage, transportation, treatment, and disposal. The 1984 Hazardous and Solid Waste Amendments (HSWA) to RCRA, Subtitle C, covers a management system that regulates hazardous wastes from the time it is generated until the ultimate disposal — the so-called "cradle-to-grave" system. Thus, under RCRA, a hazardous waste management program is based on a "cradle-to-grave" concept that allows all hazardous wastes to be traced and equitably accounted for.

Comprehensive Environmental Response, Compensation, and Liability Act (CERCLA or "Superfund") — CERCLA (or "Superfund") establishes a broad authority to deal with releases or threats of releases of hazardous substances, pollutants, or contaminants from vessels, containments, or facilities. This legislation deals with the remediation of hazardous waste sites by providing for the cleanup of inactive and abandoned hazardous waste sites. The objective is to provide a mechanism for the federal government to respond to uncontrolled releases of hazardous substances to the environment. The statute covers reporting requirements for past and present owners/operators of hazardous waste facilities, and the liability issues for owners/operators for cost of removal or remedial action and damages in case of release or threat of release of hazardous wastes. SARA strengthens and expands the cleanup program under CERCLA, focuses on the need for emergency preparedness and community right-to-know, and changes the tax structure for financing the Hazardous Substance Response Trust Fund established under CERCLA to pay for the cleanup of abandoned and uncontrolled hazardous waste sites.

Toxic Substances Control Act (TSCA) — TSCA regulates the manufacture, use, and disposal of chemical substances by authorizing the EPA to establish regulations pertaining to the testing of chemical substances and mixtures, premanufacture notification for new chemicals or significant new uses of existing substances, control of chemical substances or mixtures that pose an imminent hazard, and record-keeping and reporting requirements. It provides for a wide range of risk management actions to accommodate the variety of risk-benefit situations confronting the EPA. The risk management decisions under TSCA would consider not only the risk factors (such as probability and severity of effects), but also non-risk factors (such as potential and actual benefits derived from use of the material and availability of alternative substances).

Federal Insecticide, Fungicide and Rodenticide Act (FIFRA) — Under FIFRA, the EPA has published procedures for the disposal and storage of excess pesticides and pesticide containers in 40 CFR Part 165, Subpart C. EPA has also promulgated tolerance levels for pesticides and pesticide residues in or on raw agricultural commodities under authority of the *Federal Food, Drug, and Cosmetic Act* (40 CFR Part 180). FIFRA provides the EPA with broad authorities to regulate all pesticides, including the presence of pesticides at facilities that constitute hazardous waste sites.

   Endangered Species Act (ESA) — The ESA of 1973 (reauthorized in 1988) provides a means for conserving various species of fish, wildlife, and plants that are threatened with extinction. Also, the ESA provides for the designation of critical habitats (i.e., specific areas within the geographical area occupied by the endangered or threatened species) on which are found those physical or biological features essential to the conservation of the species in question.

   Other laws and regulations exist in the U.S. that also work towards preventing or limiting the potential impacts of hazardous wastes and other toxic materials on the environment, and on populations potentially at risk. For instance, the *Hazardous Materials Transport Act* (HMTA) provides a legislation that deals with the regulation of transport of hazardous materials, the *Fish and Wildlife Conservation Act* of 1980 requires states to identify significant habitats and develop conservation plans for these areas, the *Marine Mammal Protection Act* of 1972 protects all marine mammals — some of which are endangered species, the *Migratory Bird Treaty Act* of 1972 implements many treaties involving migratory birds — to protect most species of native birds in the U.S., and the *Wild and Scenic Rivers Act* of 1972 preserves select rivers declared as possessing outstanding remarkable scenic, recreational, geologic, fish and wildlife, historic, cultural, or other similar values.

   Requirements under several of these regulations and other federal and state environmental laws can be very important to the effective management of most contaminated site problems. Depending on the type of program under evaluation, one or more of these regulations may dominate the decision-making process. It is noteworthy that the EPA, acting for the federal government, and also several state legislators and regulators, are continually refining guidance and specifications for several programs (such as RCRA, CERCLA, TSCA, and FIFRA). Each may be carried out by a different office within the EPA, for different purposes, and under different legal authority and limitations. It is therefore important to be aware of the fact that several of the regulatory guidances can and do change periodically. One can be brought up-to-date by examining the *Federal Register* and other relevant documents issued periodically by state and local regulatory agencies. Also, it is important that the analyst has some understanding of the legal criteria specified for case-specific programs, to facilitate the process of choosing and using the appropriate guidance tools in decisions relating to the management of contaminated sites.

   Laws, legal provisions, and regulations similar to those enumerated above can be found in the legislative requirements for several other industrialized countries. In general, a "cradle-to-grave" type of system that monitors and regulates the movement of hazardous materials from manufacture, through usage, to the ultimate disposal of any associated hazardous wastes is maintained by most industrialized countries. For instance, the United Kingdom (U.K.) *Control of Pollution Act* of 1974 allows the government to restrict the production, importation, sale, or use of chemical substances in the U.K.; the Act controls the disposal of wastes on land by means of site licensing. Furthermore, the 1980 *Control of Pollution (Special Waste) Regulations* was established to ensure prenotification for a limited range of the most hazardous wastes, to keep a "cradle-to-grave" record of each disposal of special wastes, and to keep long-term records of locations of special waste disposal landfill sites. In Japan, the *Waste Management Law* of 1970 requires anyone undertaking the collection, transport, or disposal of industrial wastes to have obtained a permit; waste disposal sites are also required to be designed and operated in a manner that prevents trespassing, and the disposal sites must be isolated from both surface water and groundwater resources and measures taken to prevent leakage. In addition, the 1973 *Chemical Substances Control Act* requires a manufacturer or importer of a new chemical (i.e., one that is not enumerated in a 1973 pre-listed compilation) to provide the government with all available information concerning such a chemical; the substance can then be evaluated and classified as being dangerous or safe. Most other industrialized countries have comparable laws and regulations, and also have enforcement mechanisms in place as part of their overall environmental and contaminated sites management programs.

REFERENCES

Asante-Duah, D.K. 1993. *Hazardous Waste Risk Assessment.* CRC Press/Lewis Publishers, Boca Raton, FL.

Fredrickson, J.K., H. Bolton, Jr., and F.J. Brockman. 1993. "In Situ and On-Site Bioreclamation." *Environ. Sci. Technol.* Vol.27, No.9:1711–1716.

Gibbs, L.M. 1982. *Love Canal: My Story.* State University of New York Press, Albany, NY.

Grisham, J.W. (Ed.). 1986. *Health Aspects of the Disposal of Waste Chemicals.* Pergamon Press, Oxford.

Levine, A.G. 1982. *Love Canal: Science, Politics, and People.* Lexington Books, MA.

Schramm, G. and J.J. Warford (Eds.). 1989. *Environmental Management and Economic Development.* A World Bank Publication. Johns Hopkins University Press, Baltimore, MD.

The World Bank. 1989. *Striking a Balance — The Environmental Challenge of Development.* IBRD/The World Bank, Washington, D.C.

USEPA. 1974. *Safe Drinking Water Act.* Public Law 93-523. U.S. Environmental Protection Agency, Washington, D.C.

USEPA. 1985. National Primary Drinking Water Regulations; Volatile Synthetic Organic Chemicals; Final Rule and Proposed Rule, *Federal Register*, 50, 46830-46901. National Primary Drinking Water Regulations, Synthetic Organic Chemicals, Inorganic Chemicals and Microorganisms; Proposed Rule, *Federal Register*, 50, 56936–47025.

USEPA. 1987. The New Superfund: What It Is, How It Works. U.S. Environmental Protection Agency, Washington, D.C.

USEPA. 1988. CERCLA Compliance with Other Laws Manual (Interim Final). EPA/540/G-89/006, Office of Solid Waste and Emergency Response, U.S. Environmental Protection Agency, Washington, D.C.

USEPA. 1989. CERCLA Compliance with Other Laws Manual: Part II. *Clean Air Act* and Other Environmental Statutes and State Requirements. EPA/540/G-89/009. OSWER Directive 9234.1-02. U.S. Environmental Protection Agency, Washington, D.C.

2

DIAGNOSTIC ASSESSMENT
OF CONTAMINATED SITE PROBLEMS

The existence of contaminated sites may result in the release of chemical contaminants into various environmental compartments. Any contaminant released into the environment will be controlled by a complex set of processes (such as intermedia transfers, degradation, and biological uptake). In fact, many chemical contaminants are persistent in the environment and undergo complex interactions in more than one environmental medium. Contaminated sites should therefore be carefully and thoroughly investigated so that risks to potentially exposed populations can be determined with a reasonably high degree of accuracy. Typically, different levels of effort in the investigation will generally be required for different contaminated site problems. Ultimately, the application of a well-designed diagnostic assessment plan to a contaminated site problem will ensure that appropriate and cost-effective corrective measures are identified and implemented for the site.

2.1 INVESTIGATION OF POTENTIALLY CONTAMINATED SITES

Site investigations consist of the planned and managed sequence of activities carried out to determine the nature and distribution of contaminants at potentially contaminated sites. The activities involved usually are comprised of the identification of the principle hazards, the design of sampling and analysis programs, the collection and analysis of environmental samples, and the reporting of laboratory results for further evaluation (BSI, 1988).

The most important primary sources of contaminant release to the various environmental media are usually associated with constituents in soils at contaminated sites. The contaminated soils can subsequently impact other environmental matrices. The impacted media, having once served as "sinks", may eventually become secondary sources of contaminant releases into other environmental compartments. Table 2.1 summarizes the important sources and "sinks" or receiving media associated with typical contaminated site problems. In general, all relevant sources and impacted media should be thoroughly evaluated as part of the site investigation efforts.

In order to get the most out of a site investigation, it must be conducted in a systematic manner. Systematic methods help focus the purpose, the required level of detail, and the several topics of interest — such as physical site conditions, likely

Table 2.1   Potential Release Mechanisms from Various Contaminant Sources and
             Target Media

Primary contaminant source	Typical release causes and mechanisms	Primary receiving or impacted media
Surface impoundments (e.g., lagoons, ponds, pits)	Loading/unloading activities Overtopping dikes and surface runoff Seepage and infiltration/percolation Fugitive dust generation Volatilization	Air Soils and sediments Surface water Groundwater
Waste management units (e.g., landfill, land treatment unit, and waste pile)	Migration of releases outside unit's runoff collection and containment system Migration of releases outside the containment area from loading and unloading operations Seepage and infiltration Leachate migration Fugitive dust generation Volatilization	Air Soils and sediments Surface water Groundwater Subsurface gas (in soil pores, vents, and cracks migrating through soil)
Waste management zones (e.g., container storage area and storage tanks)	Migration of runoff outside containment area Loading/unloading area spills Leaking drums, leaks through tank shells, and leakage from cracked or corroded tanks Releases from overflows Leakage from coupling/ uncoupling operations	Air Soils and sediments Surface water Groundwater Subsurface gas (in soil pores, vents, and cracks migrating through soil)
Waste treatment plants/facilities	Effluent discharge to surface water and groundwater resources	Surface water (by dissolution, dispersion, transport, etc.) Sediments (from adsorbed chemicals) Groundwater
Incinerators	Routine releases from waste handling/preparation activities Leakage due to mechanical failure Stack emissions	Air Foliage (from particulate deposition and atmospheric fallout) Soils (from particulate deposition and atmospheric washout) Surface water (from particulate deposition and atmospheric washout)
Injection wells	Leakage from waste handling operations at the well head	Groundwater (by dissolution, diffusion, dispersion, etc.) Surface water (from groundwater recharge)

Source: Asante-Duah, 1993.

contaminants, extent and severity of contamination, effects on populations potentially at risk, potential for environmental harm, and hazards during construction activities (Cairney, 1993). In addition to establishing the concentration of contaminants at a case site, the site investigation should be designed to provide an indication of the general background or "reference" level of the target contaminants in the local environment. The systematic process required for the site investigation essentially involves the early design of a representative conceptual model of the site. This model is used to assess the physical conditions at the site as well as to identify the mechanisms and processes that could produce significant risks at the site.

When good quality assurance/quality control (QA/QC) procedures have been used in the overall process, the information derived from the investigation of a potentially contaminated site will be both reliable and of known quality. On the other hand, failure to follow good QA/QC procedures may seriously jeopardize the integrity of the data needed to make critical site restoration decisions, which could adversely impact costs of possible remediation requirements for the contaminated site.

2.1.1   A Site Investigation Strategy

Oftentimes, site investigation activities are designed and implemented in accordance with several regulatory and legal requirements of the region or area in which the potentially contaminated site or property is located. In a typical investigation, the influence of the responsible regulatory agencies may affect several operational elements, including site control measures, health and safety plans, soil borings and excavations, monitoring well permitting and specifications, excavated materials control/disposal and the management of investigation-derived wastes in general, sample collection and analytical procedures, decontamination procedures, and traffic disruption/control. Irrespective of whichever regulatory authority is involved, the basic site investigation strategy generally adopted for contaminated site problems typically will comprise of the elements summarized in Box 2.1 (Cairney, 1993).

As an important starting point in the site investigation process, the quality of data required from the study should be clearly defined. Once the level of confidence required for site data is established, strategies for sampling and analysis can be developed (USEPA, 1988). The identification of sampling requirements involves specifying the sampling design, the sampling method, sampling numbers, types, and locations, and the level of sampling quality control. In fact, sampling program designs must seriously consider the quality of data needed. If the samples are not collected, preserved, and stored correctly before they are analyzed, the analytical data may be compromised. Also, if sufficient sample amounts are not collected, the method sensitivity requirements may not be achieved.

Effective analytical protocols in the sampling and laboratory procedures are required to help minimize uncertainties in the site investigation process. In a number of situations, the laboratory designated to perform the sample analyses provides sample bottles, preservation materials, and explicit sample collection instructions because of the complexity of gathering many different samples from various matrices that may have to be analyzed using a wide range of analytical protocols. The determination of analytical requirements involves specifying the most cost-effective analytical method that, together with the sampling methods, will meet the overall data quantity and quality objectives for the site investigation.

BOX 2.1
Tasks and Elements of a Site Investigation Program

Problem Definition and Preparatory Evaluation:
- Define objectives (including the level of detail and topics of interest)
- Collect and analyze existing information (i.e., review available background information, previous reports, etc.)
- Conduct visual inspection (i.e., field reconnaissance surveys)
- Construct preliminary conceptual model of site

Sampling Design:
- Identify information required to refine conceptual model of the site
- Identify constraints and limitations (e.g., access, presence of services, financial limitations)
- Define sampling and interpretation strategy
- Determine exploratory techniques and testing program

Implementation of Sampling and Analysis Plans:
- Conduct exploratory work on site (e.g., exploratory borings, test pits, geophysical surveys, etc.)
- Perform *in situ* testing
- Carry out sampling activities
- Compile record of investigation logs, photographs, and sample details
- Perform laboratory analyses

Data Evaluation and Results Interpretation:
- Compile and present relevant data
- Carry out logical analysis of data
- Refine conceptual model for site
- Enumerate implications of results
- Report on findings

2.1.2 Selecting Target Contaminants During Site Investigations

Because of the inherent variability in the materials and the diversity of processes used in industrial activities, it is not unexpected to find a wide variety of contaminants at a contaminated site. As a consequence, there is a corresponding variability in the range and type of hazards and risks that may be anticipated from different contaminated site problems. In general, detailed background information on the critical contaminants of potential concern should be compiled as part of the site investigation program.

The investigation of a potentially contaminated site must provide information on all contaminants known, suspected, or believed to be present at the site. Thus, the investigation should cover all compounds for which the history of site activities, current visible contamination, or public concerns suggest the possibility of contamination by such compounds. Ultimately, several chemical-specific factors (such as toxicity/potency, concentration, mobility, persistence, bioaccumulative/bioconcentration potential, synergistic/antagonistic effects, potentiation/neutralizing effects, frequency of detection, and naturally occurring background thresholds) are used to further screen and select the specific target contaminants that will become the focus of the detailed site evaluation process.

2.1.3   Contaminant Fate and Transport Considerations

Environmental contamination can be transported far away from its primary source(s) of origination via natural erosional processes, resulting in the possible birth of new contaminated site problems. On the other hand, some natural processes work to lessen or attenuate contaminant concentrations in the environment through mechanisms of natural attenuation such as dispersion/dilution, ion exchange, precipitation, adsorption and absorption, filtration, gaseous exchange, photodegradation, and biodegradation. Typically, environmental fate analysis is used to assess the movement of chemicals between environmental compartments. Simple mathematical models can be used to guide the decisions involved in estimating the potential spread of contaminant plumes. Where applicable, wells or monitoring equipment can then be located in areas expected to have elevated contaminant concentrations and/or in areas considered upgradient and downgradient of a plume.

In general, as pollutants are released into various environmental media, several factors contribute to their migration and transport. A number of important physical and chemical properties affecting the environmental fate and transport of chemical contaminants are annotated in Appendix C. A more detailed discussion of the pertinent factors affecting the environmental fate and/or intermedia transfers for chemical constituents at contaminated sites can be found elsewhere in the literature (e.g., Swann and Eschenroeder, 1983; Lyman et al., 1990).

The affinity that contaminants have for soils can particularly affect their mobility by retarding transport. For instance, hydrophobic or cationic contaminants that are migrating in solution are subject to retardation effects. In fact, the hydrophobicity of a contaminant can greatly affect its fate, which explains some of the different rates of contaminant migration occurring in the subsurface environment. Also, the phenomenon of adsorption is a major reason why the sediment zones of surface water systems may become highly contaminated with specific organic and inorganic chemicals.

In the groundwater system, the solutes in the porous media will move with the mean velocity of the solvent by an advective mechanism. In addition, other mechanisms governing the spread of contaminants include hydraulic dispersion and molecular diffusion (which is caused by the random Brownian motion of molecules in solution that occurs whether the solution in the porous media is stationary or has an average motion). Furthermore, the transport and concentration of the solute(s) are affected by reversible ion exchange with soil grains, chemical degeneration with other constituents, fluid compression and expansion, and, in the case of radioactive materials, by radioactive decay.

The degree of chemical migration from a contaminated site depends on both the physical and chemical characteristics of the individual constituents at the site, and also on the physical, chemical, and biological characteristics of the site. Physical characteristics of the contaminants, such as solubility and volatility, influence the rate at which chemicals leach into groundwater or escape into the atmosphere. The characteristics of the site environment (such as the geologic or hydrogeologic features) also affect the rate of contaminant migration. In addition, under various environmental conditions some chemicals will readily degrade to substances of relatively low toxicity, while other chemicals may undergo complex reactions to become more toxic than the parent chemical constituent. All other factors being equal, the extent and rate of contaminant movement are a function of the physical containment of the chemical constituents or the contaminated zone. A classical illustration pertains to the fact that a low permeability cap over a contaminated site will minimize water percolation from the surface and therefore minimize leaching of chemicals into an underlying aquifer. Invariably, the fate of chemical compounds released into the environment forms an important basis for evaluating the exposure of biological and ecological receptors to hazardous chemicals.

2.1.4   Design of Data Collection and Evaluation Programs

The general types of site data and information required in the investigation of potentially contaminated sites relate to contaminant identities, contaminant concentrations in the key sources and media of interest, characteristics of sources and contaminant release potential, and characteristics of the physical and environmental setting that can affect the fate, transport, and persistence of the contaminants (USEPA, 1989). The design and implementation of a substantive data collection and evaluation program is vital to the effective management of contaminated site problems.

Data are generally collected at several stages of the site investigation, with initial data collection efforts usually limited to developing a general understanding of the site. Typically, a preliminary gas survey using subsurface probes and portable equipment will give an early indication of likely problem areas. Soil gas surveys are generally carried out as a precursor to exploratory excavations, in order to identify areas that warrant closer scrutiny. They can also be used to assist in the delineation of previously identified plumes of contamination. This is an important step to complete prior to the start of a full-scale site investigation.

Gases produced at contaminated sites will tend to migrate through the paths of least resistance. The presence of volatile contaminants or gas-producing materials can be determined by sampling the soil atmosphere within the ground. Installation of a gas-monitoring well network, in conjunction with sampling in buildings in the area, can be used to determine the need for corrective measures. This information can be used to determine the possibility for human exposures and to determine appropriate locations for monitoring wells and gas collection systems.

On-site vapor screening of soil samples during drilling can provide indicators of organic contamination. For example, organic vapor analyzer/gas chromatograph (OVA/GC) or gas chromatograph/photoionization detector (GC/PID) screening provides a relative measure of contamination by volatile organic chemicals. Also, predictive models can be used to estimate the extent of gas migration from a suspected subsurface source. This information can subsequently be used to identify apparent "hot spots" and to select soil samples for detailed chemical analyses. The vapor analyses on-site can also be helpful in selecting screened intervals for monitoring wells.

In areas where the contamination source is known, the sampling program should be targeted around that source. Normally sampling points should be located at regular distances along lines radiating from the contaminant source. Provisions should also be made in the investigation to collect additional samples of small, isolated pockets of material which are visually suspect.

In general, a phased sampling approach encourages the identification of key data needs as early in the site investigation process as possible. This ensures that the data collection effort is always directed toward providing adequate information that meets the data quantity and quality requirements of the study. As a basic understanding of the site characteristics is achieved, subsequent data collection efforts focus on identifying and filling in data gaps. Any additionally acquired data should be such as to further improve the understanding of site characteristics and also consolidate information necessary to effectively manage the contaminated site problem. In this way, the overall site investigation effort can be continually rescoped to minimize the collection of unnecessary data and to maximize the quality of data acquired. Overall, the data gathering process should provide a logical, objective, and quantitative balance between the time and resources available for collecting the data and the quality of data, based on the intended use of such data.

2.1.5   Analyzing Site Information

The analysis of previously acquired and newly generated data serves to provide an initial basis to understanding the nature and extent of contamination which, in turn, aids in the design of appropriate corrective action programs for contaminated sites. Consequently, at any reasonable stage of a site investigation, all available site information should be compiled and analyzed to develop a conceptual model for the site. This representation should incorporate contaminant sources and "sinks", the nature and behavior of the site contaminants, migration pathways, the affected environmental matrices, and potential receptors (Figure 2.1). In fact, the development of an adequate conceptual site model (CSM) is an important aspect of the technical evaluation scheme necessary for the successful completion of a site investigation. It integrates geologic and hydrologic information, and provides a basis for human health and ecological risk assessments. The CSM is also relevant to the development and evaluation of corrective action programs for potentially contaminated sites.

Several variables and parameters are important to the design of a realistic CSM that will meet the overall goals of the corrective action program anticipated for a potentially contaminated site (Box 2.2). In general, the CSM should be appropriately modified if the acquisition of additional data and new information necessitates a redesign. Further discussions and illustrations of the framework for developing CSMs are given in Chapter 4.

2.2   A PHASED APPROACH TO THE INVESTIGATION OF POTENTIALLY CONTAMINATED SITE PROBLEMS

Programs designed to investigate and remedy potentially contaminated site problems typically consist of a number of phases. These phases reflect the different degrees of detail in the corrective action and/or risk management decisions for the site.

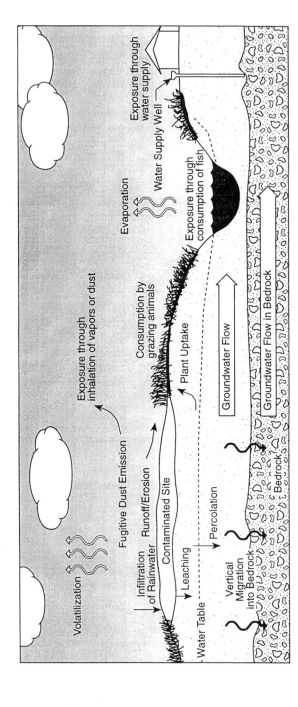

Figure 2.1 A conceptual representation of the potential migration pathways and exposure scenarios associated with a contaminated site problem.

BOX 2.2
Major Assessment Parameter Requirements for the
Construction of a Detailed Conceptual Site Model (CSM)

- Human, ecological, and welfare resources potentially at risk.
- Routes of contaminant exposure to populations potentially at risk.
- Spatial distribution of contaminants.
- Atmospheric dispersion potential and proximity of target receptors.
- Amount, concentration, hazardous nature, and environmental fate properties of the substances present.
- Site geology and hydrogeological factors.
- Area climatic regime and hydrology.
- Extent to which the sources can be adequately identified and characterized.
- Likelihood of releases if the contaminants remain on-site.
- Extent to which natural and artificial barriers currently contain contaminants, and the adequacy of the barriers.
- Assessment of the potential migration pathways.
- Extent to which site contaminants have migrated, or are expected to migrate from their source(s).
- Assessment of the likelihood of contaminant migration posing a threat to public health, welfare, or the environment.
- Extent to which contamination levels could exceed relevant regulatory standards, in relation to public health or environmental standards and criteria.

Environmental site assessments conducted as part of a corrective action investigation for potentially contaminated sites may be classified into the following general phases:

- "PHASE I" Investigation, or Preliminary Site Assessment, consisting of a reconnaissance site appraisal and reporting.
- "PHASE II" Investigation, or Comprehensive Site Assesment, comprising of a site investigation (that involves contamination and environmental damage assessment) and a preliminary feasibility study of corrective measures.
- "PHASE III" Investigation, or Remedial Measures Design and Implementation, incorporating a focused feasibility study and the detailed evaluation of site restoration measures.

In general, the objective of the initial phase (which is comprised of basic background information gathering) should be to determine the history of the site with respect to contamination sources and any relevant characteristics of the site that are readily obtainable from available records, reports, and interviews. The intermediate phase (involving a site characterization) has the primary objectives of defining the vertical and lateral extents of contamination, understanding how the contaminants are

TABLE 2.2 Summary Requirements of the Site Assessment Process

Level of investigation	Purpose of investigation	Typical "add-on" tasks performed
PHASE I: Preliminary site assessment	To provide a qualitative indication of potential contamination at a site	Records search, including geologic and hydrogeologic literature search, aerial photo reviews, review of archival and regulatory agency records (to define the historical uses of site), and anecdotal reports (on site history and practices as made available by former employees, local residents, and local historians) Site inspection/reconnaissance survey (to define present conditions at the site) Personal interviews (to supplement historical use data) Written report of findings (to document results and recommendations)
PHASE IIA: Initial site investigation	To identify the known or suspected source(s) of contamination To define the nature and extent of the contamination	Sampling of potentially impacted media at the surface Subsurface borings, well installations, and groundwater sample collection Sample analysis to identify and quantify contaminants Evaluation of sampling results and detailed report of findings, with conclusions
PHASE IIB: Expanded site investigation (or remedial investigation/ feasibility study)	To define the hydrogeology of the region and direction of contaminant plume migration To determine and evaluate those remedial options potentially applicable to the site restoration	Development of remediation goals and cleanup criteria Identification of alternative methods/ technologies for site remediation Screening of alternatives in order to select those that are most feasible option
PHASE III: Remedial measures, investigation and corrective action implementation	To recommend the site restoration method which is most feasible in terms of technical performance and cost-effectiveness	Evaluation of alternatives in terms cost, probable effectiveness, etc. Ranking of feasible alternatives and recommendation of the overall best method/technology for implementation Design and implementation of remedial options and monitoring programs

affected by hydrogeologic conditions, and providing the data necessary to design appropriate and applicable remedial measures. The final phase (that may include a risk evaluation of remedial alternatives as part of the focused feasibilty study) targets the development, selection, and implementation of appropriate corrective action plans for the site. Table 2.2 summarizes the requirements of the different levels of effort associated with the site assessment process.

2.2.1 PHASE I Investigations

A preliminary assessment (PA) is usually conducted to determine if a site is potentially contaminated as a result of past or current site activities, either due to unauthorized dumping or disposal, or from the migration of contaminants from adjacent or nearby properties. The objective of the PA is to qualitatively determine possible significant risks associated with the case-site. The PA will, at a minimum, involve record searching and a superficial physical survey that includes carrying out the following activities:

1. A review of historical records (such as site history, past operation and disposal practices, etc.).
2. A review of readily available files and databases maintained by regulatory agencies (to obtain information on site hydrogeology, characteristics of adjacent properties, known environmental problems in the area, etc.).
3. A field reconnaissance of the subject site and adjacent properties (to ascertain local soil and groundwater conditions, the proximity of the site to drinking water supplies and surface water discharges, existence of obvious contaminant sources or areas, etc.).

The PA is typically designed to document known or potential areas of concern due to the presence of contaminants, to establish the characteristics of the contaminants, to identify potential migration pathways, and to determine the potential for migration of the contaminant constituents. Data collection for the PA is accomplished by sequentially performing a records search/review, a site reconnaissance, and where practicable and/or appropriate, a soil vapor contaminant assessment. A Phase I investigation will generally conclude that either:

1. No current or historic evidence of contamination considered likely to have affected the site was found, and no further investigation is recommended; or
2. Sources of contamination that may have affected the site have been identified, and a Phase II investigation is recommended.

Thus, depending on the results of the PA, a "further-response-action" or a "no-further-response-action" may be recommended for the potentially contaminated site. Where warranted, the next level of detail in the site assessment is carried out to ascertain this initial indication of possible contamination. Whereas the PA provides some important site information, a more detailed characterization of the extent of soil and groundwater contamination may be necessary to properly assess long-term risks posed by a contaminated site.

2.2.2 PHASE II Investigations

The comprehensive site assessment typically comprises a site investigation (SI) and/or a remedial investigation/feasibility study (RI/FS). In general, comprehensive site assessments usually will involve sampling and testing to identify the types of contaminants, analyzing preselected or priority pollutants, and determining the horizontal and vertical extents of the contamination. This typically includes subsurface investigations, soil and water sampling, laboratory analyses, tank testing, and other relevant engineering investigations to quantify potential risks previously identified in the PA.

The SI is designed to verify findings from the PA, to determine the presence or absence and the extent of contamination at the site, and to identify probable remedial measures. Typically, the investigation identifies specific contaminants, their concentrations, the areal extent of contamination, the fate and transport properties of the contaminants, and the potential migration pathways of concern.

In an expanded SI, the RI strives to improve the initial site characterization. The FS investigates the most cost-effective methods of remediation that will protect public health, the environment, and public and private property under applicable and appropriate standards or regulations.

At a minimum, a comprehensive site assessment will include the collection and analysis of as many soil samples as necessary to determine the full extent of the contamination. If contamination is found to be confined to the unsaturated (vadose) zone, then no groundwater investigation may be required; otherwise, groundwater investigation is initiated. The level of detail for the data collection activities will be site-specific, and is dependent on the degree of soil and groundwater contamination found at the site.

2.2.2.1   PHASE IIA Investigations

An initial Phase II assessment will normally be used to confirm whether or not a release has occurred. This is accomplished by implementing a limited program to collect and analyze appropriate site samples. To complete this process, a sampling plan must first be developed. Subsequently, site visits are conducted during which sampling activities will be carried out. The results of this initial comprehensive site assessment will determine the need for a "further-response-action" or a "no-further-response-action". A Phase IIA investigation will generally conclude that either:

1. No evidence of contamination was discovered, and no further investigation is recommended; or
2. Contamination that may require remediation has been found, and a Phase IIB investigation is recommended.

In fact, if the site is determined to pose significant public health or environmental risks, extensive studies will typically be required to quantify the magnitude of contaminants present, delineate the limits of contamination, characterize in detail the specific chemical constituents present at the site, and assess the fate and transport properties of the specific substances at the site.

2.2.2.2   PHASE IIB Investigations

Where necessary, an expanded Phase II assessment is conducted with the main objective to characterize the contamination confirmed from the initial Phase II investigations. The characterization process involves specifying the type of contamination present, assessing the three-dimensional occurrence of the contamination, evaluating the contaminant fate and transport, determining possible human and ecological receptors potentially at risk, estimating the risks posed to the populations at risk, establishing a database to facilitate documentation of changes in the occurrences of the contamination, and conducting a preliminary screening of corrective measures. A Phase IIB investigation will generally conclude that either:

1. No evidence of extensive contamination was discovered that requires remediation, and no further investigation is recommended; or
2. Contamination that may require remediation has been found, and a Phase III investigation is recommended.

Typically, the Phase IIB assessment will generate a report made up of a site characterization, a risk assessment, and an evaluation of mitigation and remediation options with an indication of the preferred corrective action plan.

2.2.3   PHASE III Investigations

The remedial measures study involves an evaluation of corrective action programs previously identified during the expanded comprehensive site assessment, an engineering design of the selected remedial plan, and the implementation of the cleanup or mitigation measures necessary to abate public health and environmental concerns. This also includes a review of the environmental and public health risks and costs associated with a variety of proposed remedial alternatives. The documentation offered will facilitate the development of appropriate corrective action programs consistent with appropriate regulatory guidelines. The selected remedy will fulfill the jurisdictional requirements for environmental conditions at the site, and should present minimal risk to the environment and/or to public health.

The Phase III assessment will typically report on the preferred corrective action plan, incorporating the design and implementation of the remedial action plans and postremediation monitoring. This level of the site assessment will also generally include remediation cost estimates as part of the detailed evaluation process for the alternative remedial options identified for the site. This aspect of the assessment will involve evaluating the feasibility of various corrective action strategies applicable to the site scenario, and also evaluating the impact of the mitigated site on the current and future land uses at and near the site.

2.3   THE SITE ASSESSMENT DECISION PROCESS

Prior to the development of a corrective action plan for a contaminated site problem, a site assessment must be conducted to determine the true extent of contamination at the site. The use of a systematic approach will result in an optimal data gathering and evaluation process that meets uncompromising data quantity and quality objectives (Figure 2.2). Such a strategy will indeed help address potentially contaminated site problems in a cost-effective manner.

In general, once a diagnostic assessment of possible environmental contamination problems is completed for a potentially contaminated site, plans can be made towards the implementation of effectual corrective actions, where warranted. The underlying goal in conducting site assessments is to determine an appropriate level of effort in the corrective action required for a site at which contamination is suspected, or known to have occurred. The type of corrective action selected for the contaminated site problem will depend on the nature of contamination, the amount of contamination that could safely remain at the site following site restoration, and several other site-specific factors.

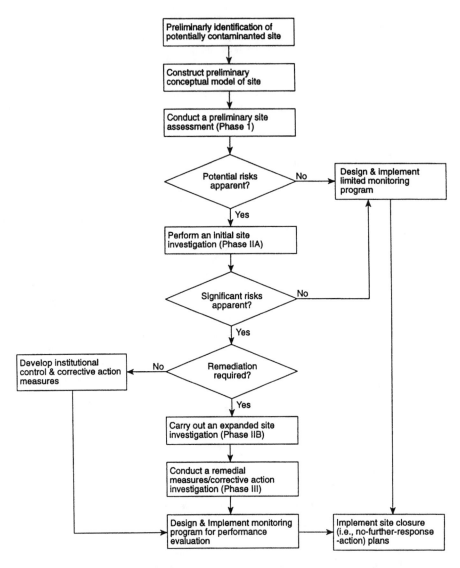

Figure 2.2 Overview of the decision process required for the systematic assessment of contaminated site problems.

A project manager needs to be cautious in accepting the results of any site investigation activity as being an absolute indicator of the true situation at a suspect site. In practice, rather than simply walk away from a site purported to be "clean", it often is a good idea to implement some form of monitoring program for the candidate site even when no evidence of contamination has been found during the site assessment. This is important if for no other reason than to account for uncertainty, because even a carefully executed site investigation program may still miss some isolated pockets of contaminants that may become long-term release sources and therefore a potential liability issue.

2.4   AN ILLUSTRATIVE EXAMPLE OF A CORRECTIVE ACTION INVESTIGATION

This section illustrates the nature of decision elements typically employed in the investigation of potentially contaminated sites. The example discusses a potential contamination problem at a small shopping center located in a rural township that depends almost exclusively on groundwater from a contiguous aquifer for its water supplies. The trigger for this investigation is the discovery that dry-cleaning solvents have spilled on concrete floor slabs at a dry-cleaning facility within this mini-mall.

Background — A preliminary environmental site assessment carried out for the Village Shopping Center (VSC) indicated the potential for soil and groundwater contamination at this facility as a result of releases from dry-cleaning and laundry activities at one section of the mall. A follow-up phase II-type assessment conducted for the site confirmed the presence of elevated levels of perchloroethylene (PCE) in the sampled soils. The soils beneath the facility location consist predominantly of fine sands, or fine sand with occasional gravels. Some plastic liner materials (likely to have been used as a plastic moisture barrier beneath the concrete floor slabs) were found in some of the exploratory borings at the site. It is believed that the plastic liner materials found beneath the concrete floor may have prevented extensive contamination of the subsurface environmental compartments.

Recommendations for the corrective action investigation — Since PCE spills would have occurred on the concrete floors at the VSC facility, and because the plastic liner material may have been serving as a "barrier" against further contaminant migration, it is very possible that any PCE encountered in the exploratory soil borings could have been introduced into the soil after the barrier was broken during the soil sample coring activities. If this hypothesis is true, then the extent of soil contamination may be even less than suspected; in addition, the possibility of any extensive groundwater contamination can also be ruled out. Under such circumstances, a more detailed assessment may indicate that no extensive and expensive remediation or cleanup program is necessary for the VSC facility. In fact, this reasoning will also support the the importance of studying complete building plans/layouts before any drilling activities that could actually facilitate the spreading of contaminants that would otherwise be sitting as a more easily removable free product.

To complete the requisite investigations for the VSC facility that will allow for appropriate corrective action decisions, a number of issues must be fully explored and evaluated, including the following:

1. Groundwater beneath the site should be investigated in order to complete the site assessment. This is because soils beneath the site have already been impacted, and PCE is reasonably mobile in the type of soil formations at this site. That is, considering the mobility of PCE and the sandy nature of soils found at the VSC facility, the possibility of a contaminated aquifer beneath the facility cannot be ignored. In particular, if it cannot be established that the liner materials may have prevented the PCE from migrating further into the subsurface environment, then the groundwater system beneath the site should be fully investigated. On the other hand, if it can be positively established that the plastic liner material did serve to prevent or minimize PCE migration into the subsurface environments, then a different set of exposure scenarios may be developed for the corrective action investigation at this facility.

2. Site conditions should be adequately characterized, in order to fully define the lateral and vertical extents of the PCE contamination. In this regard, it is noted that PCE has rather low adsorptivity to soil; consequently, if released to soils, it will generally be subject to an accelerated migration into the groundwater. In particular, PCE can move rapidly through sandy soils and may therefore reach groundwater more easily in the type of geological formations found at the VSC facility.

3. Risk assessment procedures are typically used to establish cleanup objectives for contaminated environmental media requiring corrective actions and/or for the implementation of risk management programs for contaminated sites. Such an assessment should be used to help focus corrective action assessments and risk management plans for the VSC facility. The site-specific risk assessment may also include the development of appropriate site restoration goals with reference to site conditions, land uses, and exposure scenarios pertaining specifically to this shopping center and its vicinity. It is believed that the development of site-specific cleanup levels can result in significant cost savings in this type of investigation.

A complete characterization of the "contaminated zone" at the site should facilitate the screening and selection of the best available technology for a remedial action plan that is developed for the VSC facility. Prior to implementing any remediation plan for this site, appropriate risk-based cleanup criteria should be developed and compared with the current levels of contamination present at the site (i.e., under the baseline conditions). Based on such a criteria, it may become apparent under the appropriate types of exposure scenarios that no cleanup is warranted. In fact, even if it is determined that some degree of cleanup is required, the cleanup criteria developed will aid in optimizing the efforts involved, so as to arrive at more cost-effective solutions than could otherwise have been achieved.

REFERENCES

Asante-Duah, D.K. 1993. *Hazardous Waste Risk Assessment.* CRC Press/Lewis Publishers, Boca Raton, FL.

BSI (British Standards Institution). 1988. Draft for Development, DD175: 1988 Code of Practice for the Identification of Potentially Contaminated Land and its Investigation. BSI, London, U.K.

Cairney, T. (Ed.). 1993. *Contaminated Land (Problems and Solutions).* Lewis Publishers, Boca Raton, FL.

Lyman, W.J., W.F. Reehl and D.H. Rosenblatt. 1990. *Handbook of Chemical Property Estimation Methods: Environmental Behavior of Organic Compounds.* American Chemical Society, Washington, D.C.

Swann, R.L., and A. Eschenroeder (Eds.). 1983. *Fate of Chemicals in the Environment.* ACS Symp. Ser. 225, American Chemical Society, Washington, D.C.

USEPA. 1988. Guidance for Conducting Remedial Investigations and Feasibility Studies Under CERCLA. EPA/540/G-89/004. OSWER Directive 9355.3-01, Office of Emergency and Remedial Response, U.S. Environmental Protection Agency, Washington, D.C.

USEPA. 1989. Risk Assessment Guidance for Superfund. Vol. I. Human Health Evaluation Manual (Part A). EPA/540/1-89/002. Office of Emergency and Remedial Response, U.S. Environmental Protection Agency, Washington, D.C.

3 SITE CHARACTERIZATION ACTIVITIES AT CONTAMINATED SITES

The characterization of potentially contaminated sites is a process used to establish the presence or absence of contamination at a site, to delineate the nature and extent of site contamination, and to determine possible threats posed by the site to human health and/or the environment. The scope and detail of a site characterization should generally be adequate to determine the:

- Primary and secondary sources of contamination
- Amount and extent of contamination
- Fate and transport characteristics of site contaminants
- Pathways of contaminant migration
- Types of exposure scenarios associated with the site
- Risk to human health and the environment
- Feasible solutions to mitigate receptor exposures to site contaminants

Characterization of contaminant sources, migration pathways, and potential receptors probably form the most important basis for determining the need for site remediation. The completion of an adequate site characterization is therefore considered a very important component of any corrective action program that is designed to effectively remedy a contaminated site problem.

3.1 THE SITE CHARACTERIZATION PROCESS

The site characterization process consists of the collection and analysis of a variety of environmental data necessary for the design of an effectual corrective action program. Box 3.1 enumerates several important elements of site characterization activities that are used to support corrective action decisions at contaminated sites. The implementation of these typical action items will generally help realize the overall goal of a corrective action investigation.

Credible site characterization programs generally involve a complexity of activities that require careful planning. The initial step involves a data collection activity to compile an accurate site description, history, and chronology of significant events. Subsequently, the most important functions are field sampling and laboratory analyses.

35

BOX 3.1
Elements of a Site Characterization Activity

- Investigate site physical characteristics (to include surface features, geology, soils and vadose zone properties, drainage network and surface water hydrology [with particular emphasis on an assessment of the erosion potential of the site, and the potential impacts on any surface waters in the catchment], hydrogeology, meteorological conditions, land uses, and human and ecological populations)

- Determine sources of contamination

- Design a sampling plan that will allow for a characterization of both "hot-spots" and "representative areas" through extensive sampling, field screening, visual observations, or a combination of these

- Conduct field data collection and analysis

- Determine the nature and extent of contamination (including contaminant types, contaminant concentrations, spatial and temporal distribution of contaminants, and impacted media [such as groundwater, soils, surface water, sediments, and air])

- Evaluate the contaminant fate and transport properties in the various environmental matrices

- Prepare a data summary that contains pertinent sampling information for each area of concern at the site, including the media sampled, chemicals analyzed, sampling statistics (such as frequency of detection of the contaminants of concern, range of detected concentrations, and average concentrations), background threshold concentrations in the local area, range of laboratory and/or sample quantitation limits, and sampling periods

- Identify and compile regulatory limits for site contaminants

- Develop detailed iso-concentration plots to show the distribution of the chemicals of potential concern present at the site

- Develop a site conceptual model that can be refined as additional information about the site becomes available

- Develop baseline risk screening for the site (i.e., a preliminary risk assessment for a "no-action" scenario)

- Identify risk management and site remediation requirements

The field data are collected to help define the nature and extent of contamination present at, or migrating from, a contaminated site. Consequently, it is important to use proven methods of approach for the sampling and analysis programs designed for potentially contaminated sites.

A wide variety of investigation techniques may be employed in the characterization of potentially contaminated sites. However, the appropriate methods and applicable techniques will generally be dependent on the type of contaminants, the site's geologic and hydrogeologic characteristics, site accessibility, availability of resources, and program costs. Several methods of choice used to effectively complete site characterization programs are described elsewhere in the literature of site assessment (e.g., BSI, 1988; CCME, 1993, 1994; CDHS, 1990; Driscoll, 1986; OBG, 1988; USEPA, 1985, 1987, 1988a, 1988b, 1989a, 1989b, 1989c). The literature provides technical standards for remedial investigation project scoping, data validation, surveying and mapping of contaminated site sampling locations, hydrogeologic characterization at contaminated sites, surface geophysical techniques, drilling, coring, sampling and logging at contaminated sites, borehole geophysical techniques, the design and construction of monitoring wells and piezometers, groundwater sampling, the design and construction of extraction wells at contaminated sites, etc. These types of guidance will facilitate the implementation of a worthwhile site characterization plan. Overall, the information obtained from a site characterization activity should be adequate to predict the fate and behavior of the contaminants in the environment, as well as to facilitate the design of an effectual corrective program.

3.2 DESIGNING COST-EFFECTIVE SITE CHARACTERIZATION PROGRAMS

Several types of decisions are usually made very early in the design of a site characterization program, at which time only very limited information may be available about the case site. Nonetheless, reasonable decisions can be made despite the uncertainties that may surround it. By adopting a phased approach, each phase of a site investigation should, as far as practicable, be based upon previously generated information about the site. Such an approach will help identify the specific information needed to understand the site conditions, as well as help optimize the cost of acquiring that information. A multi-layered, iterative, and flexible survey that emphasizes *in situ* measurements and focuses on mapping contaminant boundaries is critical to this process (Jolley and Wang, 1993).

The characterization of huge, complex sites may present increased logistical problems, making the design of site characterization programs for such sites more complicated. To allow for a manageable situation in the investigation of extensive or large areas, it usually is prudent to divide the site into "operable units" (OUs) or "waste management units" (WMUs) or "contaminated site zones" (CSZs). The OUs or WMUs or CSZs would define the areas of concern, which are used to guide the identification of the general sampling locations at or near the site. The areas of concern, defined by the individual OUs or WMUs or CSZs, will typically include sections or portions of a site that:

- Have different chemical constituents
- Have different anticipated concentrations or "hot spots"
- Are a major contaminant release source
- Differ from each other in terms of the anticipated spatial or temporal variability of contamination
- Must be sampled using different field procedures and/or equipment and tools

All of the areas of concern (designated as OUs or WMUs or CSZs) together should account for, or be representative of, the entire case site.

The key to a successful and cost-effective site characterization will be to optimize the number and length of field reconnaissance trips, the number of soil borings and wells, and the frequency of sampling and laboratory analyses. Opportunities to effect cost savings begin during the initial field investigation conducted as part of a preliminary assessment. In all situations, before a subsurface investigation is initiated, background information relating to the regional geology, hydrogeology, and site development history should have been fully researched. In general, minimizing of well installations and sample analyses during this phase of the characterization can greatly reduce overall costs. Indeed, well installation can even add to the spread of contamination by penetrating impermeable layers containing contaminants and then allowing such contamination to migrate into previously uncontaminated groundwater systems (Jolley and Wang, 1993). Consequently, nonintrusive and screening measurements are generally preferred modes of operations whenever possible, since these tend to minimize both sampling costs and the potential to spread contamination. Ultimately, however, sufficient information should be obtained that will reliably show the identity, areal and vertical extent, and the magnitude of contamination associated with the site.

3.2.1   Sampling and Analysis Design Considerations

A preliminary identification of the types of contaminants, the chemical release potentials, and also the potential exposure pathways should be made very early in the site characterization, because these are crucial to decisions on the number, type, and location of samples to be collected.

The type of chemicals at potentially contaminated sites may dictate the sections or areas and environmental media to be sampled. For instance, in the design of a sampling program, if it is believed that the contaminants of concern are relatively immobile (e.g., PCBs and most metals in silty or clayey materials), then sampling will initially focus on soils in the vicinity of the suspected source of contaminant releases. On the other hand, the sampling design for a site believed to be contaminated by more mobile compounds (e.g., organic solvents in sandy materials) will take account of the fact that contamination may already have migrated into groundwater systems and/or moved significant distances away from the original source area(s). Then, also, some chemicals that bioaccumulate in aquatic life will likely be present in sediments, requiring that both sediments and biota be sampled and analyzed. Consequently, knowledge on the type of contaminants will generally help focus more attention on the specific media most likely to be impacted.

Similar decisions as above will typically have to be made regarding analytical protocols. For instance, due to the differences in the relative toxicities of the different species of some chemicals (e.g., chromium may exist as trivalent chromium, Cr^{3+} or as the more toxic hexavalent chromium, Cr^{6+}), chemical speciation to differentiate between the various forms of the chemicals of potential concern present at a contaminated site may be required in the design of analytical protocols.

3.3   DEVELOPMENT OF SITE CHARACTERIZATION WORKPLANS

Workplans are generally required to specify the administrative and logistic requirements of site investigation activities. As part of the corrective action program for potentially contaminated site problems, a carefully executed investigative strategy or workplan should be developed to guide all relevant decisions. A typical workplan developed to facilitate the investigation of contaminated site problems will generally consist of the following major components:

BOX 3.2
Elements of a Typical Site Characterization Workplan

- How site mapping will be performed (including survey limits, scale of site plan to be produced, horizontal and vertical control, and significant site features)
- Number of individuals to be involved in each field sampling task and estimated duration of work
- Identification of soil boring and test pit locations on a map to be provided in a detailed workplan
- Number of samples to be obtained in the field (including blanks and duplicates), and the sampling location (illustrated on maps to be included in a detailed workplan)
- List of field and laboratory analyses to be performed
- An elaboration of how investigation-generated wastes will be handled
- A general discussion of data quality objectives (DQOs)
- Identification of pilot or bench-scale studies that will be performed, where necessary, in relationship to recommendations for remedial technologies screening, risk management strategies, and/or site stabilization processes
- A discussion of health and safety plans required for the site investigation or corrective action activities, as well as that necessary to protect populations in the general vicinity of the site

---

- A sampling and analysis plan
- A health and safety plan
- A waste management plan (for investigation-generated wastes)
- A site activity plan
- A quality assurance/quality control plan

Details of all the workplan elements, as represented in Box 3.2, should be adequately developed. In this regard, any existing information on the fate and behavior of contaminants present at a contaminated site are essential for developing the various workplan components. The major components and tasks required of most site characterization workplans are elaborated further in the following sections.

3.3.1   The Sampling and Analysis Plan

Sampling and analysis of environmental pollutants is a very important part of the decison-making process involved in the management of potentially contaminated site problems. Yet, sampling and analysis often turns out as one of the most expensive and time-consuming aspects of a site characterization project. Even of greater concern is the fact that errors in sample collection, sample handling, or laboratory analysis can invalidate projects or add to the overall project costs. All environmental samples that are intended for use in the characterization of potentially contaminated sites must therefore be collected, handled, and analyzed properly, in accordance with applicable/appropriate standards and methods.

The sampling and analysis plan (SAP) is an essential component of any environmental investigation, and more so in the planning, development, and implementation of corrective action programs for potentially contaminated sites. SAPs generally are required to specify sample types, numbers, locations, and procedures. In fact, the SAP sets the stage for developing a cost-effective and adequate corrective action plan for potentially contaminated sites. Its purpose is to ensure that sampling and data collection activities will be comparable to, and compatible with, previous data collection activities.

SAPs provide a mechanism for planning and approving field activities. The required level of detail and the scope of the planned investigation generally determines the data quality objectives (DQOs). The sampling and analysis strategies should be planned in such a manner as to minimize the costs associated with achieving the DQOs. Typically, the SAP will comprise of two major components (USEPA, 1988a, 1988b, 1989b):

1. A quality assurance project plan (QAPP) that describes the policy, organization, functional activities, and quality assurance and quality control protocols necessary to achieve the DQOs dictated by the intended use of the data.
2. A field sampling plan (FSP) that provides guidance for all fieldwork, by defining in detail the sampling and data-gathering methods to be used in a project. The FSP should be written so that a field sampling team unfamiliar with the site is able to gather the samples and any field information required.

A detailed discussion of sampling considerations and strategies for various environmental matrices can be found in the literature (e.g., CDHS, 1990; Keith, 1988, 1991; USEPA, 1988b, 1989b). Important issues to consider when one is making a decision on how to obtain reliable samples relate to the sampling objective and approach, the sample collection methods, chain-of-custody documentation, sample preservation techniques, sample shipment methods, and sample holding times. Data necessary to meet the project objectives should be specified, including the selection of sampling methods and analytical protocols for the site; this will also include an evaluation of multiple-option approaches that will ensure timely and cost-effective data collection and evaluation. Box 3.3 enumerates a checklist of items that should be reviewed in the development of a SAP (CCME, 1993; Keith, 1988, 1991).

The methods by which data of adequate quality and quantity are to be obtained to meet the overall project goals should be specified and fully documented in the SAP developed as part of a detailed site characterization workplan. It is noteworthy that the use of appropriate sample collection methods can be as important as the use of appropriate analytical methods for sample analyses. Consequently, effective analytical protocols in both the sampling and laboratory procedures should be specified by the SAP, in order to help minimize uncertainties associated with the data collection and evaluation activities.

3.3.1.1 Purpose of Sampling and Analysis

The principle objective of a sampling and analysis program is to obtain a small and informative portion of the statistical population being investigated so that contaminant levels can be established as part of a corrective action assessment program. Sampling and analysis programs are generally designed and conducted in order to (USEPA, 1989b):

BOX 3.3
Checklist for Developing Sampling and Analysis Protocols

- What observations at sampling sites are to be recorded?
- Has information concerning data quality objectives, analytical methods, limits of detection, etc., been included?
- Have instructions for modifying protocols in case of problems been specified?
- Has a list of all sampling equipment and materials been prepared?
- Are there instructions for cleaning equipment before and after sampling?
- Have instructions for each type of sample collection been prepared?
- Have instructions for completing sample labels been included?
- Have instructions for preserving each type of sample (such as maximum holding times of samples) been included?
- Have instructions for packaging, transporting, and storing samples been included?
- Have instructions for chain-of-custody procedures been included?
- Have health and safety plans been developed?
- Is there a waste management plan to deal with investigation-generated wastes?

- Determine the extent to which soils act as either contamination sources (i.e., situations when significant quantities of selected contaminants are found to be associated with soils initially and then released slowly over relatively long periods of time into other media), or sinks (i.e., situations where significant quantities of contaminants become permanently attached to soil and remain biologically unavailable)
- Determine the presence and concentration of specified contaminants in comparison to natural and/or anthropogenic background levels
- Determine the concentration of contaminants and their spatial and temporal distribution
- Obtain measurements for validation or use of transport and fate models
- Identify pollutant sources, transport mechanisms or routes, and potential receptors
- Determine the potential risks to human health and/or the environment (i.e., to flora and fauna) due to site contamination
- Identify areas of contamination where action is needed, and measure the efficacy of control or removal actions
- Contribute to research technology transfer or environmental model development study
- Meet the provisions and intent of environmental laws (such as RCRA, CERCLA, FIFRA, TSCA)

The design of a sampling and analysis program and its associated quality assurance plan takes account of the variability in the entire measurement process along with the sources and magnitude of the variation in the results generated. It also provides a means of determining whether a sampling and analysis program meets the specified DQOs.

3.3.1.2 *Sampling Requirements*

Environmental sampling at contaminated sites are conducted to characterize the site for enforcement and corrective actions. Several site-specific requirements are important in achieving the site characterization goals. Specific factors to consider when preparing the sampling plan include (WPCF, 1988):

- History of activity at the site
- Physical/chemical properties and hazardous characteristics of materials involved
- Topographic, geologic, pedologic, and hydrologic characteristics of the site
- Meteorologic conditions
- Flora and fauna of the site and vicinity (to include the identification of sensitive ecological species and systems, and the potential for bioaccumulation and biotransformation)
- Geographic and demographic information, including proximity of populations potentially at risk

A history of the site, including the sources of contaminants and a conceptual model describing the apparent migration pathways should be developed before a sampling plan is finalized. In certain situations, it may be necessary to conduct an exploratory study so that this preliminary conceptual model can be confirmed or modified, as appropriate. Thus, as additional data identify specific areas where a site conceptual model is not valid, the model should be modified to reflect the new information.

In working towards the development of a representative conceptual model for a site, it is important to be aware of the fact that some significant model features may not be very apparent, but rather will remain latent. For example, consider the case of contaminated non-soil debris present in the soil mass at a potentially contaminated site. This debris must be taken into account when evaluating the hazard posed by the site. In fact, in some cases, the debris (e.g., wood chips or shredded wood, for instance, used as absorbents for liquid wastes) rather than the soil itself may be the major source of contaminant releases. Consequently, screening the non-soil debris out of the soil material and excluding it from a hazard and/or risk analysis may bias the results and, therefore, the final conclusions reached about the site.

Elements of a Sampling Plan

An initial site evaluation should provide insight into the types of site contaminants, the populations potentially at risk, and possibly the magnitude of the site risk. These factors can be combined to design a sampling plan and to specify the size of sampling unit appropriate for the site characterization program. All sampling plans should contain the following common elements:

- Site background (that includes a description of the site and surrounding areas and a discussion of known and suspected contamination sources, probable transport pathways, and other information about the site)
- Sampling objectives (describing the intended uses of the data)
- Sampling location and frequency (that also identifies each sample matrix to be collected and the constituents to be analyzed)

BOX 3.4
Sampling Plan Checklist

- What are the DQOs, and what corrective measures are planned if DQOs are not met (e.g., resampling or revision of DQOs)?
- Do program objectives need exploratory, monitoring, or both sampling types?
- Have arrangements been made for site entry or access?
- Is specialized sampling equipment needed and/or available?
- Are field crews who are experienced in the required types of sampling available?
- Have all analytes and analytical methods been listed?
- Have required good laboratory practice and/or method QA/QC protocols been listed?
- What type of sampling approach will be used (i.e., random, systematic, judgmental, or combinations thereof)?
- What type of data analysis methods will be used (e.g., geostatistical, control charts, hypothesis testing, etc.)?
- Is the sampling approach compatible with data analysis methods?
- How many samples are needed?
- What types of QC samples are needed, and how many of each type of QC samples are needed (e.g., trip blanks, field blanks, etc.)?

- Sample designation (that establishes a sample numbering system for the specific project, and should include the sample or well number, the sampling round, the sample matrix, and the name of the site)
- Sampling equipment and procedures (including equipment to be used and material composition of equipment, along with decontamination procedures)
- Sample handling and analysis (including identification of sample preservation methods, types of sampling jars, shipping requirements, and holding times)

Regardless of the medium sampled, data variability problems may arise from temporal and spatial variations in field data. That is, sample composition may vary depending on the time of the year and weather conditions when the sample is collected. Ideally, samples from various media should be collected in a manner that accounts for temporal factors and weather conditions. If seasonal fluctuations cannot be characterized in the investigation, details of meteorological, seasonal, and climatic conditions during sampling must be well documented. Choosing an appropriate sampling interval that spans a sufficient amount of time to allow one to obtain, for example, an independent groundwater sample will generally help reduce the effects of autocorrelation. Also, sampling both background and compliance wells at the same point in time should reduce temporal effects. Consequently, the ideal sampling strategy will incorporate a full annual sampling cycle. If this strategy cannot be accommodated in an investigation, at least two sampling events should be considered that take place during opposite seasonal extremes (such as high-water/low-water, high-recharge/low-recharge, etc.).

Box 3.4 provides a convenient checklist of the issues that should be verified when planning a sampling activity for contaminated sites (CCME, 1993).

3.3.1.3 Sampling Protocols

Sampling protocols are written descriptions of the detailed procedures to be followed in collecting, packaging, labeling, preserving, transporting, storing, and documenting samples. The selection of analytical methods is also an integral part of the processes involved in the development of sampling plans, since this can strongly affect the acceptability of a sampling protocol. For example, the sensitivity of an analytical method could directly influence the amount of a sample needed in order to be able to measure analytes at prespecified minimum detection (or quantitation) limits. The analytical method may also affect the selection of storage containers and preservation techniques (Keith, 1988).

The overall sampling protocol must identify sampling locations and include all of the equipment and information needed for sampling, including the types, number and sizes of containers, labels, field logs, types of sampling devices, numbers and types of blanks, sample splits and spikes, sample volume, any composite samples, specific preservation instructions for each sample type, chain-of-custody procedures, transportation plans, field preparations (such as filter or pH adjustments), field measurements (such as pH, dissolved oxygen, etc.), and the reporting requirements (Keith, 1988). The sampling protocol should also identify those physical, meteorological, and hydrological variables to be recorded or measured at the time of sampling (Keith, 1988). In addition, information concerning the analytical methods to be used, minimum sample volumes, desired minimum levels of quantitation, and analytical bias and precision limits may help sampling personnel make better decisions when unforeseen circumstances require changes to the sampling protocol.

Table 3.1 lists the minimum documentation needed for sampling activities (CCME, 1993). The more specific a sampling protocol is, the less chance there will be for errors or erroneous assumptions. In general, the devices used to collect, store, preserve, and transport samples must *not* alter the sample in any manner. For example, special procedures may be needed to preserve samples during the period between collection and analysis.

Table 3.1   Minimum Requirements for Documenting Environmental Sampling

Sampling date	Sampling conditions or sample type
Sampling time	Sampling equipment
Sample identification number	Preservation used
Sampler's name	Time of preservation
Sampling site	Relevant sample site observations (auxiliary data)

Sampling Approach

There are three basic sampling approaches: random, systematic, and judgmental. There are also three primary combinations of each of these, i.e., stratified-(judgmental)-random, systematic-random, and systematic-judgmental (CCME, 1993; Keith, 1991). Additionally, there are further variations that can be found among the three primary approaches and the three combinations thereof. For example, the systematic grid may be square or triangular; samples may be taken at the nodes of the grid, at the center of the spaces defined by a grid, or randomly within the spaces defined by a grid

(CCME, 1993). A combination of judgmental, systematic, or random sampling is often the most feasible approach to employ in the investigation of potentially contaminated sites. However, the sampling scheme should be flexible enough to allow relevant adjustments/modifications during field activities. Several techniques and equipment that can be used in the sampling of contaminated soils, sediments, and waters are enumerated in the literature (e.g., CCME, 1993).

3.3.1.4 Laboratory and Analytical Program Requirements

Oftentimes, the initial analyses of environmental samples may be performed with a variety of field methods used for screening purposes. The purpose of using initial field screening methods is to decide if the level of pollution at a site is high enough to warrant more expensive (and more specific and accurate) laboratory analyses. Methods that screen for a wide range of compounds, even if determined as groups or homologues, are useful because they allow more samples to be measured faster and more inexpensively than with conventional laboratory analyses.

In the more detailed assessment, environmental sample analysis is generally performed by the so-called contract laboratory program (CLP) and non-CLP services. The CLP consists of routine and nonroutine standardized analytical procedures and associated quality control requirements managed under a broad quality assurance program (that includes sample projections, sample scheduling, chain-of-custody requirements, reporting and documentation requirements, audits, and data evaluations); CLP services are provided through routine analytical services and special analytical services. Non-CLP services will include field analytical support methods. The investigation of a potentially contaminated site will typically utilize both CLP and non-CLP services designed to meet the DQOs of the investigation.

Effective analytical programs and laboratory procedures are necessary to help minimize uncertainties in site investigation activities that are required to support corrective action decisions. Table 3.2 lists the minimum requirements for documenting laboratory work performed to support site characterization activities (CCME, 1993; USEPA, 1989c). The applicable analytical procedures, the details of which are outside the scope of this book, should be strictly adhered to.

Table 3.2 Minimum Requirements for Documenting Laboratory Work

Method of analysis	Method detection limits
Date of analysis	Confidence limits
Laboratory and facility carrying out analysis	Records of calculations
Analyst's name	Actual analytical results
Calibration charts and other measurement charts (e.g., spectral)	

3.3.1.5 Laboratory and Analytical Protocols

Analytical protocol and constituent parameter selection are usually carried out in a way that balances costs of analysis with adequacy of coverage. If specific chemical constituents are known to be associated with previous site activities, they should most definitely be targeted for analysis. Otherwise, a widely used and more general parameter list such as the USEPA "Priority Pollutant List" is adopted. For instance, if a metals processing and finishing plant is known to have existed at a site, then some priority pollutant metals and VOCs, both associated with cleaning and degreasing, may become the target constituents/parameters. In addition, a select but limited number of samples from the more strongly suspected areas and media may be subjected to the full prority pollutant analyses.

In general, methods such as the extraction procedure (EP) toxicity and the toxicity leaching characteristic procedure (TCLP) testing should not be used as a primary indicator of contamination. Instead, total constituent analysis should be used to indicate the magnitude and extent of contamination. If significant contamination is confirmed, then it may become necessary to conduct supplemental testing (e.g., by using EP toxicity or TCLP testing) to determine if characteristic hazardous waste definitions (based on toxicity criteria) are applicable to the particular situation.

Another noteworthy point to make relates to the analyses of groundwater samples. In this procedure, it is always important to distinguish between total (i.e., without sample filtration) and dissolved (i.e., with sample filtration) metal concentrations, since the former (i.e., the total metal analyses performed without filtration) may falsely suggest extensive groundwater contamination (possibly due to the presence of naturally occurring metals associated with suspended solids).

Guidelines for the selection of analytical methods are offered elsewhere in the literature (e.g., CCME, 1993). Usually there are several methods available for most environmental analytes of interest. Some analytes may have up to a dozen methods to select from. On the other hand, some analytes may have no proven methods available. In the latter case, it usually means that some of the specific isomers that were selected as representative compounds for environmental pollution have not been verified to perform acceptably with any of the commonly used methods (CCME, 1993).

3.3.2   The Health and Safety Plan

Contaminated sites, by their nature and definition, contain concentrations of chemicals that may be harmful to a variety of human population groups. One significant group potentially at risk from site contamination is the field crew who enters the contaminated site to collect samples and/or to monitor the extent of contamination. To minimize risks to site workers as a result of exposure to site contamination, health and safety issues must always be addressed as part of the site characterization activities. Proper planning and execution of safety protocols will help protect the site investigation team from accidents and needless exposure to hazardous or potentially hazardous chemicals. Protecting the health and safety of the field investigation team, as well as the general public, is indeed a major concern during the investigation of potentially contaminated sites.

The objective of the Health and Safety Plan (HSP) is to specify safety precautions needed to protect the populations potentially at risk during on-site activities. Consequently, a site-specific HSP should be prepared and implemented prior to the commencement of any work activities at potentially contaminated sites.

The HSP should be developed to conform with all the requirements for occupational safety and health, and also with applicable national, state, and local laws, rules, regulations, statutes, and orders as necessary to protect all populations potentially at risk. Furthermore, all personnel involved with on-site activities should have received adequate training, and there should be a contingency plan in place that meets all safety requirements. For instance, in the U.S., the HSP developed and implemented in the investigation of a potentially contaminated site should be in full compliance with all the requirements of the Occupational Saftey and Health Administration (OSHA) (i.e.,

---

BOX 3.5
Elements of a Site-Specific Health and Safety Plan

- Description of known hazards and risks associated with site activities (i.e., a health and safety risk analysis for existing site conditions, and for each site task and operation)
- Listing of key personnel and alternates responsible for site safety, response operations, and public protection
- Description of the levels of protection to be worn by investigative personnel and visitors to the site
- Delineation of work areas
- Establishment of procedures to control site access
- Description of decontamination procedures for personnel and equipment
- Establishment of site emergency procedures, including emergency medical care for injuries and toxicological problems
- Development of medical monitoring program for personnel (i.e., medical surveillance requirements)
- Establishment of procedures for protecting workers from weather-related problems
- Specification of any routine and special training required for personnel responding to environmental or health and safety emergencies
- Definition of entry procedures for confined spaces
- Description of requirements for environmental surveillance program
- Description of the frequency and types of air monitoring, personnel monitoring, and environmental sampling techniques and instrumentation to be used

---

OSHA: 29 CFR 1910.120), the requirements of USEPA (i.e., EPA: Orders 1420.2 and 1440.3), and indeed any other relevant state or local laws, rules, regulations, statutes, and orders necessary to protect the populations potentially at risk. Also, all personnel involved with on-site activities should have received the 40-hour OSHA Hazardous Waste Operations and Emergency Response Activities (HAZWOPER) training, including the 8-hour refresher course, where necessary.

Box 3.5 contains the relevant elements of a HSP that will satisfy the general requirements of a safe work activity (CDHS, 1990; USEPA, 1987). Appendix D of this book provides a generic example and format of a typical HSP, that could be tailored to the needs of a specific site.

3.3.2.1 Levels of Protection

Health and safety data are generally required to establish the level of protection needed for the site investigation crew entering potentially contaminated sites. Such data are also used to determine if there should be immediate concern for any population living in proximity of the site. Typically, protection at contaminated sites is categorized into four general levels:

1. **Level A:** Highest level of respiratory, skin, eye, and mucous membrane protection. This level is used if a chemical substance has been identified at concentrations that warrant using fully encapsulating equipment, or the chemical substance presents a high degree of contact hazard that warrants using fully encapsulating equipment. Work performed in a confined or poorly ventilated area also requires Level A protection until conditions change and a lower level of protection is appropriate.
2. **Level B:** Highest level of respiratory protection, but lesser level of skin and eye protection. This is the minimum level recommended during initial visits to a site until the nature of the site hazards are determined to demand less protection.
3. **Level C:** Appropriate for situations where criteria for using air-purifying respirators are met, but skin and eye exposure is unlikely. Generally, hazardous airborne substances are known and their concentrations have been measured.
4. **Level D:** No special safety equipment required other than those typically used at any construction site.

In general, safety plans should include requirements for hard hats, safety boots, safety glasses, respirators, self-contained breathing apparatus, gloves, and hazardous materials suits, if needed. In addition, personal exposure monitoring and/or monitoring ambient air concentrations of some chemicals may be necessary to meet safety regulations. Details of specific items of required safety equipment are discussed elsewhere (e.g., OBG, 1988).

The health and safety officer establishes the level of protection required and determines whether the level should be advanced or reduced. For most of the typical site activities conducted in the U.S., direct worker contact with hazardous materials in soil can be mitigated by using Health and Safety "Level-D" personal protective equipment — consisting of coveralls, safety boots, glasses, and a hard hat. To protect workers from unacceptable levels of airborne materials, at least "Level-C" equipment that includes a full-facepiece air-purifying respirator will be required. In certain other cases, worker exposure to toxic materials will be such as to warrant "Level-B" or even "Level-A" equipment, in order to provide yet greater levels of protection against exposure.

3.3.3   The Investigation-Derived Waste Management Plan

Investigation-derived wastes (IDWs) are those wastes generated during site characterization and remedial activities. There are several ways by which IDWs may be produced, including drill cuttings or core samples from soil boring or monitoring-well installations, drilling muds, purge water removed from sampling wells before groundwater samples are collected, water, solvents or other fluids used to decontaminate field equipment, goundwater and surface water samples that must be disposed of after analysis, and waste produced by on-site pilot-scale facilities constructed to test technologies best suited for remediation of a contaminated site. Other IDWs may result from disposable sampling equipment (DE), and disposable personal protective equipment (PPE).

The objective of an IDW management plan is to specify procedures needed to address the handling of both hazardous and nonhazardous IDWs. The site-specific procedures should prevent contamination of clean areas and should comply with existing regional and/or local regulations. Specifically, the IDW plan should include the characterization of IDW, delineation of any areas of contamination, and the

---

BOX 3.6
Requirements for the IDW Management Process

- Characterize IDW through the use of existing information (manifests, Material Safety Data Sheets [MSDS], previous test results, knowledge of the waste generation process, and other relevant records) and best professional judgment
- Leave a site in no worse condition than existed prior to the investigation
- Remove those wastes that pose an immediate threat to human health or the environment
- Delineate an "area of contamination" unit for leaving on-site wastes that do not require off-site disposal or extended above-ground containerization (e.g., RCRA hazardous soil cuttings)
- Comply with all regulatory requirements (e.g., federal and state ARARs) to the extent practicable
- Carefully plan and coordinate the IDW management program (e.g., containerize and dispose of RCRA hazardous groundwater, decontamination fluids, and PPE or DE [if generated in excess of 100 kg/month] at RCRA Subtitle C facilities; *but* leave RCRA nonhazardous soil cuttings, groundwater, and decontamination fluids — preferably without containerization and testing — on-site)
- Minimize the quantity of wastes generated

---

identification of waste disposal methods. Ultimately, the site manager should select investigation methods that minimize the generation of IDWs.

An IDW management plan describing the storage, treatment, transportation, and disposal of any materials (both hazardous and nonhazardous) generated during a site characterization activity should be included in the project workplan. The most important elements of the IDW management approach are summarized in Box 3.6 (USEPA, 1991). To the extent practicable, the handling, storage, treatment, or disposal of any IDWs produced during site characterization and remedial activities must satisfy all regulatory requirements and stipulations that are applicable or relevant and appropriate to the site location (e.g., federal and state ARARs under EPA programs). The procedures must also satisfy any limit requirements on the amount and concentration of the hazardous substances, pollutants, or contaminants involved.

To handle IDWs properly, the site manager must, among other things, determine the waste types (e.g., soil cuttings, groundwater, decon fluids, PPE, or DE), the waste characteristics, and the quantities of anticipated wastes. Minimizing the amount of wastes generated during a site characterization activity ultimately reduces the number of IDW handling problems and costs for disposal. Insofar as possible, provisions should be made for the proper handling and disposal of IDWs on-site. In general, most regulatory agencies do not recommend removal of IDWs from the site of origin, especially in situations where the wastes do not pose any immediate threat to human health or the environment. This is because removing wastes from such sites usually would not benefit human health and the environment, and could result in an inefficient spending of a significant portion of the total funds available for the site characterization and corrective action programs.

3.3.3.1 An Illustrative Decision Process for Screening IDWs

For illustrative purposes, consider an inactive site located in Southern California — at which site characterization activities are underway. Initial sampling and analysis of some drummed IDWs indicated several contaminants in both soil and groundwater samples. The question has been raised as to whether these IDWs should be put back on the site, or if they should be trucked out for off-site disposal.

The goal here is to present a decision process that will help manage the IDWs in a cost-effective manner. The process involved should ensure that the resulting decision or action does not increase site risks, and that it minimizes the quantity of IDWs that require off-site handling.

The Decision Process

The following discussion represents a screening procedure that will facilitate decisions on whether or not leaving soil cuttings and purged groundwater on-site will present significantly increased site risks. The rationale and justification for the proposed decision process include the following:

- Several important elements of an effective IDW management approach can be fulfilled by adopting a systematic process that allows cost-effective decisions to be made without compromising the technical effectiveness of disposal options selected to address the IDWs from the site.
- Typically, leaving IDWs on-site results in a more cost-effective way to manage such wastes without increasing risks. Based on this premise, appropriate soil IDWs can be backfilled into shallow pits or spread around the locations where the wastes had come from; appropriate liquid IDWs can be discharged and allowed to infiltrate into soils at the site.
- Should a particular site be considered a candidate for "no further action" (NFA) or closure, the proposed methodology will not reduce the likelihood of an NFA or closure decision. Conversely, returning IDWs to a site for which remediation is a likely future activity will neither affect the feasibility study program nor the remedial action selected.

Based on these rationales, one can use the proposed approach as a guide to screen and categorize the IDWs so that appropriate (i.e., technically justifiable and cost-effective) disposal practices can be selected. The methodologies involved will help determine if liquid IDWs can be discharged at the site to grade or into unlined impoundments, or if such wastes need special treatment and handling. The process should also help determine whether solid IDWs (i.e., soil cuttings) can be put back and spread out at the site of origination, or if such wastes should be disposed of in a Class III, Class II, or Class I landfill.

Figure 3.1 shows a flow chart for the decision process proposed for use as a guide in the screening of IDWs containing several chemical constituents. Foremost, it should be determined whether or not the IDWs constitute hazardous wastes under prevailing local regulations — in this case the *California Code of Regulations* (CCR). Title 22 of the CCR establishes compound-specific contentration limits (i.e., the Soluble Threshold Limit Concentrations [STLCs] and the Total Threshold Limit Concentrations [TTLCs]) for selected toxic substances, whereas Title 23 CCR contains regulations related to discharges of wastes to land. In general, the material tested is of concern if any of the "extractable" concentrations (as determined by the Waste

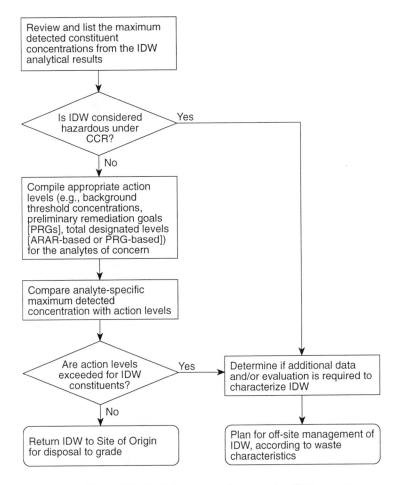

Figure 3.1   Decision process for screening IDWs.

Extraction Test [WET]) of its toxic constituents (in mg/L) equals or exceeds the STLC and/or if any of the total concentrations of its toxic constituents (in mg/kg) equals or exceeds the TTLC (CRWQCB, 1989). As appropriate, IDWs determined to be hazardous wastes may be transferred to a RCRA waste-handling facility for further management actions, or such wastes may be handled at other TSDFs.

For IDWs that are not considered hazardous following the CCR screening, further screening is conducted to determine if the wastes can be left at the site, based on appropriate regulatory limits or action levels (such as preliminary remediation goals [PRGs], background threshold concentrations, etc.). Consequently, for the chemicals that remain after the initial screening, analytical levels can be compared to local background threshold concentrations (where available), or to appropriate PRGs (e.g., the U.S. EPA Region IX PRGs), or to "total designated levels" (derived from ARARs or comparable regulatory limits, environmental attenuation factors, and leachability factors for the waste constituents [CRWQCB, 1989]). In general, where no standards are available, contaminant action levels can be established using the best scientifically justifiable information and professional judgment made on a site- and matrix-specific basis. If the contaminant concentrations in the IDWs fall within the range provided by the action levels (viz., background thresholds, PRGs, total designated levels, etc.), then an "acceptable" contaminant concentration is indicated, which should allow the IDWs to be put back on-site.

Use of the proposed approach will enable logical and consistent waste management decisions to be made, and will help determine which categories of IDWs can be disposed of at the site of origination. The overall decision process will also allow a systematic determination to be made as to whether an IDW contains hazardous substances, and whether the hazardous substances are present in such amounts as to be of significant concern, or to constitute RCRA hazardous wastes, or to be considered as Titles 22 and 23 CCR wastes, or to be perceived as wastes requiring regulation under other statutes. It is expected that this type of site-specific and matrix-specific decision process used to screen IDWs for on-site disposal will result in a more cost-effective way to manage such wastes than would otherwise be the case.

Proposed Contaminant Limits for an Example Problem

Tables 3.3A and 3.3B show the contaminant screening limits proposed for use as a guide in the screening of IDWs from the hypothetical site contaminated by a suite of chemicals shown in these tables. By comparing analytical results of the IDWs with the background thresholds, PRGs, and/or total designated levels shown in these tables, a decision can be made on the "status" of specific IDWs. If the contaminant concentrations in the IDWs fall within the range provided by the action levels, then, in general, the IDWs may be put back on to the site.

3.3.4 The Site Activity Plan

The objective of the site activity plan is to review and critically evaluate previous site investigation activities, establish goals for proposed new activities, describe the proposed site activities (to include their relevance, possible impact, execution criteria, and associated logistical requirements), establish procedures/methods to be followed during the execution of site activities, and to confirm the format for reporting the results of site activities. Box 3.7 lists several specific elements required as part of a detailed site activity workplan (USEPA, 1985, 1989b).

3.3.5 The Quality Assurance/Quality Control Plan

Quality assurance (QA) refers to a system for ensuring that all information, data, and resulting decisions compiled from an investigation (e.g., monitoring and sampling tasks) are technically sound, statistically valid, and properly documented. The QA program consists of a system of documented checks used to validate the reliability of a data set.

Quality control (QC) is the mechanism through which quality assurance achieves its goals. Quality-control programs define the frequency and methods of checks, audits, and reviews necessary to identify problems and corrective actions, thus verifying product quality. All QC measures should be performed for at least the most sensitive chemical constituents from each sampling event/date.

A detailed quality assurance/quality control (QA/QC) plan, describing specific requirements for QA and QC of both laboratory analysis and field sampling/analysis,

Table 3.3A Proposed Contaminant Limits for Screening Soil Cutting Investigation-Derived Wastes (IDWs) for On-Site Disposal to a Hypothetical Site Located in Southern California

	Screening Levels for Solid Wastes (Soil Cuttings)		
Analyte/Compound	Action Level Based on the U.S. EPA Region IX Preliminary Remediation Goal (mg/kg)	Total Designated Level Based on ARARs (as an Action Level) (mg/kg)	Local Background Threshold Concentration (mg/kg)
Metals			
Aluminum	100,000	10,000	25,000
Antimony	820	60	0.4
Arsenic	3.3	500	30
Barium	100,000	10,000	500
Beryllium	1.3	40	5
Boron	92,000	NA	25
Cadmium	490	100	20
Calcium	NA	NA	33,000
Chromium (total)	1,600	500	140
Chromium (VI)	NA	500	NA
Cobalt	NA	8,000	20
Copper	76,000	10,000	180
Iron	NA	3,000	100,000
Lead	NA	500	22
Magnesium	NA	NA	14,000
Manganese	1,000	500	1,900
Mercury	610	20	1.5
Molybdenum	1,000	35,000	80
Nickel	41,000	1,000	500
Potassium	NA	NA	10,000
Selenium	1,000	100	0.4
Silver	1,000	500	2
Sodium	NA	20,000	11,000
Thallium	NA	20	3.5
Vanadium	14,000	2,400	70
Zinc	100,000	50,000	2,000

Table 3.3A (continued) Proposed Contaminant Limits for Screening Soil Cutting Investigation-Derived Wastes (IDWs) for On-Site Disposal to a Hypothetical Site Located in Southern California

Analyte/Compound	Screening Levels for Solid Wastes (Soil Cuttings)		
	Action Level Based on the U.S. EPA Region IX Preliminary Remediation Goal (mg/kg)	Total Designated Level Based on ARARs (as an Action Level) (mg/kg)	Local Background Threshold Concentration (mg/kg)
Volatiles			
1,1,1-Trichloroethane	300	20	NA
1,1,2-Trichloroethane	5.1	3.2	NA
2-Butanone (methyl ethyl ketone)	5,200	NA	NA
4 - Methyl-2-pentanone (MIBK)	1,590 (estimate)	NA	NA
Acetone	13,000	70	NA
Benzene	4.6	0.1	NA
Carbon disulfide	74	NA	NA
cis-1,2-Dichloroethene	300	0.6	NA
Dichloromethane (methylene chloride)	39	10	NA
Ethyl Benzene	310	68	NA
Tetrachloroethylene(PCE)	58	0.5	NA
Toluene	280	100	NA
Trichloroethylene(TCE)	25	0.5	NA
Trichloromethane(Chloroform)	1.6	10	NA
Xylenes (total)	99	175	NA
Semivolatiles			
2-Methylnaphthalene	NA	0.7 (based on PQL)	NA
4-Methylphenol	5,100	NA	NA
Anthracene	1.9	960	NA
Benzo(a)anthracene	3.9	0.01	NA
Benzo(a)pyrene	0.39	0.02	NA
Benzo(b)fluoranthene	3.9	0.02	NA
Benzo(g,h,i)perylene	NA	0.00028	NA

Benzo(k)fluoranthene	3.9	0.00028	NA
Benzoic acid	100,000	NA	NA
Bis(2-ethylhexyl)phthalate	200	0.4	NA
Butylbenzylphthalate	100,000	10	NA
Carbazole	140	NA	NA
Chrysene	390	0.02	NA
Dibenzo(a,h)anthracene	0.39	0.03	NA
Dibenzofuran	NA	0.001 (estimate)	NA
Diethylphthalate	100,000	2,300	NA
Di-n-butylphthalate	204,000 (estimate)	3,000 (estimate)	NA
Di-n-octylphthalate	20,000	NA	NA
Fluoranthene	41,000	30	NA
Fluorene	28	130	NA
Indeno(1,2,3-c,d)pyrene	3.9	0.04	NA
Naphthalene	80	NA	NA
Pentachlorophenol	24	0.1	NA
Phenanthrene	NA	10 (estimate)	NA
Pyrene	31,000	96	NA
Pesticides/Herbicides/PCBs			
4,4'- DDT	8.4	0.001	NA
Aldrin	0.17	0.005	NA
alpha - BHC (HCH alpha)	0.45	0.07	NA
gamma - BHC (Lindane)	2.2	0.4	NA
Polychlorinated Biphenyls (PCBs)	0.37	0.05	NA

Notes: This list represents analytical parameters with results above their detection limits.

NA = not available/applicable.

PQL = sample/practical quantitation limit.

USEPA Region IX Preliminary Remediation Goals (PRGs) are reported for industrial soils.

Total Designated Levels are estimated by applying an attenuation factor (of 100 for inorganics and 10 for organics) and a leachability factor of 1–100 to appropriate regulatory limits.

Table 3.3B   Proposed Contaminant Limits for Screening Goundwater Investigation-Derived Wastes (IDWs) for On-Site Disposal to a Hypothetical Site Located in Southern California

	Screening Levels for Liquid Wastes (Groundwater)		
Analyte/Compound	Action Level Based on the U.S. EPA Region IX Preliminary Remediation Goals (mg/L)	Total Designated Level Based on ARARs and Regulatory Limits (mg/L)	Local Background Threshold Concentration (mg/L)
Metals			
Aluminum	3,700	100	1.2
Antimony	1.5	0.6	0.026
Beryllium	0.002	0.4	0.0003
Cadmium	1.8	1	0.01
Chromium (total)	3,700	5	0.06
Chromium (VI)	18	NA	NA
Iron	NA	30	15
Lead	0.4	5	0.01
Manganese	18	5	2.6
Nickel	73	10	9.3
Selenium	18	1	0.05
Thallium	0.29 (estimate)	0.2	0.005
Volatiles			
1,1-Dichloroethane	10	0.05	NA
1,1-Dichloroethene	0.00068	0.06	NA
2-Butanone (MEK)	25	NA	NA
Acetone	7.7	NA	NA
cis-1,2-Dichloroethene	0.77	0.06	NA
Dichloromethane (methylene chloride)	0.062	0.05	NA

Toluene	9.3	10	NA
trans-1,2-Dichloroethene	1.5	0.1	NA
Trichloroethylene(TCE)	0.025	0.05	NA
Trichlorofluoromethane	17	1.5	NA
Vinyl Chloride	0.00028	0.005	NA
Semivolatiles			
4-Chloro-3-methylphenol	NA	0.02 (based on PQL)	NA
4-Methylphenol	1.8	NA	NA
Benzoic acid	1,500	NA	NA
Benzo(a)pyrene	0.00012	0.002	NA
Benzo(b)fluoranthene	0.0012	0.002	NA
Benzo(g,h,i)perylene	NA	0.00003	NA
Bis(2-ethylhexyl)phthalate (DEHP)	0.061	0.04	NA
Di-n-butylphthalate	37 (estimate)	NA	NA
Fluoranthene	15	3	NA
Fluorene	3.1	13	NA
Pyrene	11	9.6	NA

Notes: This list represents analytical parameters that exceeded state and/or federal MCLs for groundwater IDWs, or where the analytical results are above their detection limits where no MCLs exist.

NA = not available/applicable.

PQL = sample/practical quantitation limit.

USEPA Region IX Preliminary Remediation Goals (PRGs) are reported for tap water; the groundwater action levels are estimated by applying an attenuation factor (of 100 for inorganics and 10 for organics) to the PRG.

Total Designated Levels are estimated by applying an attenuation factor (of 100 for inorganics and 10 for organics) to an ARAR or other regulatory limit.

BOX 3.7
Elements of a Site Activity Plan

- Site background information summary
- Workplan objectives
- Personnel required (including training, organization, and equipment)
- Nonstandard equipment description and contract services
- Hazards expected (physical/chemical) and project impact
- Site location/accessibility (including effect on sampling and remediation plans)
- Project schedule and budget
- Equipment and personnel mobilization/demobilization

---

should be part of the site characterization project workplan. The plan requirements will typically relate to, but not be limited to the following: the use of blanks, spikes, and duplicates; sampling procedures, cleaning of sampling equipment, storage, transportation, DQOs, chain-of-custody, and methods of analysis. The practices to be followed by the site investigation team and the oversight review which will ensure that DQOs are met, must be clearly described in the QA/QC plan.

Some aspects of the field program can and should be subjected to a quality assessment survey. This is accomplished by submitting sample blanks (alongside the environmental samples) for analysis on a regular basis. The various blanks and checks that are recommended as part of the quality assurance plan include the following (CCME, 1994):

- *Trip Blank:* required to identify contamination of bottles and samples during travel and storage. To prepare the trip blank, the laboratory fills containers with contaminant-free water and delivers them to the sampling crew; the field sampling crew subsequently ship and store these containers with the actual samples obtained from the site characterization activities. It is recommended to include one trip blank per shipment, especially where volatile contaminants are involved.
- *Field Blank:* required to identify contamination of samples during collection. This is prepared in the same manner as the trip blank (i.e., the laboratory fills containers with contaminant-free water and delivers them to the sampling crew); subsequently, however, the field sampling crew expose this water to site air (just like the actual samples obtained from the site characterization activities). It is recommended to include one field blank per site or sampling event/day.
- *Equipment Blanks:* required to identify contamination from well and sampling equipment. To obtain an equipment blank, casing materials and sampling devices are flushed with contaminant-free water, which is then analyzed. Typically, equipment blanks become important only if a problem is suspected, such as using a bailer to sample from multiple wells.
- *Blind Replicates:* required to identify laboratory variability. To prepare the blind replicate, a field sample is split into three containers and labeled as

different samples before shipment to the laboratory for analyses. It is recommended to include one blind replicate per day, or an average of one per 10 to 25 samples where large numbers of samples are involved.

- *Spiked Samples:* required to identify errors due to sample storage and analysis. To obtain the spiked sample, known concentration(s) are added to the sample bottle and then analysed. It is recommended to include one spiked sample per site, or an average of one per 25 samples where large numbers of samples are involved.

Data generated during a site characterization will provide a basis for site restoration decisions. The data should therefore give a valid representation of the true site conditions. The development and implementation of a QA/QC program during a sampling and analysis program is critical to obtaining reliable analytical results. The soundness of the QA/QC program has a particularly direct bearing on the integrity of the environmental sampling and also the laboratory work. Thus, the general design process for an adequate QA/QC program, as discussed elsewhere in the literature (e.g., CCME, 1994; USEPA, 1987b, 1987c), should be followed religiously.

3.4   THE IMPLEMENTATION OF A SITE INVESTIGATION PROGRAM

Site investigations are conducted in order to characterize site conditions and environmental samples collected from the site. Collected samples will generally be submitted to a certified analytical laboratory for analysis. The sample characterization considers detailed information in relation to the following:

- Field operations (including the role of individuals, chain-of-custody procedures, maintaining field log book, and site monitoring)
- Sampling locations and rationale
- Field sampling and mapping
- Field quality control samples
- Decontamination procedures
- Analytical requirements and sample handling
- Sample delivery
- Data compilation and analyses
- Summary site evaluations

Essentially, the characterization of environmental samples helps determine the specific type(s) of contaminants, their abundance/concentrations, the lateral and vertical extent of contamination, the volume of materials involved, and the background contaminant levels for native soils and water resources in the vicinity of the site.

3.4.1   Initial Site Inspection

Geophysical surveys, limited field screening, or limited field analyses may be performed during an initial site inspection. These types of preliminary screening activities may help determine the variability of the media, provide general interest

---

BOX 3.8
Tasks Required of an Initial Site Inspection

- Utilizing field analytical procedures, compile data on volatile chemical contaminants, radioactivity, and explosivity hazards in order to determine appropriate health and safety levels
- Determine if any site condition could pose an imminent danger to public health and safety
- Confirm the information contained in previous documents
- Record apparent discrepancies and observable data that may be missing in previous documents
- Update site conditions if undocumented changes have occurred
- Perform an inventory of possible off-site sources of contamination
- Obtain information on location of access routes, sampling points, and site organizational requirements for the field investigation

---

background information, or determine if the site conditions have changed in comparison to what may have been reported in previous investigations.

Typically, the goals of the initial site inspection include the accomplishment of several tasks, as indicated in Box 3.8. Available information should be carefully reviewed and evaluated to provide the foundation for executing additional on-site activities. The preliminary data receive confirmation from observations made during site visits. The types of information generated serve as a useful database for project scoping. The review and initial site visit are used in a preliminary interpretation of site conditions.

3.4.2 Sampling and Sample Handling Procedures

The collection of representative samples generally involves different procedures for different situations. USEPA (1987), among others, discusses several sampling methods that can be used in various types of situations. In every situation, all sampling equipment is cleaned using a nonphosphate detergent, a tap-water rinse, and a final rinse with distilled water prior to a sampling activity. Decontamination of equipment is necessary so that sample results do not show false positives. Decontamination water generated from the site activities (e.g., during decontamination of hand-auger and soil-sampling equipment) is transferred into containers and sampled for analysis.

All sampling should be conducted in a manner that maintains sample integrity and encompasses adequate quality assurance and control. Specific sample locations should be chosen such that representative samples can be collected. Also, samples should be collected from locations with visual observations of surface contamination, so that possible worst-case conditions may be identified. The use of field blanks and standards, and also spiked samples, can account for changes in samples which occur after sample collection. The following general statements apply to most sampling efforts:

- Refrigeration and protection of samples should minimize the chemical alteration of samples prior to analysis.
- Field analyses of samples will effectively avoid biases in determinations of parameters/constituents which do not store well (e.g., gases, alkalinity, pH).
- Field blanks and standards will permit the correction of analytical results for changes which may occur after sample collection (i.e., during preservation, storage, and transport). That is, field blanks and standards enable quantitative correction for biases due to systematic errors arising from handling, storage, transport, and laboratory procedures.
- Spiked samples and blind controls provide the means to correct combined sampling and analytical accuracy or recoveries for the actual conditions to which the samples have been exposed.

In general, sampling equipment should be constructed of inert materials. Collected soil samples are placed in resealable plastic bags; fluid samples are placed in air-tight glass or plastic containers. When samples are to be analyzed for organic constituents, glass containers are required. The samples are then labeled with an indelible marker. Each sample bag or container is labeled with a sample identification number, sample depth (where applicable), sample location, date and time of sample collection, preservation, and possibly a project number and the sampler's initials. A chain-of-custody form listing the sample number, date and time of sample collection, analyses requested, a project number, and persons responsible for handling the samples is then completed. Samples are generally kept on ice prior to and during transport/shipment to a federal- or state-certified laboratory for analysis; completed chain-of-custody records should accompany the samples to the laboratory. Further details of the appropriate technical standards for sampling and sample handling procedures can be found in the literature (e.g., CCME, 1994; CDHS, 1990).

3.4.3 Monitoring Programs

Decisions regarding monitoring network design and operation are generally made in the light of available data. To a great extent, monitoring can be considered as an evolutionary process that should be refined as more relevant information is obtained. In fact, effective monitoring efforts are both dynamic and flexible, and this should be explicitly indicated in the site characterization plan. Overall, it is prudent to specify monitoring programs that will permit the collection of high-quality, representative data for the most sensitive chemical constituents of interest.

Most monitoring programs designed for contaminated site problems are directed at groundwater investigations and, to some extent, surface water quality assessment. The practical elements of a viable long-term groundwater monitoring effort typically will consist of an evaluation of the hydrogeologic setting, proper well placement and construction, evaluation of well performance and purging strategies, and the execution of effective sampling protocols (to include the selection of appropriate sampling mechanisms and materials, as well as sample collection and handling procedures) (USEPA, 1985). Most of these elements, or variations thereof, are also applicable to monitoring programs in other environmental media.

REFERENCES

BSI. 1988. Draft for Development, DD175: 1988. *Code of Practice for the Identification of Potentially Contaminated Land and its Investigation.* British Standards Institution, London, U.K.

CCME (Canadian Council of Ministers of the Environment). 1993. "Guidance Manual on Sampling, Analysis, and Data Management for Contaminated Sites." Volume I: Main Report (Report CCME EPC-NCS62E), and Volume II: Analytical Method Summaries (Report CCME EPC-NCS66E). The National Contaminated Sites Remediation Program, Canadian Council of Ministers of the Environment, Winnipeg, Manitoba.

CCME. 1994. Subsurface Assessment Handbook for Contaminated Sites. Canadian Council of Ministers of the Environment (CCME), The National Contaminated Sites Remediation Program (NCSRP), Report No. CCME-EPC-NCSRP-48E. Canadian Council of Ministers of the Environment, Ottawa, Canada.

CDHS. 1990. Scientific and Technical Standards for Hazardous Waste Sites. Prepared by the California Department of Health Services, Toxic Substances Control Program, Technical Services Branch, Sacramento.

CRWQCB. 1989. The Designated Level Methodology for Waste Classification and Cleanup Level Determination. Staff Report, Central Coast Region, California Regional Water Quality Control Board (June 1989). Sacramento.

Driscoll, F.G., 1986. *Groundwater and Wells.* Johnson Division, St. Paul, MN.

Jolley, R.L. and R.G.M. Wang (Eds.). 1993. *Effective and Safe Waste Management: Interfacing Sciences and Engineering With Monitoring and Risk Analysis.* Lewis Publishers, Boca Raton, FL.

Keith, L.H., (Ed.). 1988. *Principles of Environmental Sampling.* American Chemical Society, Washington, D.C.

Keith, L.H., 1991. *Environmental Sampling and Analysis—A Practical Guide.* Lewis Publishers, Boca Raton, FL.

OBG (O'Brien and Gere Engineers, Inc.). 1988. *Hazardous Waste Site Remediation: The Engineer's Perspective.* Van Nostrand Reinhold, New York.

USEPA. 1985. Characterization of Hazardous Waste Sites: A Methods Manual, Volume 1, Site Investigations. U.S. Environmental Protection Agency, Environmental Monitoring Systems Laboratory, EPA-600/4-84-075. Las Vegas, NV.

USEPA. 1987. RCRA Facility Investigation (RFI) Guidance, EPA/530/SW-87/001, U.S. Environmental Protection Agency, Washington, D.C.

USEPA. 1988a. Guidance for Conducting Remedial Investigations and Feasibility Studies Under CERCLA. EPA/540/G-89/004. OSWER Directive 9355.3-01, Office of Emergency and Remedial Response, U.S. Environmental Protection Agency, Washington, D.C.

USEPA. 1988b. Interim Report on Sampling Design Methodology. Environmental Monitoring Support Laboratory, EPA/600/X-88/408. U.S. Environmental Protection Agency, Las Vegas, NV.

USEPA (U.S. Environmental Protection Agency). 1989a. Ground-Water Sampling for Metals Analyses. Office of Solid Waste and Emergency Response. EPA/540/4-89-001.

USEPA (U.S. Environmental Protection Agency). 1989b. Soil Sampling Quality Assurance User's Guide. 2nd ed., EPA/600/8-89/046, Experimental Monitoring Support Laboratory (EMSL), ORD, U.S. Environmental Protection Agency, Las Vegas, NV.

USEPA. 1989c. User's Guide to the Contract Laboratory Program. Office of Emergency and Remedial Response, OSWER Dir. 9240.0-1. U.S. Environmental Protection Agency, Washington, D.C.

USEPA. 1991. Management of Investigation-Derived Wastes During Site Inspections. Office of Emergency and Remedial Response, EPA/540/G-91/009, May 1991, U.S. Environmental Protection Agency, Washington, D.C.

WPCF. 1988. *Hazardous Waste Site Remediation: Assessment and Characterization.* Spec. Publ. Technical Practice Committee, Water Pollution Control Federation, Alexandria, VA.

4   CORRECTIVE ACTION ASSESSMENT OF CONTAMINATED SITE PROBLEMS

The primary objective of any corrective action response program is to ensure public safety and welfare by protecting human health, the environment, and public and private properties. Corrective action programs for contaminated sites may vary greatly, ranging from a "no-action/no-(immediate)-costs" alternative to a variety of extensive and costly remediation or site restoration options. The ability to select an appropriate and cost-effective course of action will generally depend on a careful assessment of both short- and long-term risks potentially posed by the site; it also depends on the site-specific restoration goal that is established and accepted for the site.

Typically, several pertinent questions are asked during the planning, development, and implementation of corrective action programs that are directed at restoring contaminated sites (Box 4.1). These, and indeed several other issues, are relevant to making risk reduction and risk management decisions for potentially contaminated sites (BSI, 1988; Cairney, 1993; Jolley and Wang, 1993; USEPA, 1985a, 1987a, 1987b, 1988b, 1989a, 1991; WPCF, 1988). This chapter enumerates several significant elements and factors that will help answer these questions and which will ultimately affect the type of corrective action decision accepted for a contaminated site problem.

4.1 THE CORRECTIVE ACTION ASSESSMENT PROCESS

Corrective action assessments are generally designed with the goal to minimize potential negative impacts associated with contaminated site problems. Invariably, site assessments are a primary activity in the overall corrective action assessment process. The objective of the site assessment is to determine the nature and extent of potential impacts from the release or threat of release of hazardous substances.

A site investigation effort, which is a major component of the site assessment activity, aims at collecting representative samples from the potentially contaminated site. Depending on the adequacy of historical data and the sufficiency of details about the likely contaminants at a site, sampling programs can be designed to search for specific chemical constituents that become indicator parameters for the sample analyses. Where specific information about historical uses of the site is lacking, a more comprehensive sampling and analytical program will generally be required; in this case, the sampling and analysis program may be carried out in phases — moving from

BOX 4.1
Questionnaire Chart for Corrective Action Assessments

- Do contaminant sources exist?
- Are there visible sources of contamination?
- What are the sources of the contaminants?
- Is the site potentially contaminated with hazardous or toxic chemicals? What are the toxic agents involved?
- Are there confining layers or porous layers in the soil horizon?
- Is there soil erosion or recent cuts or fills on site?
- What is the nature of drainage and surface flow patterns at the site and immediate vicinity?
- What are the site characteristics, hydrological features, meteorological or climatic factors, land-use patterns, and agricultural practices affecting the transport and distribution of site contaminants?
- What is the distribution of the chemicals over the site and vicinity?
- Are there known "hot-spots" at the site?
- What is an appropriate background or control region to use for corrective action investigations?
- Is there any area that poses an immediate and life-threatening exposure?
- What are the important transport processes and migration pathways that contribute to intermedia transfers and the spread of contamination?
- Are there present or future potential receptors that could be adversely affected by the contaminant sources? In particular, are there sensitive ecosystems or residences located downgradient, downstream, or downwind from the site?
- Are there one or more pathways through which site contaminants might migrate from the source and reach potential receptors? What are the dominant routes of exposure at the site?
- Have populations already been impacted, and/or are populations potentially at risk? What are the potential risks posed to human health and the envirmonment if no further response action is taken at the site?
- Does the risk level exceed benchmark levels specified by environmental compliance regulations? What site-specific cleanup criteria will be appropriate for the site?
- At the given concentrations at the site, which areas are considered as posing risks to the environment or surrounding populations? Which areas must therefore be remediated in order to reduce risks to an "acceptable" level?
- Are estimated risk levels low enough such that a "no-action" alternative is still protective of public health and the environment?
- What contaminants and environmental media should become the target for site remediation?

- How much contaminated material should be remediated to achieve an acceptable site restoration goal?

- Which remedial alternatives can be applied at the contaminated site in order to achieve adequate cleanup?

- Will exposure pathways be interrupted or will receptors be protected as a result of removal or remedial actions?

- What institutional control measures and risk management strategies are required in the overall corrective action decision?

---

a more general scope to specificity as adequate information becomes available on the site. The results from these activities will facilitate a complete analysis of possible contaminants present at the site. The information so obtained is used to determine current and potential future risks to human health and the environment. Ultimately, corrective actions are developed and implemented with the principal objective to protect public health and the environment.

The design of an effectual corrective action program for contaminated site problems should consider several important decision elements (Box 4.2). These basic elements should be completely evaluated and the relevant information used to support appropriate corrective action decisions (Figure 4.1). Although different levels of effort may be required for different sites, a comprehensive site assessment program is always essential for making good corrective action decisions. An adequately conducted site assessment will indeed allow a project manager to develop and select a final corrective action plan or cleanup solution that best suits the conditions at a particular site. The nature of this assessment will generally depend on the stage of investigation.

Corrective action assessment programs are generally designed to facilitate the development of scientifically based methods necessary to support corrective action decisions. The process allows a multimedia approach to site characterization and the establishment of DQOs, identifies the specific parameters for which data must be collected, identifies the preferred data gathering, handling and analytical techniques to be used, and allows project managers to develop site-specific and cost-effective cleanup solutions. In general, once quality-assured information has been compiled for a potentially contaminated site and a benchmark risk criteria established, an acceptable cleanup criteria can be determined that will be used to guide further site restoration decisions. This will subsequently form an important basis for developing a corrective action plan (CAP) for the site.

4.2   DEVELOPMENT OF CONTAMINATED SITE CONCEPTUAL MODELS

A conceptual site model (CSM) provides a structured framework for characterizing possible threats posed by potentially contaminated sites. The CSM aids in the organization and analysis of basic information relevant to the corrective action decisions about a site. Thus, the development of a comprehensive CSM is a generally recommended and vital part of the corrective action assessment for contaminated site

---

BOX 4.2
Elements of a Corrective Action Assessment Process

- Characterization of the contaminated site, including the physical setting, site geology, topography, hydrogeology, and meteorological conditions
- Identification of contaminant types and their characteristics
- Assessment of the fate and transport characteristics of site contaminants, including the identification of anticipated degradation, reaction, and/or decomposition by-products
- Determination of the critical environmental media of concern (such as air, surface water, groundwater, soils and sediments, and terrestrial and aquatic biota)
- Delineation of potential migration pathways
- Identification and characterization of potential human and ecological receptors
- Development of a conceptual representation or model of the site
- Evaluation of potential exposure scenarios, and the potential for human and ecosystem exposures
- Assessment of the environmental and health impacts of the site contaminants if they should reach critical human and ecological receptors
- Decision-making on the corrective action needs for the site
- Development of a risk management strategy for the site
- Design of effective long-term monitoring and surveillance programs as a necessary part of an overall corrective action plan

---

problems. As site characterization activities progress, the CSM may be revised as necessary and used to direct the next iteration of sampling activities.

The conceptual evaluation model is used to facilitate the assessment of the nature and extent of contamination. It also identifies potential contamination sources, potential exposure pathways, and potential human and ecological receptors. The development of CSMs typically involves the following evaluations:

- A contaminant release analysis (to determine contaminant release rates into specific environmental media over time)
- A contaminant transport and fate analysis (to provide guidance for evaluating the transport, transformation, and fate of contaminants in the environment following their release, to identify off-site areas affected by contaminant migration, and to determine contaminant concentrations in these areas)
- An exposed population analysis (to determine the likelihood of human and ecological receptors coming into contact with the contaminants of concern)
- An integrated exposure analysis (to provide guidance for calculating and integrating exposures to all populations affected by the various exposure scenarios associated with the contaminated site)

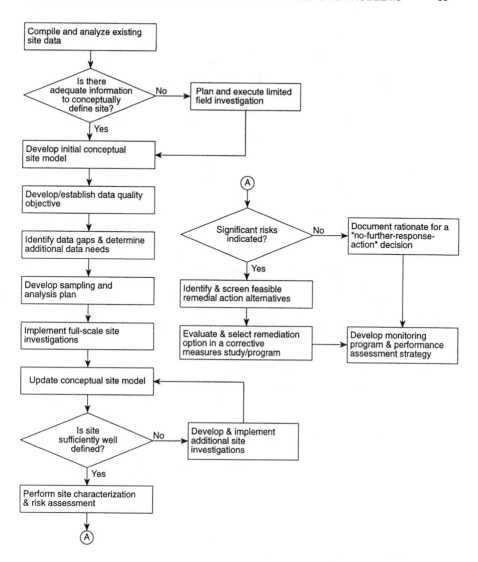

Figure 4.1   The corrective action assessment process used in the investigation of contaminated site problems.

   Site history and preliminary assessment or site inspection data generally are very useful sources of information for developing the conceptual evaluation model. In general, the CSM should include known and suspected sources of contamination, types of contaminants and affected media, known and potential migration pathways, potential exposure routes, target receiving media, and known or potential human and ecological receptors. Such information can be used to develop a conceptual understanding of the contaminated site problem, so that potential risks to human health and the environment can be evaluated more completely. Ultimately, the CSM helps to identify data gaps and assists in developing strategies for data collection. This effort, in addition to assisting in the identification of appropriate sampling locations, will also assist in the screening of potential remedial options or technologies.

In a typical scenario in which there is a source release, contaminants may be transported from the contaminated media by several processes into other environmental compartments (as depicted by the diagrammatic representation in Figure 4.2). For instance, precipitation may infiltrate into soils at a contaminated site and leach contaminants from the soil as it migrates through the contaminated material and the unsaturated soil zone. Infiltrating water may continue its downward migration until it encounters the water table at the top of the saturated zone; the mobilized contaminants may be diluted by the available groundwater flow. Once a contaminant enters the groundwater system, it is possible for it to be transported to a discharge point or to a water supply well location. There also is the possibility of continued downward migration of contaminants into a bedrock aquifer system. Contaminants may additionally be carried by surface runoff into surface water bodies. Air releases of particulates and vapors present additional migration pathways for the contaminants. Consequently, all these types of scenarios will typically be evaluated as part of the corrective action assessment process for a contaminated site problem.

The information contained in a CSM is developed during the various stages of a corrective action process, and also from controlled field and laboratory experiments that are conducted in studies pertaining to the potentially contaminated site. Once synthesized, this information may be presented in several different forms such as map views (showing sources, pathways, receptors, and the distribution of contamination), cross-sectional views (illustrating sectional component hydrogeology), and tabular forms (summarizing and comparing contaminant concentrations against background thresholds, ARARs, or risk-based standards).

4.2.1   Design Requirements for the Conceptual Site Model

CSMs offer an integrated approach for assessing human and ecological population exposures to contaminants released from potentially contaminated sites. A CSM is comprised of the following basic elements:

- Identification of site contaminants
- Characterization of the source(s) of contamination
- Delineation of potential migration pathways
- Identification and characterization of both human and ecological receptors
- Determination of the type of interconnections between contaminant sources, migration pathways, and potential receptors

Relationships among these elements, as illustrated in Figure 4.3, provide a basis for testing a range of exposure hypotheses for a given site or area.

In general, the complexity and degree of sophistication of a CSM should be consistent with the complexity of the site and the amount of data available. For instance, if in the characterization of the source(s) of contamination there are multiple sites in proximity to one another such that it is almost impossible to differentiate between the individual source(s), the affected sites may be clustered together into a single zone. Migration pathways and receptors can then be determined for the zone

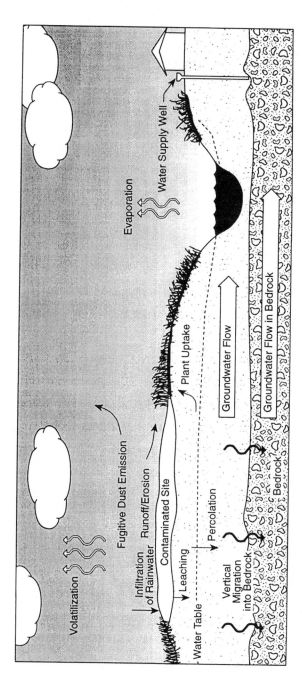

Figure 4.2   Diagrammatic representation of a contaminated site conceptual model.

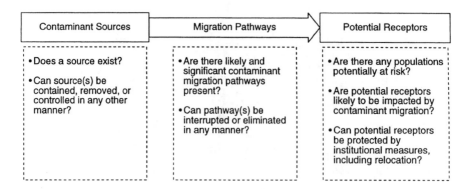

Figure 4.3 Hypothesis to test in the evaluation of the various elements of a conceptual site model.

rather than for individual sites. All assumptions and limitations in this type of situation should be clearly identifiable and justifiable.

4.2.2 Contaminant Fate and Transport Assessment

Chemicals released into the environment are affected by several complex processes and phenomena that facilitate intermedia transfers. The processes that affect the fate and transport of contaminants are identified as an important part of a corrective action assessment. The physical and chemical characteristics of constituents present at potentially contaminated sites determine the fate and transport properties, and thus the degree of migration through the environment. Some of the important properties affecting the fate and transport of chemicals in the environment are (Grisham, 1986):

- Solubility in water (which relates to leaching, partitioning, and mobility in the environment)
- Partition coefficient (relating to intermedia transfers, bioaccumulation potential, and sorption by organic matter)
- Hydrolysis (which relates to persistence in the environment or biota)
- Vapor pressure and Henry's Law constant (relating to atmospheric mobility and the rate of vaporization)
- Photolysis (which relates to persistence as a function of exposure to light)
- Half-life (relating to the degradation potential and the transformation products)

These parameters are elaborated further in Appendix C. Further details and additional parameters of possible interest are discussed elsewhere in the literature (e.g., Lyman et al., 1990; Swann and Eschenroeder, 1983).

Several site characteristics may also influence the environmental fate of chemicals, including the amount of ambient moisture, humidity levels, temperatures and wind speed; geologic, hydrologic, pedologic, and watershed characteristics; topographic features of the site and vicinity; vegetative cover of site and surrounding area; and land-use characteristics. Other factors such as soil temperature, soil moisture content, initial contaminant concentration in the impacted media, and media pH may additionally affect the release of a chemical constituent from the environmental matrix in which it is found.

4.2.2.1 Contaminant Fate and Transport Modeling

Chemical contaminants entering the environment tend to be partitioned or distributed across various environmental compartments. A good prediction of chemical concentrations in the various environmental media is essential to adequately characterize contaminated sites, the results of which can be used to support exposure and risk assessments and, therefore, corrective action decisions. Mathematical algorithms can be used to predict the potential for contaminants to migrate from a contaminated site to potential receptors, using pathways analyses concepts.

Mathematical models often serve as valuable tools for evaluating the behavior and fate of chemical constituents in the various environmental media. Transport and fate of contaminants can be predicted through the use of various methods, ranging from simple mass-balance and analytical procedures to multidimensional numerical solution of coupled differential equations. A summary listing of selected models potentially applicable to contaminant fate and transport modeling is included in Appendix E of this book; several additional models of interest, together with model selection criteria and limitations, are discussed elsewhere in the literature of exposure modeling (e.g., CCME, 1994; CDHS, 1986; USEPA, 1985b, 1987c, 1988c, 1988d). In general, the appropriateness of a particular model depends on the characteristics of the problem. The screening of models should be tied to the study goals. Indeed, the wrong choice of model could result in the generation of false information, with consequential negative impacts in any decision made thereof.

Due to the heterogeneity in environmental compartments and natural systems, models used for exposure assessments should be adequately tested, and sensitivity runs should be carried out to help determine the most sensitive and/or critical parameters considered in the evaluation. Assistance on specific questions relating to exposure modeling (such as data requirements, appropriate models, etc.) can be obtained from such oufits as the EPA's Center for Exposure Assessment (Athens, Georgia). By the use of the appropriate mathematical models, a comprehensive exposure assessment can be developed as input to corrective action assessment programs. Ultimately, the choice of appropriate models that will give reasonable indications of the contaminant behavior will help produce a representative CSM that is important to the corrective action assessment process.

4.2.2.2 Modeling the Erosion of Contaminated Soils into Surface Waters as an Example

Contaminated runoff and overland flow of toxic contaminants constitutes one source of concern for surface water contamination at uncontrolled hazardous waste sites. Surface runoff release estimation procedures (e.g., USEPA, 1988d) can be applied to such uncontrolled sites.

A hypothetical problem is discussed to demonstrate the potential impacts that contaminated sites may have on surface water resources, and to illustrate the applicability of simple contaminant transport models to support a corrective action investigation at such sites.

Background to the Hypothetical Problem

This example problem involves an abandoned industrial process facility used for handling heavy/waste oils at the XYZ site located in the eastern U.S. Petroleum products are believed to have been released from this facility, leading to surface soil contamination. An environmental site assessment conducted for the XYZ site indicated the presence of the following organic constituents in the soils: benzene, benzo(*a*)anthracene, benzo(*a*)pyrene, ethylbenzene, naphthalene, phenanthrene, phenol, and toluene.

It is apparent from the physical setting of the site that contaminated soils may be carried off-site from the XYZ property almost exclusively by erosional/overland transport flow processes. A creek that adjoins the site is potentially threatened by possible contaminant loadings. It is required by environmental laws/regulations to protect the surface water quality (from the migration of contaminated soil that may be washed into the nearby creek). However, current levels of chemicals in soil at the site would likely result in the exceedence of water quality criteria to be met for this creek, following loadings from erosion runoff. The PRPs are therefore required to clean up the site to such levels that will not significantly impact the creek after receiving runoff carrying contaminated soils from the XYZ site.

Modeling the Migration of Contaminated Soil to Surface Runoff:
Sediment Loss and Contaminant Flux Calculations

Assuming that all contamination is from adsorbed waste oil constituents present at the XYZ site, surface runoff release of chemicals can be estimated by using the modified universal soil loss equation (MUSLE) and sorption partition coefficients derived from the compounds' octanol-water partition coefficients. MUSLE allows the estimation of the amount of surface soil eroded in a storm event of a given intensity, while sorption coefficients allow the projection of the amount of contaminant carried along with the soil and the amount carried in dissolved form. The procedures used here are fully described in the literature elsewhere (e.g., DOE, 1987; USEPA, 1988d).

Soil Loss Calculation

The soil loss caused by a storm event is given by the modified universal soil loss equation (Williams, 1975; DOE, 1987):

$$Y \ (S)_E = a \ (V_r \ q_p)^{0.56} \ KLSCP$$

where:

$Y \ (S)_E$	=	sediment yield (tons per event)
a	=	conversion constant
V_r	=	volume of runoff (acre-ft)
q_p	=	peak flow rate (ft^3/s)
K	=	soil erodibility factor (commonly expressed in tons/acre/dimensionless rainfall erodibility unit)
L	=	the slope-length factor (dimensionless ratio)
S	=	the slope-steepness factor (dimensionless ratio)
C	=	the cover factor (dimensionless ratio)
P	=	the erosion control practice factor (dimensionless ratio)

Computation of V_r: The storm runoff volume generated at the case site is calculated by the following equation (Mills et al., 1982):

$$V_r = 0.083 \ A \ Q_r$$

where:

A = contaminated area ≈ 6 acres (estimated during remedial investigation)
Q_r = depth of runoff from site (in.)

The depth of runoff, Q_r, is determined by (Mockus, 1972):

$$Q_r = \frac{\left(R_t - 0.2 \ S_w\right)^2}{\left(R_t + 0.8 \ S_w\right)}$$

where:

Q_r = the depth of runoff from the site (in.)
R_t = the total storm rainfall (in.)
S_w = water retention factor (in.)

The Soil Conservation District may be consulted on a value for the total storm rainfall (R_t) to use. In this case, the 2-year 24-h event represents a typical annual storm; use of this suggested storm event yields an R_t value of approximately 3.5 in. Also, an average annual storm of between 44 and 48 in. can be expected for this area. Thus,

$$R_t = 3.5 \text{ in. and average annual rainfall} \approx 46 \text{ in.}$$

This means that the average number of average rainfall events per year $\approx \frac{46}{3.5} \approx 13$ events, yielding 920 storm events over a 70-year period.

The value of the water retention factor, S_w, is obtained as follows (Mockus, 1972):

$$S_w = \left\{ \frac{1000}{CN} - 10 \right\}$$

where:

S_w = water retention factor (in.)
CN = the Soil Conservation Services (SCS) runoff curve number (dimensionless)

The CN factor for the site is determined by the soil type at the XYZ site, its condition, and other parameters that establish a value indicative of the tendency of the soil to absorb and hold precipitation or to allow precipitation to run off the site. CN values for uncontrolled hazardous waste sites can be estimated from tables in the literature (e.g., USEPA, 1988d; Schwab et al., 1966); charts have also been developed for determining CN values (e.g., USBR, 1977) and these may be used for estimating the CN. Based on existing site conditions of fill material that is very heterogeneous, higher infiltration and relatively low to moderate runoff is anticipated at the site. An estimate of CN = 74 for the soil type is taken as conservative enough for our purposes (USEPA, 1988d). This represents a moderately low runoff potential and an above average infiltration rate of 4 to 8 mm/hr seems reasonable for the site. In this case, S_w, Q_r, and V_r are estimated to be

$$S_w = \left\{ \frac{1000}{74} - 10 \right\} = 3.51 \text{ in.}$$

$$Q_r = \frac{\left\{ 3.5 - (0.2)(S_w) \right\}^2}{\left\{ 3.5 + (0.8) S_w \right\}} = 1.24 \text{ in.}$$

$$V_r = (0.083)(5.9)(Q_r) = 0.61 \text{ acre} - \text{ft}$$

Computation of q_p — The peak runoff rate, q_p, is determined by (Haith, 1980):

$$q_p = \frac{1.01 \text{ A } R_t Q_r}{T_r \left(R_t - 0.2 S_w \right)}$$

where:

q_p = the peak runoff rate (ft^3/s)
A = contaminated area (acres)
R_t = the total storm rainfall (in.)
Q_r = the depth of runoff from the watershed area (in.)
T_r = storm duration (hr)
S_w = water retention factor (in.)

For the typical storm represented by the 2-year 24-h rainfall event suggested for this scenario, and given the following parameters:

A = 5.9 acres
R_t = 3.5 in.
Q_r = 1.24 in.
T_r = 24 h
S_w = 3.51 in.

and therefore

$$q_p = 0.38 \text{ cfs}$$

Estimation of K — The soil erodibility factors are indicators of the erosion potential of given soil types and are therefore site specific. The value of the soil erodibility factor, K, as obtained from the Soil Conservation Service, is 0.32. This compares reasonably well with estimates given by charts and nomographs found in the literature (e.g., DOE, 1987; Erickson, 1977; Goldman et al., 1986). That is,

$$K = 0.32 \text{ tons/acre/rainfall erodibility unit}$$

Estimation of LS — The product of the slope-length and slope-steepness factors, LS, is determined from charts/nomographs given in the literature (e.g., USEPA, 1988d; DOE, 1987; Mitchell and Bubenzer, 1980). This is based on a slope length of 570 ft and a slope of 0.5%, obtained from information generated during the remedial investigations, yielding:

$$LS = 0.14$$

C and P Factors — A value of C equal to 1.0 is used here, assuming no vegetative cover exists at this site. This will help simulate a worst-case scenario. Similarly, a worst-case (conservative) P value of 1 for uncontrolled sites is used.

Soil Loss Computation — The following is a summary of estimated parameters used as input to the model calculations:

$$
\begin{aligned}
a &= 95 \\
V_r &= 0.61 \text{ acre–ft} \\
q_p &= 0.38 \text{ cfs} \\
K &= 0.32 \text{ tons/acre/rainfall erodibility unit} \\
LS &= 0.14 \\
C &= 1.0 \\
P &= 1.0
\end{aligned}
$$

Thus, for a typical rainstorm event of 2-year 24-h magnitude, substitution of these estimated parameters into the MUSLE yields

$$Y \ (S)_E = 1.89 \text{ tons/event}$$

Dissolved/Sorbed Contaminant Loading
After computing the soil loss during a storm event, the amounts of adsorbed and dissolved substance loadings on the creek can be calculated. The amounts of adsorbed and dissolved substances are determined by using the following equations (Haith, 1980; DOE, 1987):

$$S_s = \left[\frac{1}{\left(1 + q_c / K_d \beta\right)} \right] C_i A$$

$$D_s = \left[\frac{1}{\left(1 + K_d \beta / q_c\right)} \right] C_i A$$

where:

$$
\begin{aligned}
S_s &= \text{sorbed substance quantity (kg)} \\
q_c &= \text{available water capacity of the top cm of soil (dimensionless)} \\
K_d &= \text{sorption partition coefficient } (\text{cm}^3/\text{gm})
\end{aligned}
$$

$$= f_{oc} \times K_{oc},$$ where f_{oc} is the organic carbon content/fraction of the soil and K_{oc} is the soil/water distribution coefficient, normalized for organic content

β = soil bulk density (g/cm^3)

C_i = total substance concentration (kg/ha-cm)

= $C_s \times \beta \times \phi$, where C_s is the chemical concentration in soil (mg/kg), β is the soil bulk density (g/cm^3), and ϕ is a conversion factor

A = contaminated area (ha-cm)

D_s = dissolved substance quantity (kg)

The model assumes that only the contaminant in the top 1 cm of soil is available for release via runoff.

Estimating K_d — The soil sorption partition coefficient for a given chemical can be determined from known values of certain other physical/chemical parameters, primarily the chemical's octanol-water partition coefficient, solubility in water, or bioconcentration factor. The sorption partition coefficient, K_d, is given by the following relationship:

$$K_d = f_{oc} \, K_{oc}$$

where:

f_{oc} = the organic carbon content/fraction of the soil

K_{oc} = the soil/water distribution coefficient, normalized for organic content

An organic carbon fraction/content of 0.1% is assumed for this site.

Estimating β, θ_c — The contaminated site area, A, is 6 acres. The soil bulk density, β, is estimated to be about 1.4 gm/cm^3 (Walton, 1984). The available water capacity of the top 1 cm of soil is estimated at about $\theta_c \approx 120$ mm/m, i.e., $\theta_c \approx 0.12$ (Walton, 1984).

Estimation of C_i — The total substance concentration, C_i [kg/ha-cm], is obtainable by multiplying the chemical concentration in soil, C_s [mg/kg] by the soil bulk density, β [g/cm^3] and an appropriate conversion factor, ϕ:

$$C_i \, [\text{kg/ha-cm}] = C_s \, \beta \, \phi$$

where ϕ is a conversion factor equal to 0.1. For the estimated value of $\beta = 1.40$ g/cm^3, this results in:

$$C_i \, [\text{kg/ha-cm}] = 0.1 \, \beta \, C_s \quad \text{or} \quad C_i = 0.14 \, C_s$$

Maximum concentrations found at the XYZ site are used in this computation to help simulate the possible worst-case scenario.

Computation of Dissolved/Sorbed Contaminant Loading — The following parameters are used to estimate the dissolved and sorbed quantities of contaminants reaching the creek:

A = 5.9 acres

θ = 0.12

β = 1.4 g/cm^3

f_{oc} = 0.1%

Consequently, the equation for sorbed substance quantity, S_s [kg] is simplified into the following:

$$S_s = \left\{ \frac{1}{\left(1 + \frac{0.12}{[1.4\,K_d]} \right)} \right\} (0.14\,C_s)(2.388) = \left\{ \frac{1.4\,K_d}{1.4\,K_d + 0.12} \right\} (0.334\,C_s)$$

$$= \frac{0.468\,K_d\,C}{(0.12 + 0.028\,K_{oc})} = \frac{0.009\,K_{oc}\,C_s}{(0.12 + 1.4\,K_d)}$$

Similarly, the equation for dissolved substance quantity, D_s [kg] is simplified into:

$$D_s = \left\{ \frac{1}{1 + [K_d\,\beta/\theta_c]} \right\} (0.334\,C_s) = \left\{ \frac{1}{1 + [1.4\,K_d/0.12]} \right\} (0.334\,C_s)$$

$$= \left(\frac{0.12}{1 + \{1.4\,K_d / [0.12 + 1.4\,K_d]\}} \right)(0.334\,C_s) = \frac{0.04\,C_s}{[0.12 + 1.4\,K_d]}$$

$$= \frac{0.04\,C_s}{[0.12 + 0.028\,K_{oc}]}$$

Computation of Total Contaminant Loading on Creek
    After calculating the amount of sorbed and dissolved contaminant, the total loading to the receiving waterbody is calculated as follows (Haith, 1980; DOE, 1987):

$$PX_i = \left[\frac{Y(S)_E}{100\,\beta} \right] S_s$$

$$PQ_i = \left[\frac{Q_r}{R_t} \right] D_s$$

where:

PX_i	=	sorbed substance loss per event (kg)
$Y(S)_E$	=	sediment yield (tons per event, metric tons)
β	=	soil bulk density (g/cm³)
S_s	=	sorbed substance quantity (kg)
PQ_i	=	dissolved substance loss per event (kg)
Q_r	=	total storm runoff depth (cm)
R_t	=	total storm rainfall (cm)
D_s	=	dissolved substance quantity (kg)

PX_i and PQ_i can be converted to mass per volume terms for use in estimating contaminant concentration in the receiving waterbody, if divided by the site storm runoff volume, V_r (where $V_r = a\ A\ Q_r$). The contaminant concentrations in the surface runoff, C_{sr}, are then given by:

$$C_{sr} = \frac{(PS_i + PQ_i)}{V_r}$$

Next, the contaminant concentrations in the creek are computed by a mass balance analysis according to the following relationship:

$$C_{cr} = \frac{(C_{sr} \times q_p)}{[q_p + Q_{cr}]}$$

where:

C_{cr} = concentration of contaminant in creek (mg/L)
C_{sr} = concentration of contaminant in surface runoff (mg/L)
q_p = peak runoff rate (cfs)
Q_{cr} = volumetric flow rate of creek (cfs)

Recommended Soil Cleanup Levels for Surface Water Protection
By performing back-calculations based on contaminant concentrations in the creek as a result of the current constituent loadings from the XYZ site, a conservative estimate is made as to what the maximum acceptable contaminant concentration should be on the site so as *not* to adversely impact the creek. The back-modeling is carried out as follows:

$$C_{max} = \left(\frac{C_{std}}{C_{cr}}\right) \times C_s$$

where:

C_{max} = maximum acceptable soil concentration on site (mg/kg)
C_{std} = applicable surface water quality criteria (mg/L)
C_{cr} = constituent concentration in creek under current loading conditions (mg/L)
C_s = soil chemical concentration prior to cleanup (mg/kg)

Based on the appropriate maximum acceptable soil concentration value, C_{max}, the site may be cleaned up to such levels as *not* to impact the surface water quality. The overall computational process for the analysis of surface water contamination is presented in Table 4.1, with the last column showing the restoration goals established for the cleanup of soils at the XYZ site. No significant/adverse impacts to the creek will be expected if site remediation is carried out to residual contaminant levels corresponding to the maximum acceptable soil concentrations estimated for this site.

Table 4.1 Development of Site Restoration Goals (Modeling the Migration of Contaminated Soils Via Overland Flow into a Creek)

K = 3.15E − 01	R_t = 3.50E + 00	# Storm Events/year = 1.31E + 01
L = 5.75E + 02	T_r = 2.40E + 01	# Storm Events in 70 years = 9.20E + 02
S = 5.00E − 03	Average Annual Storm = 4.60E + 01	S_w = 3.51E + 00
LS = 1.40E − 01	River Flow Rate, QV_r = 8.45E − 01	Q_r = 1.24E + 00
C = 1.00E + 00	CN = 7.40E + 01	V_r = 6.12E − 01
P = 1.00E + 00	Bulk Density = 1.44E + 00	Peak Flow, Q_p = 3.88E − 01
A = 5.95E + 00	Available Water Capacity = 1.20E − 01	$Y(S)e$ = 1.87E + 00
	Organic Fraction, Foc = 1.00E − 03	

Chemical Constituent	K_{oc} (m³/kg)	K_{oc} (cm³/g)	K_d^a (cm³/g)	Soil Chemical Concentration C_s (mg/kg)	Sorbed Quantity S_s (kg)	Dissolved Quantity D_s (kg)	Sorbed Loss PX_s (kg)	Dissolved Loss PQ_s (kg)	Total Loss (kg)	Concentration in Runoff (mg/L)	Concentration in Surface Water (mg/L)	Chronic Water Quality Standard (mg/L)	Attenuation Factor	Maximum Acceptable Soil Concentration (mg/kg)
Benzene	8.30E − 02	8.30E + 01	8.30E − 02	5.39E + 01	9.33E + 01	4.99E − 03	1.21E − 01	1.77E − 03	1.23E − 01	1.63E − 01	5.13E − 02	5.10E + 00	9.52E − 04	5,354
Benz(a)anthracene	1.38E + 03	1.38E + 06	1.38E + 03	8.42E + 01	2.92E + 01	1.00E − 02	3.50E + 02	3.55E − 03	3.50E + 02	4.63E + 02	1.46E + 02	3.00E − 01	1.73E + 00	0.17
Benzo(a)pyrene	5.50E + 03	5.50E + 06	5.50E + 03	1.08E + 02	3.74E + 01	1.00E − 02	4.48E + 02	3.55E − 03	4.49E + 02	5.94E + 02	1.87E + 02	3.00E − 01	1.73E + 00	0.17
Ethylbenzene	1.10E + 00	1.10E + 03	1.10E + 00	6.52E + 02	2.10E + 02	9.31E − 02	2.74E + 00	3.30E − 03	2.74E + 00	3.63E + 00	1.14E + 00	4.30E − 01	1.75E − 03	246
Naphthalene	1.45E + 00	1.45E + 03	1.45E + 00	4.80E + 02	1.57E + 02	9.47E − 03	2.05E + 00	3.36E − 03	2.05E + 00	2.72E + 00	1.25E + 00	3.00E − 01	2.60E − 03	115
Phenanthrene	1.40E + 01	1.40E + 04	1.40E + 01	2.68E + 02	9.24E + 01	9.96E − 03	1.20E + 00	3.53E − 03	1.21E + 00	1.60E + 00	7.34E − 01	3.00E − 01	2.74E − 03	110
Phenol	1.42E − 02	1.42E + 01	1.42E − 02	7.90E + 02	3.99E + 01	1.46E − 03	5.19E − 01	5.17E − 04	5.20E − 01	6.88E − 01	2.17E − 01	5.80E + 00	2.74E − 04	21,154
Toluene	3.00E − 01	3.00E + 02	3.00E − 01	3.38E + 03	9.17E + 02	7.84E − 03	1.19E + 01	2.78E − 03	1.19E + 01	1.58E + 01	4.98E + 00	5.00E + 00	1.47E − 03	3,395

a A soil organic carbon fraction of 0.1% is assumed; $K_d = f_{oc} \times K_{oc}$.

4.2.3   Development of Exposure Scenarios

Exposure scenarios are derived and modeled based on the movement of chemicals in various environmental compartments. The exposure scenario associated with a given hazardous situation may be well defined if the exposure is known to have already occurred. In most cases associated with the investigation of potentially contaminated sites, however, corrective action assessments are carried out to determine potential risks due to exposures that may not yet have occurred. Consequently, hypothetical exposure scenarios are generally developed for such applications.

Several tasks are undertaken to facilitate the development of complete exposure scenarios. The relevant tasks involve determining the sources of site contamination, identifying constituents of concern, identifying affected environmental media, delineating contaminant migration pathways, identifying potential receptors, determining potential exposure routes, constructing a conceptual model for the site, and delineating likely and significant exposure pathways. At all times, it is necessary to develop as realistic an exposure scenario as possible. This can then be used to support an evaluation of the risks posed by the potentially contaminated site; it will also allow appropriate decisions to be made regarding the need for, and extent of, remediation.

Box 4.3 illustrates the typical range and variety of *potential* exposure patterns that can be anticipated from contaminated sites (Asante-Duah, 1993). This is by no means complete, but does demonstrate the multiplicity and interconnections of numerous pathways through which populations may be exposed to contamination. In fact, the listed exposure scenarios may not all be relevant for every contaminated site problem; on the other hand, a number of other exposure scenarios may have to be evaluated due to site-specific conditions.

If numerous potential exposure scenarios exist, or if a complex exposure scenario has to be evaluated, it usually is helpful to use an event-tree model to clarify potential outcomes and/or consequences. The event tree concept, as exemplified by Figure 4.4, offers an efficient way to develop exposure scenarios. By using such an approach, the various exposure contingencies can be identified and organized in a systematic manner. Once developed, priorities can be established to help focus the available effort on the aspects of greatest need. Table 4.2 illustrates an equivalent analytical protocol for developing the set of exposure scenarios; it is noteworthy from the elementary similarities that this representation is analogous to the event tree structure shown in Figure 4.4.

4.3   DEVELOPMENT OF DATA QUALITY OBJECTIVES

Data quality objectives (DQOs) are statements that specify the data needed to support corrective action decisions. They are described to establish the desired degree of data reliability, the specific data requirements and considerations, and an assessment of the data applications as determined by the overall study objective(s). The DQO represents the full set of qualitative and quantitative constraints needed to specify the level of uncertainty that an analyst can accept when making a decision based on a particular set of data.

DQOs are an important aspect of the quality assurance requirements in the entire corrective action assessment process. Despite the fact that the DQO process is considered flexible and iterative, it generally follows a well-defined sequence of stages that will allow effective and efficient data management. It is apparent that, as the quantity and quality of data increases, the risk of making a wrong decision generally decreases

---

BOX 4.3
Example Types of General Exposure Scenarios
Associated with Contaminated Site Problems

- Receptor exposure via the ingestion of dirt, inhalation of air-borne contaminants, and/or absorption through the skin after dermal contact with contaminated soil

- Receptor exposure to indoor air from soil gas

- Receptor exposure resulting from the use of contaminated groundwater that is used for municipal or local water supplies. Exposure may be via ingestion (from culinary uses), inhalation (e.g., during showering activities), and/or dermal contact (through washing and showering activities)

- Receptor exposure resulting from use of contaminated surface water. Exposure may be via ingestion, inhalation, and/or dermal contact that occurs from the culinary uses of the surface water, or from recreational uses of the surface water

- Receptor exposure via the food chain, due to consumption of aquatic animals. Typically, aquatic life may have been exposed to contaminants as a result of contaminated runoff and/or contaminated groundwater discharging into surface water bodies; this can result in bioaccumulation of some chemicals in fish exposed to the contaminants. The contaminated fish may subsequently be consumed by humans

- Receptor exposure through the food chain due to consumption of game and livestock. Typically, game or livestock may have been exposed by ingestion and/or contacts to contaminated materials. Humans can subsequently be exposed by consuming food that has become contaminated as a result of bioaccumulation through the food chain

- Receptor exposure resulting from the consumption of dairy products from cattle that consumed feed containing surface residues of chemicals and/or contaminated water

- Receptor exposure resulting from the consumption of crops with bioaccumulated chemicals deposited onto soils, directly onto edible portions of plants, or accumulated through plant root uptake

- Interreceptor transfers, such as ingestion of mammalian breast milk containing chemicals absorbed by the feeding mother

---

(USEPA, 1987a). Consequently, the DQO process should be carefully developed to allow for a responsible program that will produce adequate quantity and quality of information required for corrective action decisions.

4.3.1 Attributes of the DQO Development Process

The DQO process consists of a planning tool that enables an investigator to specify the quality of the data required to support the objectives of a corrective action assessment program. In the investigation of potentially contaminated sites, DQOs are used as qualitative and quantitative statements that specify the quality of data required

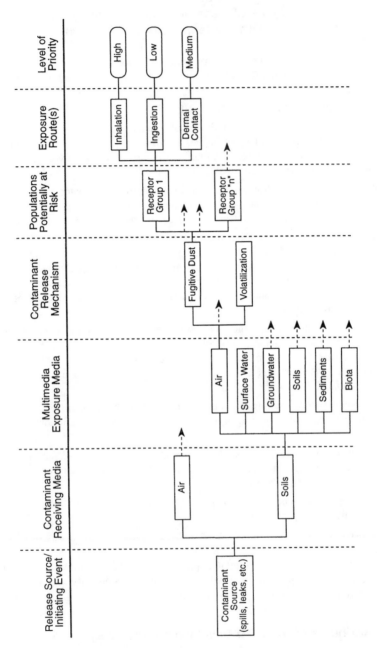

Figure 4.4 An event tree representation of exposure scenarios.

Table 4.2 Illustrative Tabular Analysis Chart for Developing Exposure Scenarios at Potentially Contaminated Sites

Primary Contaminant Source(s)	Primary Release Mechanism(s)	Contaminated Exposure Medium	Contaminant Release Source(s)	Contaminant Release Mechanism(s)	Potential Receptor Location	Receptor Groups Potentially at Risk	Potential Exposure Routes	Pathway Potentially Complete and Significant?
Underground storage tanks Fuel pipelines Wash pads in spill areas	Spills and leakages	Air	Contaminated surface soils	Fugitive dust generation	On-site	On-site facility worker	Inhalation	No
							Incidental ingestion	No
							Dermal absorption	No
						Construction worker	Inhalation	Yes
							Incidental ingestion	No
							Dermal absorption	No
					Off-site	Downwind worker	Inhalation	Yes
							Incidental ingestion	No
							Dermal absorption	No
						Downwind resident	Inhalation	No
							Incidental ingestion	No
							Dermal absorption	No
				Contaminated surface soils and waters	Volatilization	On-site	On-site facility worker	Inhalation	No
							Dermal absorption	No
						Construction workers	Inhalation	No
							Dermal absorption	No
					Off-site	Nearest downwind worker	Inhalation	No
							Dermal absorption	No
						Nearest downwind resident	Inhalation	No
							Dermal absorption	No
		Soils	Contaminated soils and/or buried wastes	Direct contacting	On-site	On-site facility worker	Incidental ingestion	Yes
							Dermal absorption	Yes
						Construction worker	Incidental ingestion	Yes
							Dermal absorption	Yes
			Redeposited contaminated soils from fugitive dust	Direct contacting	Off-site	Downwind worker	Incidental ingestion	Yes
							Dermal absorption	Yes
						Downwind resident	Incidental ingestion	Yes
							Dermal absorption	Yes

Table 4.2 (continued) Illustrative Tabular Analysis Chart for Developing Exposure Scenarios at Potentially Contaminated Sites

Primary Contaminant Source(s)	Primary Release Mechanism(s)	Contaminated Exposure Medium	Contaminant Release Source(s)	Contaminant Release Mechanism(s)	Potential Receptor Location	Receptor Groups Potentially at Risk	Potential Exposure Routes	Pathway Potentially Complete and Significant?
		Surface water	Contaminated surface soils	Surface runoff to surface impoundments	On-site	On-site facility worker	Inhalation	No
							Incidental ingestion	No
							Dermal absorption	No
						Construction worker	Inhalation	No
							Incidental ingestion	No
							Dermal absorption	No
				Erosional runoff	Off-site	Downslope resident	Inhalation	No
							Incidental ingestion	No
							Dermal absorption	No
						Recreational population	Inhalation	Yes
							Incidental ingestion	Yes
							Dermal absorption	Yes
			Contaminated groundwater	Groundwater discharge	Off-site	Downslope resident	Inhalation	No
							Incidental ingestion	No
							Dermal absorption	No
						Recreational population	Inhalation	Yes
							Incidental ingestion	Yes
							Dermal absorption	Yes

Affected medium	Contaminated source	Release/transport mechanism	Location	Potential receptor	Exposure route	Potentially complete pathway
Groundwater	Contaminated soils	Leaching/percolation	On-site	On-site facility worker	Inhalation	No
					Incidental ingestion	No
					Dermal absorption	No
				Construction worker	Inhalation	No
					Incidental ingestion	No
					Dermal absorption	No
			Off-site	Downgradient resident	Inhalation	Yes
					Incidental ingestion	Yes
					Dermal absorption	Yes
Drainage sediments	Contaminated surface soils	Surface runoff/episodic overland flow	On-site	On-site facility worker	Inhalation	Yes
					Incidental ingestion	Yes
					Dermal absorption	Yes
				Construction worker	Inhalation	No
					Incidental ingestion	No
					Dermal absorption	No
			Off-site	Nearest downgradient resident	Inhalation	No
					Incidental ingestion	No
					Dermal absorption	No
Biota	Contaminated environmental media	Consumption of, and/or contact with contaminated media	On-site	On-site facility worker	Incidental ingestion	No
					Dermal absorption	No
				Construction worker	Incidental ingestion	No
					Dermal absorption	No
			Off-site	Nearest resident	Incidental ingestion	No
					Dermal absorption	No
				Recreational population	Incidental ingestion	No
					Dermal absorption	No

to support corrective action decisions. These are determined based on the end uses of the data to be collected.

There are several benefits to establishing DQOs for corrective action assessment programs, including the following particularly important ones (CCME, 1993):

- The DQO process ensures that the data generated are of known quality
- All projects have some inherent degree of uncertainty, and DQOs help data users plan for uncertainty. By establishing DQOs, data users evaluate the consequences of uncertainty and specify constraints on the amount of uncertainty they can tolerate in the expected study results. Thus, the likelihood of an incorrect decision is gaged *a priori*
- The DQO process facilitates communication among data users, data collectors, managers, and other technical staff before time and money are spent collecting data. This will result in increased cost-efficiency in the data collection program
- The DQO process provides a logistical structure for project planning that is iterative and that encourages the data users to narrow many vague objectives to one or a few critical questions
- The structure of the DQO process provides a convenient way to document activities and decisions that can prove useful in litigation or administrative procedures
- The DQO process establishes quantitative criteria for use as cut-off line, as to when to stop sampling

Overall, the DQO process results in a well thought out sampling and analysis plan. Consequently, DQOs should be established prior to data collection activities, to ensure that data collected are sufficient and of adequate quality for their intended uses. In fact, the DQO should be integrated with development of the sampling and analysis plan, and should be revised as needed based on the results of each data collection activity.

4.4 ESTABLISHING BACKGROUND THRESHOLDS FOR ENVIRONMENTAL CONTAMINANTS

During corrective action assessments, it often becomes necessary to establish media-specific background thresholds for selected chemical constituents present at contaminated sites. The background threshold is meant to give an indication of the level of contamination in the environment that is not necessarily attributable to the project site under investigation. This serves to provide a "reference point-of-departure" that can be used to determine the magnitude of contamination in other environmental samples obtained from the "culprit" site. In general, background samples (or control site samples) are preferably collected near the time and place of the environmental samples of interest.

4.4.1 Selection of "Control" or "Reference" Sites for Background Sampling

Control (or reference) sites are important for understanding the significance of environmental sampling and monitoring data. Such sites should be selected that have common characteristics (i.e., identical in their physical and environmental settings) with the contaminated areas except for the pollution source. There are two types of control sites, whose differentiation is based primarily on the closeness of the control site to the environmental sampling site: local and area.

Local control sites are usually adjacent or very near the project sampling sites. In selecting and working with local control sites, the following principles apply (Keith, 1991; CCME, 1993):

- Local control sites generally should be upwind, upstream, or upgradient from the environmental sampling site.
- When possible, local control site samples should be taken first to avoid contamination from the environmental sampling site.
- Travel between local control sites and environmental sampling areas should be minimized because of potential cross contamination caused by humans, equipment, and/or vehicles.

In contrast to a local control site, an area control site is in the same general area (e.g., a city or region) as the sampling site, *but not adjacent to it.* The factors to be considered in area control site selection are similar to those for local control sites.

In general, local control sites are preferable to area control sites because they are physically closer. However, when a suitable local control site cannot be found, an area control site will still allow important background samples to be obtained (Keith, 1991; CCME, 1993).

4.4.2   Background Sampling Criteria

Background samples are typically collected and evaluated to determine the possibility of a potentially contaminated site contributing to off-site contamination levels in the vicinity of the site. The background samples would not have been significantly influenced by contamination from the subject site. However, these samples must be obtained from an environmental matrix that has the same basic characteristics as the matrix at the subject site, in order to provide a justifiable basis for comparison.

A number of investigations have reported the significance of matrix effects on environmental contaminant levels. For example, the analysis of data from different soil types in the same background area revealed levels of select inorganic constituents that were over twice as high in silt/clay as in sand (LaGoy and Schulz, 1993). This type of situation illustrates the importance of giving adequate consideration to the effects of natural variations in soil composition when one is designing a field sampling program. In fact, unless background samples are collected and analyzed under the same conditions as the environmental test samples, the presence and/or levels of the analytes of interest and the effects of the matrix on their analysis cannot be known or estimated with any acceptable degree of certainty. Therefore, background samples of each significantly different matrix must always be collected when different types of matrices are involved, such as various types of water, sediments, and soils in or near a sampling site area.

A minimum of one background sample per medium will usually be collected, although more may be desired especially in complex environmental settings. In general, if the natural variability of a particular constituent present at a site is relatively large, the sampling plan should reflect this site-specific characteristic.

In typical sampling programs, background air samples would consist of upwind air samples and, perhaps, different height samples. Background soil samples would be collected near a site in areas upwind of the site, the upwind locations having been determined from a wind rose (i.e., a diagram or pictorial representation that summarizes pertinent statistical information about wind speed and direction at a specified location) for the geographical region in which the site is located. Background groundwater samples generally come from upgradient wells, in relation to groundwater flow direction at the site. Background surface water and sediment samples may be collected under both high and low flow conditions; as far as practicable, sample collection from nearby lakes and wetlands should comprise shallow and deep samples (when sufficient water depth allows), to account for such differences potentially due to stratification or incomplete mixing. More detailed sampling considerations and strategies for the various environmental media of general interest can be found in the literature (e.g., Keith, 1991; USEPA, 1988e, 1989b).

4.4.3   Uses of Background Sampling Results

Background samples are used to demonstrate whether or not a site is contaminated by determining if the site samples are truly different from the background in the geographical area. In fact, some sort of background sampling is usually required as part of corrective action assessments. This will then allow a valid scientific comparison to be made between environmental samples (suspected of containing chemical contaminants) and control site samples (possibly containing only naturally low or anthropogenic levels of contaminants).

Typically, the statistical evaluation of background sampling results consist of the following one-tailed test of significance: the null hypothesis that there is no difference between contaminant concentrations in the background areas and the on-site chemical concentrations vs. the alternative hypothesis that concentrations are higher on-site. Broadly speaking, however, a statistically significant difference between background samples and site-related contamination should not, by itself, be a cause for alarm nor should it trigger off a cleanup action; further evaluations (such as conducting a risk assessment) will ascertain the significance of the contamination. In cases where remediation is required, the established background thresholds may be used as part of the process for setting realistic cleanup goals. The ability to estimate baseline or threshold concentrations can indeed become a critical factor in the formulation of reasonable cleanup objectives at potentially contaminated sites.

In general, background sampling is conducted to distinguish site-related contamination from naturally occurring or other nonsite-related levels of the chemicals of concern. Anthropogenic levels (which are concentrations of chemicals that are present in the environment due to human-made, non-site sources, such as industry and automobiles), rather than naturally occurring levels, are preferably used as a basis for evaluating background sampling data. It is noteworthy that, even at background concentration levels, a site may still be posing significant risks to human health and the environment. Whereas a PRP may not be responsible for cleanup under such circumstances, it is important to evaluate and document this situation so that at least some institutional control measures can be implemented to protect populations potentially at risk.

4.5   "NO-FURTHER-RESPONSE-ACTION" DECISIONS

A major reason for conducting corrective action assessments for potentially contaminated site problems is to be able to make informed decisions regarding site restoration programs. The fundamental purpose of a site restoration program is to protect human health and the environment from the unintended consequences of environmental contamination problems. In some situations, a decision that "no further action" (NFA) is required for a site may be appropriate for a contaminated site problem.

The process involved in a NFA decision is intended to indicate that, based on the best available information, no further response action is necessary to ensure that the case site does not pose significant risks to human health or the environment. Under such circumstances, a NFA closure document will usually be prepared for the site or group of sites, in accordance with applicable regulatory requirements.

The NFA document usually is a stand-alone report, containing sufficient information to support the "no-further-response-action" decision (NFRAD). It should therefore include site-specific evidence along with well-supported technical reasoning for the NFRAD. Additionally, the NFA document should clearly address specific CSM hypotheses tested to verify likely and complete exposure pathways or scenarios.

The appropriate criteria for establishing NFRADs are derived from statutory and regulatory provisions (such as CERCLA and RCRA under the U.S. federal statutes and regulations), requiring that human health and the environment are adequately protected for any corrective action decision on potentially contaminated sites. Consequently, the NFRAD will generally consider contaminant cleanup standards specified by several regulations (such as ARARs, action levels, site-specific risk-based criteria, and local or area background threshold levels).

4.5.1   Site Designations

A broad categorization scheme for a variety of site designations may be used to facilitate the planning, development and implementation of appropriate NFRADs. For instance, sites may be designated as belonging to a number of different groups depending on the level of effort required to implement the appropriate response actions. Typical designations, which are by no means exhaustive, are enumerated below.

- *Areas of No Confirmed Contamination (ANCCs).* These consist of areas (e.g., suspected sources) where the results of records search and site investigations show that no hazardous substances were stored for any substantial period of time, released into the environment or site structures, or disposed of on the site property; or areas where the occurrence of such storage, release, or disposal is not considered to have been probable, so that no further response action is necessary. The ANCC determination can be made at any of several investigative decision stages. Consequently, NFA decisions under this category can be made if, after a preliminary site assessment (or an equivalent effort), it can be concluded that a source or suspected release of contamination does not exist and the site can be described as part of an ANCC.
- *Areas Below Action Levels (ABALs).* These consist of impacted areas where no response or remedial action is required to ensure protection of human health and the environment. ABALs include areas where an environmental investigation (such as a Phase II-type assessment) has demonstrated that hazardous materials have been released, stored, or disposed of, but are present in quantities that require no response or remedial action to protect human health and the environment. An ABAL designation means that levels of hazardous substances

detected in a given area *do not* exceed media-specific action levels (e.g., chemical-specific ARARs or risk-based concentrations); *do not* result in significant carcinogenic or noncarcinogenic risks, nor otherwise exceed applicable federal or state and local requirements. Consequently, NFA decisions under this category can be made if, after a baseline risk assessment, it can be concluded that the site is part of an ABAL, or otherwise poses no significant risk to human health and the environment.

- *Areas Where Remedies Have Been Implemented/Completed (ARICs).* These consist of areas where the site records indicate that hazardous materials are known to have been released or disposed of, but where all remedial actions necessary to protect human health and the environment with respect to any hazardous substances remaining on the site have been taken. Consequently, NFA decisions under this category can be made if it can be demonstrated from the available evidence that the selected remedy is complete, that remediation goals have been met, and that the remaining contamination at the site does not pose a significant threat to human health or the environment.

Ultimately, the NFA determination rests on whether or not complete exposure pathways exist at a site, and whether or not any of the complete pathways are significant. It is therefore logical to infer that the NFA decision criteria are indeed linked to the use of the CSM as a decision tool. An evaluation of the elements or components of a CSM will generally serve as an important basis for many NFRADs. This involves a demonstration that the source-pathway-receptor linkage cannot be completed at the site or, that if the linkage can be completed, the risk posed by the contamination present does not exceed "acceptable" reference standards established for the site or area.

4.6 CORRECTIVE MEASURE EVALUATION TOOLS AND LOGISTICS

Corrective measure evaluation typically is comprised of a feasibility study, the purpose of which is to examine site characteristics, cleanup goals, and the performance of alternative remedial technologies so that the most effective approach for the restoration of a contaminated site can be identified. A well-designed feasibility study should address every contaminant migration pathway and environmental medium that poses, or could pose, unacceptable risks to human health or the environment. The ultimate goal of corrective action assessments is to select and implement the most cost-effective mitigation strategy that provides adequate protection to public health, welfare, and the environment.

Corrective measure studies are generally designed to identify and evaluate remedial alternatives that are potentially suitable for addressing contaminated site problems. Oftentimes, a variety of decision tools may be employed to assist the decision maker with the choice of the optimum corrective measure. Table 4.3 lists some example application tools appropriate for such purposes. Several other logistical tools can be used to support corrective action assessment programs, in order to arrive at informed decisions on potentially contaminated site problems. The use of mathematical models, decision analysis techniques, and statistical methods are of particular interest in corrective measure evaluations; these support tools are elaborated below.

Table 4.3   Example Decision Tools Potentially Useful to Corrective Action Assessment Programs

Name of Decision Tool	Description and Uses of Decision Tool	Sources of Information on Decision Tool
AERIS (Aid for Evaluating the Redevelopment of Industrial Sites)	AERIS is an expert system consisting of a multimedia risk assessment model. It serves as a useful remediation model used to identify cleanup objectives. AERIS consists of a computer program capable of deriving cleanup guidelines for industrial sites where redevelopment is being considered. It is used to generate site-specific cleanup guidelines. AERIS can be used to identify the factors that are likely to be major contributors to potential exposure concerns at sites, and those aspects of a redevelopment scenario with the greatest need for better site-specific information. AERIS is designed to evaluate situations where the soil had been contaminated sufficiently long enough to establish equilibrium or near-equilibrium conditions. Thus, it is not suitable for evaluating recent spill sites or locations	Decommissioning Steering Committee, Canadian Council of Resource and Environment Ministers (CCREM), Ottawa, Canada SENES Consultants Ltd., Richmond Hill, Ontario, Canada
AIR3D	The American Petroleum Institute (API) AIR3D model is software designed to simulate air flow in the vadose zone during soil vapor extraction or venting. It is a powerful tool to assist site engineers in the efficient design of vapor extraction systems. AIR3D is both a deterministic model and an optimization model. Linear programming techniques are incorporated to determine the optimum number and location of venting wells. Optimization can be performed based upon minimizing the number of wells, or on minimizing the cost of installing the system	American Petroleum Institute (API), Washington, D.C. Geraghty and Miller, Inc., Reston, VA
AIRTOX (Air Toxics Risk Management Framework)	AIRTOX is a decision analysis model for air toxics risk management. The framework consists of a structural model that relates emissions of air toxics to potential health effects, and a decision tree model that organizes scenarios evaluated by the structural model. AIRTOX evaluates the magnitude of health risks to a population, a specific source's contribution to the total health risk, and the cost-effectiveness of current and future emission control measures	EPRI (Electric Power Research Institute), Palo Alto, CA

Table 4.3 (continued)   Example Decision Tools Potentially Useful to Corrective Action Assessment Programs

Name of Decision Tool	Description and Uses of Decision Tool	Sources of Information on Decision Tool
DSS (Exposure and Risk Assessment Decision Support System)	The American Petroleum Institute (API) exposure and risk assessment Decision Support System (DSS) is a software system designed to assist environmental professionals in estimating human exposure and risk from sites contaminated with petroleum products. It estimates receptor point concentrations by executing fully incorporated unsaturated zone, saturated zone, air emission, air dispersion, and particulate emission models The computational modules of the DSS can be implemented in either a deterministic or Monte Carlo mode. The latter is used to quantify the uncertainty in the exposure and risk results due to uncertainty in the input parameters From physical, chemical, and toxicological property data provided in the DSS databases, risk assessments can be conducted for 16 hydrocarbons, 6 petroleum additives, and 3 metals. The databases can also be expanded to include up to 100 other constituents Overall, the DSS is a user-friendly tool that can be used to estimate site-specific exposures and risks, identify the need for site remediation, develop and negotiate site-specific cleanup levels with regulatory agencies, and efficiently and effectively evaluate the effect of uncertainty in the input parameters on estimated risks using Monte Carlo techniques	American Petroleum Institute (API), Washington, D.C. Geraghty and Miller, Inc., Reston, VA
GEMS (Graphical Exposure Modeling System) and PCGEMS (Personal Computer version of the Graphical Exposure Modeling System)	(PC)GEMS is an interactive management tool that allows quick and meaningful analysis of environmental problems. It consists of an interactive computer system for environmental modeling, physicochemical property estimation, and statistical analysis. The environmental modeling programs allows for the simulation of the migration and transformation of chemicals through the air, surface water, soil, and groundwater subsystems (PC)GEMS is a complete information management tool designed to help exposure assessment studies. It allows users to estimate chemical properties,	EPA, Research Triangle Park, NC Office of Pesticides and Toxic Substances, Exposure Evaluation Division, USEPA, Washington, D.C. General Sciences Corporation (GSC), Laurel, MD

assess fate of chemicals in receiving environments, model resulting chemical concentrations, determine the number of people potentially exposed, and estimate the resultant human exposure and risk

LEADSPREAD	LEADSPREAD consists of a mathematical model for estimating blood lead concentrations as a result of contacts with lead-contaminated environmental media. A distributional approach is used, allowing estimation of various percentiles of blood lead concentration associated with a given set of inputs. The method is adapted to a computer spreadsheet	Office of Scientific Affairs, Department of Toxic Substances Control (DTSC), California EPA, Sacramento, CA
	LEADSPREAD provides a methodology for evaluating exposure and the potential for adverse health effects resulting from multipathway exposure to inorganic lead in the environment. It can be used to determine blood levels associated with multiple pathway exposures to lead at potentially contaminated sites	
RAPS (Remedial Action Priority System)	RAPS was developed for use by the U.S. Department of Energy (DOE) in order to set priorities for the investigation and possible cleanup of chemical and radioactive waste disposal sites. It is intended to be used in a comparative rather than predictive mode	Battelle Pacific Northwest Laboratory, Richland, WA
	The RAPS methodology considers four major pathways of contaminant migration: groundwater, surface water, overland flow, and atmospheric. Estimated concentrations in the air, soil, sediments, and water media are used to assess exposure to neighboring populations. The estimated environmental concentrations form the basis of subsequent human exposure calculation and determination of a Hazard Potential Index (HPI)	
	The RAPS methodology is not truly multimedia since it is based on use of independent modules which do not interact spatially or temporally; that is, transfer of pollutant is in one direction only. This modular framework permits updating of, or inclusion of, additional components with advancing technology	
ReOpt	ReOpt is remediation software that can be used in the selection of suitable technologies for the cleanup of contaminated sites. It speeds up site cleanup decisions	Sierra Geophysics, Inc., Seattle, WA Battelle Pacific Northwest Laboratory, Richland, WA
	ReOpt contains information about technologies that might potentially be used for cleanup at contaminated	

Table 4.3 (continued)   Example Decision Tools Potentially Useful to Corrective Action Assessment Programs

Name of Decision Tool	Description and Uses of Decision Tool	Sources of Information on Decision Tool
	sites, auxiliary information about possible hazardous or radioactive contaminants at such sites, and selected pertinent regulations that govern disposal of wastes containing these contaminants The ReOpt software enables a user to quickly and easily review a variety of remediation options and determine their effectiveness for the particular site under investigation. The user provides/specifies a series of conditions, and ReOpt provides a short list of choices ReOpt will enable engineers and planners involved in environmental restoration efforts to quickly identify potentially applicable environmental restoration technologies and access corresponding information required to select cleanup activities for contaminated sites. The analyst can automatically select potentially appropriate technologies by simply specifying the contaminants or contaminated medium of interest. The analyst can also select any technology and review the technical description of the process with accompanying schematic diagrams, as well as examine the technical and regulatory constraints which govern the technology	
RISK*ASSISTANT	RISK*ASSISTANT provides an array of analytical tools, databases, and information-handling capabilities for risk assessment. It has the ability to tailor exposure and risk assessments to local conditions RISK*ASSISTANT is designed to assist the user in rapidly evaluating exposures and human health risks from chemicals in the environment at a particular site. The user need only provide measurements or estimates of the concentrations of chemicals in the air, surface water, groundwater, soil, sediment, and/or biota	Hampshire Research Institute, Alexandria, VA USEPA, Research Triangle Park, NC California EPA, Sacramento, CA New Jersey Department of Environmental Protection, Trenton, NJ
RISKPRO	RISKPRO is a complete software system designed to predict the environmental risks and effects of a wide range of human health-threatening situations. It consists of a multimedia/multipathway environmental pollution modeling system. It provides modeling tools to	General Sciences Corporation (GSC), Laurel, MD

	predict exposure from pollutants in the air, soil, and water	
	RISKPRO graphically represents its results through maps, bar charts, wind-rose diagrams, isopleth diagrams, pie charts, and distributional charts. Its mapping capabilities can also allow the user to create custom maps showing data and locations	
SITES (The Contaminated Sites Risk Management System)	SITES is a flexible interactive computerized decision-support tool for organizing relevant information and conducting risk management analyses of contaminated sites. It has the dimensionality to model multiple chemicals, pathways, population groups, health effects, and remedial actions. The user completely defines the scope of the analyses. The model uses information from diverse sources, such as site investigations, transport and fate modeling, behavioral and exposure estimates, and toxicology. It explicitly addresses the many uncertainties and allows for quick sensitivity analyses. Both deterministic and probabilistic analyses are possible	EPRI (Electric Power Research Institute), Palo Alto, CA
	The decision-tree structure in SITES allows for explicit examination of key uncertainties and the efficient evaluation of numerous scenarios. The model's design and computer implementation facilitates extensive sensitivity analysis	
	SITES is a PC computer-based integrating framework used to help evaluate and compare site investigation and remedial action alternatives in terms of health and environmental effects and total economic costs/impacts	
WET (Wastes-Environments-Technologies model)	WET is a RCRA risk/cost policy model that establishes a system to allow users to investigate how trade-offs of costs and risks can be made among wastes, environments, and technologies (W-E-Ts) in order to arrive at feasible regulatory alternatives. The system assesses waste streams in terms of likelihood and severity of human exposure to their hazardous constituents and models their behavior in three media — air, surface water, and groundwater. The exposure/risk score is tallied based on key flow and transport parameters	Office of Health and Environmental Assessment, USEPA, Washington, D.C.
	WET is used to assist policymakers in identifying cost-effective options that minimize risks to health and the environment	

4.6.1   Use of Mathematical Models

Mathematical models can be used to perform several functions that will assist in the effective management of environmental contamination problems, as long as they are used properly and with an understanding of their limitations. In particular, mathematical models find several specific applications in corrective action assessments, ranging from problem definition and system conceptualization, to exposure and risk assessments, to the development and evaluation of remedial options.

Modeling can be used in the optimization of remedial action designs because a number of alternative designs can be evaluated rapidly and quantitatively. They can assist in the development of a conceptual design for the most cost-effective action, by simulating different configurations of the selected action. For example, a groundwater pumping action may be conceptually designed by evaluating pumping rates, number, and spatial locations of wells, and location of screened intervals. Optimizing a conceptual design for the groundwater pumping action may involve evaluating alternative locations for wells, pumping rates, and remedial action configurations to identify which specific combination will be most effective (USEPA, 1985b). In addition, models are frequently used to help assess cleanup levels, determine the levels of required source removal, project the performance characteristics for remedial action designs, as well as formulate postremediation and closure requirements.

In planning for corrective action programs, one of the major benefits associated with using mathematical models is that environmental concentrations useful for exposure assessment and risk characterization can be estimated for several locations and time periods of interest. Since field data frequently are limited and insufficient to accurately and completely characterize a site and nearby conditions, models can be particularly useful for studying spatial and temporal variabilities, together with potential uncertainties. In addition, sensitivity analyses can be performed by varying specific parameters and using models to explore the ramifications (as reflected by changes in model outputs). Models can indeed be used for several purposes in the investigation of potentially contaminated site problems. The most common application of mathematical models in corrective action assessments relate to the following:

- To predict contaminant fate and transport (e.g., in the prediction of contaminant migration in various environmental compartments, or in the prediction of future concentrations of contaminants at a water supply or compliance boundary well).
- The screening of remedial alternatives (e.g., screening of alternatives is performed to eliminate those remedial actions deemed infeasible due to technical, public health, institutional and/or cost reasons).
- Analysis of possible remediation or corrective action alternatives (e.g., in the design of monitoring and corrective action plans, or in the simulation of several scenarios during the design of groundwater extraction-injection well networks).
- Conceptual design tasks. Models can be used to refine and, in some cases, optimize conceptual designs prior to their implementation.

Models (analytical or numerical) may also be used for the detailed analysis of corrective action alternatives. Appropriate models may be used to obtain information on the effectiveness, durability (i.e., design life), and expected exposures and risks as a result of implementing different remedial actions. In a corrective action evaluation program, models may be used to determine the general technical feasibility, and any potential environmental impacts arising from implementation of different remedial actions.

4.6.1.1 Model Selection for Corrective Action Evaluations

Numerous model classification systems with different complexities exist. The choice of which model to use for specific applications is subject to numerous factors. Simply choosing a more complicated model over a simple one will not necessarily ensure a better solution to the problem at hand. Since a model is a mathematical representation of a complex system, some degree of mathematical simplification usually must be made about the system being modeled. Data limitations must be weighted appropriately, since it usually is not possible to obtain all of the input parameters due to the complexity (e.g., anisotropy and nonhomogeneity) of natural systems.

The effective use of models in corrective action assessment programs depends greatly on the selection of models most suitable for this purpose. Model selection is dependent on the overall goal of the investigation, the complexity of the site, and the type of corrective actions being considered. Guidance for effective selection of models in corrective action assessments is provided in other sources (e.g., CCME, 1994; CDHS, 1990; USEPA, 1985b, 1987c, 1988c, 1988d). In several situations, a "ballpark" or "order-of-magnitude" estimate of effectiveness is usually all that is required for screening analyses. Thus, simple analytical models (i.e., models with simplified underlying assumptions) are often sufficient for this application. In general, analytical models are appropriate for well-defined systems for which extensive data are available, and/or for which the limiting assumptions are valid. Whereas analytical models may suffice for some situations, numerical models may be required for more complex configurations and complicated systems.

4.6.2 Application of Environmental Decision Analysis Methods

Decision analysis is a management tool consisting of a conceptual and systematic procedure for rationally analyzing complex sets of alternative solutions to a problem, in order to improve the overall performance of the decision-making process. Decision theory provides a logical and systematic framework to structure the problem objectives and to evaluate and rank alternative potential solutions to the problem. Environmental decision analyses typically involve the use of a series of techniques to comprehensively develop corrective action plans, and to evaluate appropriate mitigative alternatives in a technically defensible manner.

Multi-attribute decision analysis and utility theory have been suggested (e.g., Keeney and Raiffa, 1976; Lifson, 1972) for the evaluation of problems involving multiple conflicting objectives, such as is the case for decisions on contaminated site management programs. In such situations, the decision maker is faced with the problem of having to trade off the performance of one objective for another. In addressing these types of problem, a mathematical structure may be developed around utility theory that presents a deductive philosophy for risk-based decisions (Keeney, 1984; Keeney and Raiffa, 1976; Lifson, 1972; Starr and Whipple, 1980).

The use of structured decision support systems have proven to be efficient and cost effective in making sound environmental decisions. Such tools can indeed play vital roles in improving the environmental decision-making process. It should be

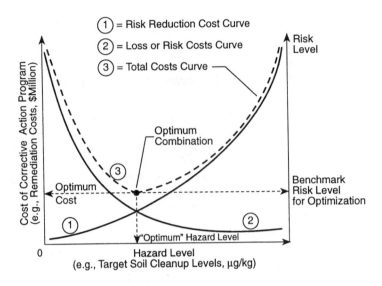

Figure 4.5 A schematic representation of the relationships between corrective action costs (e.g., cleanup or remediation costs), hazard levels (e.g., post-remediation contaminant cleanup levels), and risk levels (e.g., risks from exposure to site contamination).

acknowledged, however, that despite the fact that decision analysis presents a systematic and flexible technique that incorporates the decision-maker's judgment, it does not necessarily provide a complete analysis of the public's perception of risk.

The management of contaminated site problems generally involve competing objectives, with the prime objective to minimize both hazards and corrective action costs under multiple constraints. Once a minimum acceptable and achievable level of protection has been established via hazard assessment, alternative courses of action can be developed that weigh the magnitude of adverse consequences against the cost of corrective measures. In general, reducing hazards would require increasing costs, and cost minimization during hazard abatement will likely leave higher degrees of unmitigated hazards (Figure 4.5). Typically, a decision is made based on the alternative that accomplishes the desired objectives at the least total cost (total cost here being the sum of hazard cost and remedial cost).

As part of a corrective action assessment program, it is almost inevitable that the environmental analyst will often have to make choices between alternative remedial methods. These are based on an evaluation of risk trade-offs and relative risks among available decision alternatives, evaluation of the cost effectiveness of corrective action plans, or a risk-cost-benefit comparison of several management options. In fact, comparison between risks, benefits, and costs in various corrective action strategies is an important aspect of risk management programs. A number of the alternative methods that are potentially applicable to corrective action assessment programs are discussed below.

4.6.2.1 Cost-Effectiveness Analysis

Cost-effectiveness analysis involves a comparison of the costs of alternative methods to achieve some set goal(s) of risk reduction, such as an established benchmark risk or cleanup criteria. The process compares the costs associated with different methods of achieving a specific corrective action goal. The analysis can be used to allocate limited resources among several risk abatement programs, aimed at achieving the maximum positive results per unit cost. The procedure may also be used to project and compare total costs of several corrective action plans.

In the application of cost-effective analyses to corrective action assessments, a fixed goal is established, and then policy options are evaluated on the ability to achieve that goal in a most cost-effective manner. The goal generally consists of a specified level of "acceptable" risk, and the remedial options are compared on the basis of the monetary costs necessary to reach the benchmark risk. Cost constraints can also be imposed so that the options are assessed on their ability to control the risk most effectively for a fixed cost. The efficacy of the corrective action alternatives in the hazard reduction process can subsequently be assessed, and the most cost-effective course of action (i.e., the one with minimum cost meeting the constraint of a bench-mark risk/hazard level) can then be implemented. This would then guarantee the objective of meeting the constraints at the lowest feasible cost.

4.6.2.2 *Risk-Cost-Benefit Optimization*

Subjective and controversial as it might appear to express certain hazards in terms of cost, especially where public health and/or safety is concerned, it nevertheless has been used to provide an objective way of evaluating corrective action problems. This is particularly true where risk factors are considered in the overall study.

Risk-cost-benefit analysis is a generic term for techniques encompassing risk assessment and the inclusive evaluation of risks, costs, and benefits of alternative projects or policies. In performing risk-cost-benefit analysis, one attempts to measure risks, costs, and benefits, to identify uncertainties and potential trade-offs, and then to present this information coherently to decision makers. A general form of objective function for use in a risk-cost-benefit analysis that treats the stream of benefits, costs, and risks in a net present value calculation is given by (Crouch and Wilson, 1982; Massmann and Freeze, 1987):

$$\Phi = \sum_{t=0}^{T} \frac{1}{(1+r)^t} \left\{ B(t) - C(t) - R(t) \right\}$$

where:

Φ	=	objective function (\$)
t	=	time, spanning 0 to T (years)
T	=	time horizon (years)
r	=	discount rate
B(t)	=	benefits in year t (\$)
C(t)	=	costs in year t (\$)
R(t)	=	risks in year t (\$)

The risk term is defined as the expected cost associated with the probability of significant impacts or failure, and is a function of the costs due to the consequences of failure in year t. In general, trade-off decisions made in the process will be directed at improving both short- and long-term benefits of the program.

4.6.2.3 *Utility Theory Applications*

Risk trade-offs between increased expenditure of a remedial action and the hazard reduction achieved upon implementation may be assessed by the use of multi-attribute decision analysis and utility theory methods. Multi-attribute decision analysis and utility theory can indeed be applied in the investigation and management of contaminated site problems, in order to determine whether one set of remedial alternatives is more or less desirable than another set. With such a formulation, an explicitly logical and justifiable solution can be assessed for the complex decisions involved in corrective action assessment programs. In using expected utility maximization, the preferred alternative will be the one that maximizes the expected utility — or equivalently, the one that minimizes the loss of expected utility. In a way, this is a nonlinear generalization of cost-benefit or risk-benefit analysis.

Even though utility theory offers a rational procedure for corrective measure studies, it may transfer the burden of decision to the assessment of utility functions. Also, several subjective assumptions are used in the application of utility functions that are a subject of debate. The details of the paradoxes surrounding conclusions from expected utility applications are beyond the scope of this elaboration and are not discussed here.

Utility-Attribute Analysis

Attributes measure how well a set of objectives is being achieved. Through the use of multiple attributes scaled in the form of utilities, and weighted according to their relative importance, a decision analyst can describe an expanded set of consequences associated with a corrective action assessment program. Adopting utility as the criterion of choice among alternatives allows a multifaceted representation of each possible consequence. Hence, in its application to contaminated site management problems, both hazards and costs can be converted to utility values, as measured by the relative importance that the decision maker attaches to either attribute.

The utility function need not be linear since the utility is not necessarily proportional to the attribute. Thus, curves of the forms shown in Figure 4.6 can be generated for the utility function. An arbitrary value (e.g., **0** or **1** or **100**) of **1** can be assigned to the "ideal" situation (i.e., a "no hazard/no cost scenario"), and the "doom" scenario (i.e., "high hazard/high cost") is then assigned a corresponding relative value (e.g., **–1** or **0** or **1**) of **0**. The shape of the curves is determined by the relative value given each attribute. The range in utilities is the same for each attribute, and attributes should, strictly speaking, be expressed as specific functions of system characteristics.

In assigning utility value to hazard, it is commonplace to rely on various social and environmental goals which can help determine the threats posed by the hazard, rather than use the direct concept of hazard. These utility values can then be used as the basis for selection among remedial alternatives.

Preferences and Evaluation of Utility Functions

Evaluation of utility functions require skill, and when the utility function represents the preferences of a particular interest group, additional difficulties arise. Nonetheless, risk trade-offs may be determined by applying weighting factors of preferences in a utility-attribute analysis.

Preferences are directly incorporated in the utility functions by assigning an appropriate weighting factor to each utility term. The weighting factors are changed to reflect varying trade-off values associated with alternative decisions. If minimizing

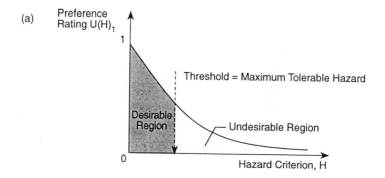

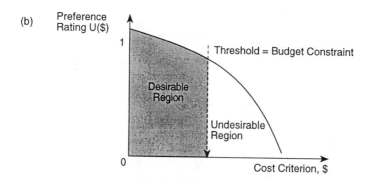

Figure 4.6   Utility functions giving the relative values of: (a) hazards and (b) costs in similar (dimensionless) terms.

hazards is **k** times as important as minimizing costs, then weighting factors of **k/(k+1)** and **1/(k+1)** would be assigned to the hazard utility and the cost utility, respectively. These weighting factors would reflect, or give a measure of, the preferences for a given utility function. Past decisions can help provide empirical data that can be used for quantifying the trade-offs and therefore the **k** values. The given utilities are weighted by their preferences, and are summed over all the objectives. For **n** alternatives, the value of the i-th alternative would be determined according to:

$$V_i = \frac{\mathbf{k}}{(k+1)}\,U\!\left(\mathbf{H}_i\right) + \frac{1}{(k+1)}\,U\!\left(\$_i\right)$$

where:

V_i = the total relative value for the **i**-th alternative
$U(H_i)$ = the hazard utility, H, for the **i**-th alternative
$U(\$_i)$ = the cost utility, $, associated with alternative **i**

In general, the largest total relative value would ultimately be selected as the best alternative.

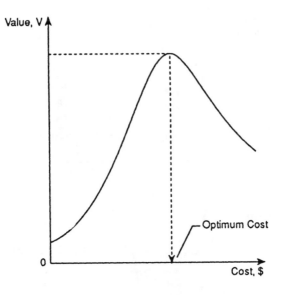

Figure 4.7   Value function for costs.

Utility Optimization

   To facilitate the development of an optimum corrective action assessment program, the total relative value can be plotted against the cost (Figure 4.7). From this
plot, the optimum cost is that cost value which corresponds to the maximum total
relative value. The optimum cost is equivalently obtained, mathematically, as follows:

$$\frac{\delta V}{(\delta \$)} = \frac{\delta}{(\delta \$)}\left[\frac{k}{(k+1)}U(H) + \frac{1}{(k+1)}U(\$)\right] = 0$$

or,

$$k\frac{\delta U(H)}{(\delta \$)} = -\frac{\delta U(\$)}{(\delta \$)}$$

where $\frac{\delta U(H)}{(\delta \$)}$ is the derivative of hazard utility relative to cost, and $\frac{\delta U(\$)}{(\delta \$)}$ is the
derivative of cost utility relative to cost. The optimum cost is obtained by solving this
equation for \$; this would represent the most cost-effective option for project execution.

   In an evaluation similar to the one presented above, a plot of total relative value
against hazard provides a representation of the 'optimum hazard' (Figure 4.8). Again,
this result can be evaluated in an analytical manner similar to that presented above for
cost; the 'optimum hazard' is given mathematically by:

$$\frac{\delta V}{(\delta H)} = \frac{\delta}{(\delta H)}\left[\frac{k}{(k+1)}U(H) + \frac{1}{(k+1)}U(\$)\right] = 0$$

$$k\frac{\delta U(H)}{(\delta H)} = -\frac{\delta U(\$)}{(\delta H)}$$

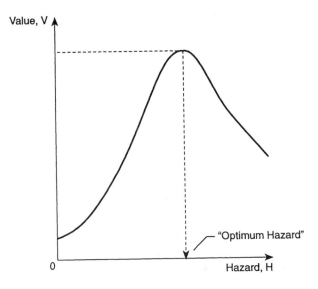

Figure 4.8 Value function for hazards.

where $\dfrac{\delta U(H)}{(\delta H)}$ is the derivative of hazard utility relative to hazard, and $\dfrac{\delta U(\$)}{(\delta H)}$ is the derivative of cost utility relative to hazard. Solving for **H** yields the "optimum" value for the hazard.

4.6.3   Statistical Analysis of Environmental Data

Several statistical methods of approach can serve to facilitate the evaluation of corrective action programs. For instance, statistical procedures can aid in the determination of sampling size requirements that are necessary to produce a statistically reliable data set required for corrective action evaluations. Also, statistical procedures used for data evaluation can significantly affect corrective action response decisions.

Over the years, extensive technical literature has been developed regarding the "best" probability distribution to utilize in different scientific applications. Of the many statistical distributions available, the Gaussian (or normal) distribution has been widely utilized to describe environmental data. However, there is considerable support for the use of the lognormal distribution in describing environmental data; the use of lognormal statistics for the data set X_1, X_2, X_3, ..., X_n requires that the logarithmic transform of these data, $\ln(X_1)$, $\ln(X_2)$, $\ln(X_3)$, ..., $\ln(X_n)$, can be expected to be normally distributed. Consequently, chemical concentration data in air, water, and soil have been described by the lognormal distribution, rather than being defined by a normal distribution (Gilbert, 1987; Leidel and Busch, 1985; Rappaport and Selvin, 1987). In fact, the use of a normal distribution (whose central tendency is measured by the arithmetic mean) to describe environmental contaminant distribution, rather than lognormal statistics (whose central tendency is defined by the geometric mean), will often result in significant overestimation, and may be overly conservative.

To demonstrate the possible significance of the choice of statistical distributions on the analysis of environmental data, consider a case involving the estimation of the mean, standard deviation, and confidence limits of monthly groundwater sampling

Table 4.4   Example Statistical Analysis of Environmental Data

| Sampling Event/ Parameter | Chemical Contamination Levels (μg/L) | |
	Original "Raw" Data $\{ X \}$	Log-Transformed Data $\{ Y = \ln(X) \}$
January	0.049	−3.016
February	0.056	−2.882
March	0.085	−2.465
April	1.200	0.182
May	0.810	−0.211
June	0.056	−2.882
July	0.049	−3.016
August	0.048	−3.037
September	0.062	−2.781
October	0.039	−3.244
November	0.045	−3.101
December	0.056	−2.882
Mean	$X_{a\text{-}mean} = 0.213$	$Y_{a\text{-}mean} = -2.445$ $(X_{g\text{-}mean} = 0.087)^a$
Standard Deviation	$SD_x = 0.379$	$SD_y = 1.154$
95% Confidence Interval (95% CI)	0.213 ± 0.834 $(CL_x$ is −0.621 to 1.047 μg/L$)^b$	-2.445 ± 2.539 $(CL_y$ is 0.0068 to 1.099 μg/L$)^c$
95% Upper Confidence Level (95% UCL)^d	1.047 μg/L	1.099 μg/L

[a] Transforming the average of the Y values back to arithmetic values yields the geometric mean value of $X_{g\text{-}mean} = e^{-2.445} = 0.087$. It is recognized that the arithmetic mean of $X_{a\text{-}mean} = 0.213$ μg/L is substantially larger than the geometric mean of 0.087 μg/L. The two large sample values in the data set very strongly bias the arithmetic mean; the logarithmic transform acts to suppress the extreme values.

[b] The development of a 95% confidence limit (refer to standard statistics texts for details of procedures) for the untransformed data gives a confidence interval of $0.213 \pm 0.379\ t = 0.213 \pm 0.834$ (where t = 2.20 obtained from the student t-distribution for n = 12−1 = 11 degrees of freedom). This indicates a nonzero probability of a negative concentration, making it nonsensical.

[c] The development of a 95% confidence limit (refer to standard statistics texts for details of procedures) for the log-transformed data gives a confidence interval of $-2.445 \pm 1.154\ t = -2.445 \pm 2.539$ (where t = 2.20 obtained from the student t-distribution for n = 12−1 = 11 degrees of freedom). Transforming these values back to the arithmetic realm gives the 95% confidence interval from 0.0068 to 1.099 μg/L, which makes more sense than provided under (b) above.

[d] The 95% UCL for both normally and lognormally distributed data sets are similar. Hence, the 95% UCL is a preferred statistical parameter to use in corrective action assessments rather than the mean values which can be significantly different for the two types of distribution.

data (Table 4.4). The results from this example analysis illustrate the potential effects that could result from the choice of one distribution type over another, and also the implications of selecting specific statistical parameters in the evaluation of environmental sampling data. Note, for instance, that whereas the mean values from the normally and lognormally distributed data are significantly different, their 95% upper confidence levels (UCLs) are relatively closer in values. In general, the use of arithmetic or geometric mean values for estimating average concentrations tends to bias this central tendency statistic. Furthermore, sparse sampling data usually are all that is available from previous site investigations; these tend to be highly variable, and arithmetic or geometric averaging would not produce representative concentration estimates.

Geostatistical techniques that account for spatial variations in concentrations may be employed for estimating the average concentrations at a site, as required for long-term exposure assessments (e.g., USEPA, 1988a; Zirschy and Harris, 1986). A technique called block kriging is frequently used to estimate soil chemical concentrations in sections of contaminated sites in which sparse sampling data do exist. In this case, the site is divided into blocks (or grids), and concentrations are determined within blocks by using interpolation procedures that incorporate sampling data in the vicinity of the block. The sampling data are weighted in proportion to the distance of the sampling location from the block. Also, weighted moving-average estimation techniques based on geostatistics are applicable for estimating average contaminant levels present at a site.

In general, statistical analysis procedures used in the evaluation of environmental data should reflect the character of the underlying distribution of the data set. The appropriateness of any distribution assumed or used to fit a given data set should preferably be checked prior to its application; this can be accomplished by using such procedures as the Chi-Square Test for Goodness of Fit, as described in standard textbooks of statistics (e.g., Fruend and Walpole, 1987; Gilbert, 1987; Miller and Freund, 1985; Sharp, 1979; Wonnacott and Wonnacott, 1972). The choice of statistical parameters for corrective action decisions is also critical to the effectiveness of the overall corrective action plan. Because of the uncertainty associated with any estimate of exposure concentration, the upper confidence interval (i.e., the 95% upper confidence limit), on the average, is frequently used in corrective action assessment programs and indeed most environmental impact evaluation programs.

4.6.3.1 Statistical Methods of Analysis

Several of the available statistical methods and procedures finding potential use in corrective action assessment programs can be found in the literature on statistics (e.g., Freund and Walpole, 1987; Gilbert, 1987; Hipel, 1988; Miller and Freund, 1985; Sharp, 1979; Wonnacott and Wonnacott, 1972; Zirschy and Harris, 1986). The broad types of commonly used methods that find general application in the evaluation of environmental data are briefly discussed below.

Parametric Vs. Nonparametric Statistics

There are several statistical techniques available for analyzing data that are not dependent on the assumption that the data follow any particular statistical distribution. These distribution-free methods, referred to as *nonparametric* statistical tests, have fewer and less stringent assumptions. Conversely, several assumptions have to be met before one can use a *parametric* test. In fact, whenever the set of requisite assumptions are met, it is preferred to use a parametric test, because it is more powerful than the nonparametric test. However, to reduce the number of underlying assumptions required to test a hypothesis (such as the presence of specific trends in a data set), nonparametric tests are generally employed. Nonparametric techniques are generally selected when the sample sizes are small and the statistical assumptions of normality and homogeneity of variance are tenuous.

Nonparametric tests are usually adopted for use in environmental impact assessments because the statistical characteristics of the often messy environmental data make it difficult, or even unwise, to use many of the available parametric methods. It is noteworthy, however, that nonparametric tests tend to ignore the magnitude of the observations in favor of the relative values or ranks of the data. Consequently, as Hipel (1988) notes, a given nonparametric test with few underlying assumptions, that is designed, for instance, to test for the presence of a trend, may only provide a "yes" or "no" answer as to whether or not a trend may be contained in the data. The output from the nonparametric test may not give an indication of the type or magnitude of the trend. To have a more powerful test about what is occurring, many assumptions must be made, and as more and more assumptions are formulated, a nonparametric test begins to look more like a parametric test (Hipel, 1988).

Hypothesis Testing

A typical and common example of statistical applications in corrective action assessment programs relates to the use of the analysis-of-variance (ANOVA) model in hypotheses testing of environmental variables. ANOVA is a method for partitioning the total variation in a set of data into the different sources of variation that are present. ANOVA models are also used to analyze the effects of the independent variable(s) on the dependent variable(s).

The application of ANOVA to environmental impact evaluations will generally result in a summary table that provides a convenient form for presenting information contained in the environmental data sets. For instance, ANOVA can be used to study data from groundwater monitoring wells as an integral part of a corrective action program. In the context of groundwater monitoring, wells or groups of wells represent the independent variables and the groundwater contaminant concentrations represent the dependent variable. The ANOVA will help determine if different wells (or group of wells) have significantly different concentrations of the contaminants of concern, by comparing site or compliance monitoring wells with background wells. The contrasts of interest may involve a comparison between the mean concentration of the background wells and the mean concentration of each compliance well.

4.6.3.2   An Illustrative Statistical Evaluation to Aid a Corrective Action Decision

This section consists of a simple statistical correlation analysis that is used to support a corrective action decision for a PCB-oil contaminated site located within the rural and predominantly agricultural community of NK4. This site has an area of about 30 acres, about half of which is occupied by a 6-m-deep lagoon. The lagoon contains oily liquids and PCBs in excess of 7000 ppm in certain sediment pockets. A thick clay unit at the lagoon bottom provides protection of a water supply aquifer at the NK4 site. The lagoon water level is up to 3 m higher than the groundwater table, causing a groundwater mound in the vicinity of the lagoon. Nonetheless, there does not appear to be significant hydraulic flows of lagoon water into the aquifer, apparently because of possible sealing/ barrier effects created by the oily bottom sludge together with the clay unit.

Study Objective

A simple statistical evaluation is being considered to facilitate cost-effective corrective action decisions for the NK4 site. It is generally believed that the PCBs will tend to be concentrated in the oils. The specific objective of this evaluation is to confirm if any reasonable correlation exists between the occurrence of oil/grease and PCB levels in the lagoon sediments. The rationale for this evaluation is that if it is determined that high levels of oil/grease are associated with high levels of PCBs and vice versa, then a remedial design can be tailored to extract PCB-laden oils first, followed by the dredging of PCB-contaminated "hot-spots" in the sediment zones. The oils and sediments can then be treated by separate incineration processes.

Choice of Statistical Tests

Nonparametric statistical tests are employed in this evaluation since these have fewer and less stringent assumptions in comparison with parametric analyses. The nonparametric correlational technique selected for this analysis is the Spearman rank coefficient of correlation. This coefficient or number indicates the exact strength and direction of the relationship between the two sets of variables being compared — in this case the PCB and oil/grease levels found in sediment samples taken from the lagoon at the NK4 site. A rough gage for interpreting the Spearman rank coefficient of correlation is given as follows (Sharp, 1979):

HIGH: 0.85 to 1.0 (or –0.85 to –1.0)
MODERATE: 0.50 to 0.84 (or –0.50 to –0.84)
LOW: 0 to 0.49 (or 0 to –0.49)

Subsequently, a two-tailed hypothesis test is performed to determine the level of significance of the estimated correlation coefficient.

Statistical Evaluation

The statistical evaluation process employed in this study is described below.

- The Spearman Rank Correlation Coefficient (or the Spearman rho), ρ, used in this evaluation is given by:

$$\rho = 1 - \frac{\left(6 \sum D^2\right)}{N\left(N^2 - 1\right)}$$

where N is the number of individual points in a group and D represents the difference between the ranks for the groups. Results from the application of this technique to the PCB and oil/grease levels in the lagoon sediment samples is presented in Table 4.5. A Spearman rho of 61% (i.e., $\rho = 0.61$) is indicated, which represents a moderate positive correlation between the levels of PCB and oil/grease found in the sediment samples taken from the lagoon.

- To test the level of significance of this correlation involves testing the null hypothesis, H_o: *no correlation exists between PCB and oil/grease levels in sediment*, against the corresponding alternative hypothesis, H_a: *PCB and oil/grease levels in lagoon sediments are correlated*. This is a two-tailed test, and is performed at a significance level of $\alpha = 0.01$. Using standard statistical tables (e.g., Sharp, 1979), $\rho_{\alpha,n}$ from the tables is compared against the computed ρ value, and H_o is rejected if $|\rho| > |\rho_{\alpha,n}|$, otherwise H_o is accepted.

Table 4.5   Statistical Correlation Analysis for PCB
             and Oil/Grease Contamination at the NK4 Site

Sampling Event/Group	PCB concentration in soils/sediments (ppm)	Oil/Grease concentration in soils/sediments (ppm)
1	14.6	9,630
2	1.8	507
3	8.3	2,230
4	3.3	2,870
5	67.7	98,900
6	20.9	71,600
7	25.6	71,600
8	31.1	43,700
9	46	81,600
10	194	188,000
11	1.5	2,020
12	3.9	91,200
13	3.8	16,600
14	3.8	14,300
15	951	230,000
16	826	25,000
17	22.7	73
18	7,241	130,000
19	5,639	1,700
20	840	110
21	2.5	620
22	0.7	100
23	5.2	110
24	57.6	150,000
25	6.1	270
26	13.4	7,400
27	8.9	15,000
28	0.6	1,900
29	9.0	4,800
30	3.0	15,000
31	13.1	85,000
32	2.0	1,300
33	2.3	2,080
34	78	130,000
35	16.9	77,000
36	203	109,000
37	119	56,000
38	25.6	25,000
39	88	107,000
40	7.2	6,800
41	114	107,000
42	142	60,000
43	99	170,000
44	34	3,700
45	9.9	12,000
46	6.3	7,200
47	4.3	750

Note: The Spearman rank correlation coefficient for the data set is 0.61.

- At a level of significance of $\alpha = 0.01$ (which corresponds to 99% confidence level), H_o is compared against H_a to arrive at a statistical decision. This helps determine whether or not the observed value of $\rho = 0.61$ differs from zero only by chance. The following steps are used to help arrive at this statistical decision (e.g., Sharp, 1979):

1. Compute $Z = \rho \{(N - 1)^{0.5}\}$,
   i.e. $Z = 0.61\{(47 - 1)^{0.5}\} = 4.14$
2. Determine the probability of Z from standard statistical tables as follows:
   Prob $\{Z = 4.14\} = 2 \times (0.00003) = 0.00006$
   i.e. $P = 0.00006$
3. If the probability value, P, obtained from the tables is less than or equal to α, then H_o is rejected,
   i.e.   reject H_o if $P \leq \alpha$

Now, since P $(= 0.00006) < \alpha$ $(= 0.01)$, H_o is rejected and H_a accepted. This means that there is a reasonable degree of correlation between the occurrence of oil/grease and PCB in the lagoon sediment samples.
• The statistical conclusion, made at a 99% level of confidence, is that high levels of PCBs are likely to occur where high levels of oil/grease exist in the lagoon at NK4.

The Decision Process
Results obtained from this statistical analysis indicate a good chance that high levels of PCBs will likely be found concentrated in the oil/grease. In fact, PCBs are not soluble in water and are generally immobile in the solid matrix. Consequently, if the oil can be extracted from the lagoon sediments and/or water, then PCB levels in these matrices may be greatly reduced. This would, in turn, reduce the volumes of high PCB-contaminated sediments and/or water that have to be treated. A recommended remedial strategy for the NK4 site will consist of the dredging and on-site incineration of the oil/PCB sludge and sediments, together with an on-site treatment of the lagoon water.

In general, it is expected that the costs to incinerate PCB-contaminated oil will be considerably less than the incineration of PCB-contaminated sediments. Thus, the knowledge gained from the statistical evaluation may be used to support a decision to initially extract PCB-laden oils, followed by a sediments cleanup process. An effective restoration process may be dry excavation to remove sediments, after dewatering the lagoon. The lagoon can be dewatered by using partially penetrating barrier walls in conjunction with a system of pumping wells. The aqueous phase (to be treated and reused or discharged) can then be separated from the PCB-laden oil phase (to be incinerated). The barrier wall used in the lagoon dewatering system can also serve as a containment system that will minimize potential migration of contaminants; the wells may also aid further groundwater pump and treat remedial action that may be required for groundwater remediation.

4.7 THE CORRECTIVE ACTION RESPONSE

When there is a source release, contaminants may be transported to different potential receptors via several environmental media (such as air, soils, groundwater, and surface water). A complexity of processes may affect contaminant migration at contaminated sites, resulting in human and ecological receptors outside the source area potentially being threatened. Consequently, it is imperative to adequately characterize a site and its surroundings through a well-designed site investigation program (in which all contaminant sources and impacted media are thoroughly investigated), in order to arrive at appropriate and cost-effective corrective action decisions.

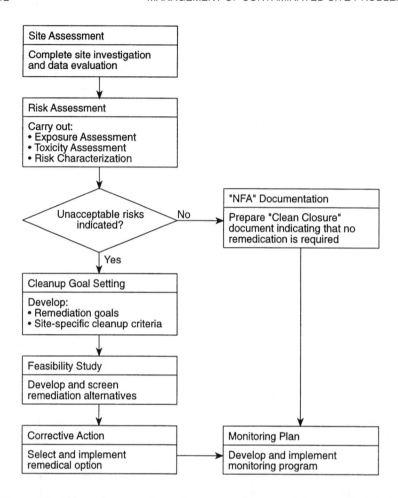

Figure 4.9   Steps in the corrective action program for a potentially contaminated site.

A multimedia approach to site characterization is generally adopted, so that the significance of possible air, water, soil, and biota contamination can be established through appropriate field sampling and analytical procedures that will yield the high-quality environmental data needed to support the corrective action decision. To accomplish this, samples are gathered and analyzed for the chemicals of potential concern in the appropriate media of interest. Effective analytical protocols in the sampling and laboratory procedures are used to minimize uncertainties associated with the data collection and evaluation activities. A quality assurance program (consisting of a system of documented checks which validate the reliability of acquired data sets) is an important part of this whole process. Indeed, the development and implementation of a firm QA/QC program during a sampling and analysis program is critical to obtaining reliable analytical results which become the basis for the corrective action decision process.

Figure 4.9 summarizes the major tasks involved in a corrective action response program for potentially contaminated site problems. Ultimately, the information gathered during remedial investigation activities is used to map out the extent of contamination, and to evaluate the potential risks associated with the subject site so that an appropriate corrective action decision can be made. Invariably, the health and environmental risk estimates become an important element in the corrective action decision process. This is because the risk assessment provides the decision maker with technically defensible and scientifically valid procedures for determining whether or not a potentially contaminated site represents significant adverse human health or ecological risks that warrants consideration as a candidate for mitigation.

In general, risk assessment becomes an integral part of the remedial investigation process. This is used to facilitate decisions on whether or not remedial actions are needed to abate site-related risks, and also in the enforcement of regulatory decisions. A more elaborate discussion of risk assessment applications to corrective action assessment programs is presented in the next chapter.

REFERENCES

Asante-Duah, D.K. 1993. *Hazardous Waste Risk Assessment.* CRC Press/Lewis Publishers, Boca Raton, FL.

Ashby, E. 1980. "What Price the Furbish Lousewort"? Proc. 4th Conf. on Environ. Engr. Educ., Toronto, Canada.

BSI (British Standards Institution). 1988. Draft for Development, DD175: 1988 Code of Practice for the Identification of Potentially Contaminated Land and its Investigation. BSI, London, U.K.

Cairney, T. (Ed.). 1993. *Contaminated Land (Problems and Solutions).* Lewis Publishers, Boca Raton, FL.

CCME. 1993. "Guidance Manual on Sampling, Analysis, and Data Management for Contaminated Sites." Vol. I. Main Report (Report CCME EPC-NCS62E), and Vol. II. Analytical Method Summaries (Report CCME EPC-NCS66E). The National Contaminated Sites Remediation Program, Canadian Council of Ministers of the Environment, Winnipeg, Manitoba.

CCME. 1994. Subsurface Assessment Handbook for Contaminated Sites. Canadian Council of Ministers of the Environment (CCME), The National Contaminated Sites Remediation Program (NCSRP), Report No. CCME-EPC-NCSRP-48E. Ottawa, Canada.

CDHS. 1986. The California Site Mitigation Decision Tree Manual. California Department of Health Services, Toxic Substances Control Division, Sacramento.

CDHS. 1990. Scientific and Technical Standards for Hazardous Waste Sites. Prepared by the California Department of Health Services, Toxic Substances Control Program, Technical Services Branch, Sacramento.

Crouch, E. A. C. and R. Wilson. 1982. *Risk/Benefit Analysis.* Ballinger, Boston, MA.

DOE. 1987. The Remedial Action Priority System (RAPS): Mathematical Formulations. U.S. Department of Energy, Office of Environmental Safety and Health, Washington, D.C.

Erickson, A.J. 1977. Aids for Estimating Soil Erodibility — "K" Value Class and Tolerance. U.S. Department of Agriculture, Soil Conservation Service, Salt Lake City, UT.

Freund, J.E. and R.E. Walpole (Eds.). 1987. *Mathematical Statistics.* Prentice-Hall, Englewood Cliffs, NJ.

Gilbert, R.O. 1987. *Statistical Methods for Environmental Pollution Monitoring.* Van Nostrand Reinhold, New York.

Goldman, S.J., K. Jackson and T.A. Bursztynsky. 1986. *Erosion and Sediment Control Handbook.* McGraw-Hill, New York.

Grisham, J.W. (Ed.). 1986. *Health Aspects of the Disposal of Waste Chemicals.* Pergamon Press, Oxford.

Haith, D.A. 1980. A mathematical model for estimating pesticide losses in runoff. *J. Environ. Qual.* 9(3):428–433.

Hipel, K.W. 1988. Nonparametric Approaches to Environmental Impact Assessment. *Water Resour. Bull. AWRA,* Vol. 24, No.3: 487–491.

Jolley, R.L. and R.G.M. Wang (Eds.). 1993. *Effective and Safe Waste Management: Interfacing Sciences and Engineering With Monitoring and Risk Analysis.* Lewis Publishers, Boca Raton, FL.

Keeney, R.L. 1984. Ethics, Decision Analysis, and Public Risk. *Risk Anal.,* 4, 117–129.

Keeney, R.D. and H. Raiffa. 1976. *Decisions with Multiple Objectives: Preferences and Value Tradeoffs.* John Wiley & Sons, New York.

Keith, L.H. 1991. *Environmental Sampling and Analysis — A Practical Guide.* Lewis Publishers, Boca Raton, FL.

LaGoy, P.K. and C.O. Schulz. 1993. "Background Sampling: An Example of the Need for Reasonableness in Risk Assessment." *Risk Anal.* Vol. 13, No. 5, 483–484.

Leidel, N. and K. A. Busch. 1985. Statistical Design and Data Analysis Requirements. In, *Patty's Industrial Hygiene and Toxicology,* Vol. IIIa, 2nd ed., John Wiley & Sons, New York.

Lifson, M.W. 1972. *Decision and Risk Analysis for Practicing Engineers.* Barnes and Noble, Cahners Bks, Boston, MA.

Lyman, W.J., W.F. Reehl and D.H. Rosenblatt. 1990. *Handbook of Chemical Property Estimation Methods: Environmental Behavior of Organic Compounds.* American Chemical Society, Washington, D.C.

Massmann, J. and R.A. Freeze. 1987. Groundwater Contamination from Waste Management Sites: The Interaction between risk-based engineering design and regulatory policy, I. Methodology; II. Results. *Water Resour. Res.,* Vol. 23, No. 2, 351–380.

Miller, I. and J.E. Freund. 1985. *Probability and Statistics for Engineers.* 3rd ed. Prentice-Hall, Englewood Cliffs, NJ.

Mills, W.B., Dean, J.D., Porcella, D.B., et al. 1982. Water Quality Assessment: A Screening Procedure for Toxic and Conventional Pollutants: Parts 1, 2, and 3. U.S. Environmental Protection Agency. Environmental Research Laboratory. Office of Research and Development. EPA-600/6-82/004 a.b.c., Athens, GA.

Mitchell, J.K. and G.D. Bubenzer. 1980. Soil Loss Estimation. In *Soil Erosion,* M.J. Kirby and R.P.C. Morgan, Eds., John Wiley & Sons, New York.

Mockus, J. 1972. Estimation of direct runoff from storm rainfall. In National Engineering Handbook. Section 4: Hydrology. U.S. Department of Agriculture, Soil Conservation Service. Washington, D.C.

Rappaport, S.M. and J. Selvin. 1987. "A Method for Evaluating the Mean Exposure from a LogNormal Distribution." *J. Am. Ind. Hyg. Assoc.* 48:374–379.

Schwab, G.O., R.K. Frevert, T.W. Edminster and K.K. Barnes. 1966. *Soil and Water Conservation Engineering,* 2nd ed. John Wiley & Sons. New York.

Sharp, V.F. 1979. *Statistics for the Social Sciences.* Little, Brown, Boston, MA.

Starr C. and C. Whipple. 1980. Risks of Risk Decisions. *Science* 208, 1114.

Swann, R.L. and A. Eschenroeder (Eds.). 1983. *Fate of Chemicals in the Environment.* ACS Symp. Ser. 225, American Chemical Society, Washington, D.C.

USBR. 1977. Design of Small Dams. U.S. Department of the Interior, Bureau of Reclamation. U.S. Government Printing Office, Washington, D.C.

USEPA. 1985a. Characterization of Hazardous Waste Sites: A Methods Manual, Vol. 1, Site Investigations. U.S. Environmental Protection Agency, Environmental Monitoring Systems Laboratory, EPA-600/4-84-075. Las Vegas, NV.

USEPA. 1985b. Modeling Remedial Actions at Uncontrolled Hazardous Waste Sites. EPA/540/2-85/001 (April, 1985). Office of Emergency and Remedial Response, U.S. Environmental Protection Agency, Washington, D.C.

USEPA. 1987a. Data Quality Objectives for Remedial Response Activities. EPA/540/ G-87/003, Office of Emergency and Remedial Response, U.S. Environmental Protection Agency, Washington, D.C.

USEPA. 1987b. RCRA Facility Investigation (RFI) Guidance, EPA/530/SW-87/001, U.S. Environmental Protection Agency, Washington, D.C.

USEPA (U.S. Environmental Protection Agency). 1987c. Selection Criteria for Mathematical Models Used in Exposure Assessments: Surface Water Models. EPA-600/8-87/042. Office of Health and Environmental Assessment, Washington, D.C.

USEPA. 1988a. GEO-EAS (Geostatistical Environmental Assessment Software) User's Guide. Environmental Monitoring Systems Laboratory, Office of R&D, EPA/600/ 4-88/033a. U.S. Environmental Protection Agency, Las Vegas, NV.

USEPA. 1988b. Guidance for Conducting Remedial Investigations and Feasibility Studies Under CERCLA. EPA/540/G-89/004. OSWER Directive 9355.3-01, Office of Emergency and Remedial Response, U.S. Environmental Protection Agency, Washington, D.C.

USEPA. 1988c. Selection Criteria for Mathematical Models Used in Exposure Assessments: Ground-Water Models. EPA-600/8-88/075. Office of Health and Environmental Assessment, U.S. Environmental Protection Agency, Washington, D.C.

USEPA. 1988d. Superfund Exposure Assessment Manual, Report No. EPA/540/1-88/001, OSWER Directive 9285.5-1, Office of Remedial Response, U.S. Environmental Protection Agency, Washington, D.C.

USEPA. 1988e. Interim Report on Sampling Design Methodology. Environmental Monitoring Support Laboratory, EPA/600/X-88/408. U.S. Environmental Protection Agency, Las Vegas, NV.

USEPA. 1989a. Risk Assessment Guidance for Superfund. Vol. I. Human Health Evaluation Manual (Part A). EPA/540/1-89/002. Office of Emergency and Remedial Response, U.S. Environmental Protection Agency, Washington, D.C.

USEPA (U.S. Environmental Protection Agency). 1989b. Risk Assessment Guidance for Superfund. Vol. II. Environmental Evaluation Manual. EPA/540/1-89/001. Office of Emergency and Remedial Response, U.S. Environmental Protection Agency, Washington, D.C.

USEPA. 1989c. Soil Sampling Quality Assurance User's Guide. 2nd ed. Environmental Monitoring Systems Laboratory, EPA/600/8-89/046. U.S. Environmental Protection Agency, Las Vegas, NV.

USEPA. 1990. Guidance for Data Useability in Risk Assessment, Interim Final. Office of Emergency and Remedial Response. EPA/540/G-90/008. U.S. Environmental Protection Agency, Washington, D.C.

USEPA. 1991a. Risk Assessment Guidance for Superfund: Vol. I. Human Health Evaluation Manual (Part B, Development of Risk-based Preliminary Remediation Goals). PB92-963333. Office of Emergency and Remedial Response, U.S. Environmental Protection Agency, Washington, D.C.

USEPA. 1991b. Risk Assessment Guidance for Superfund: Vol. I. Human Health Evaluation Manual (Part C, Risk Evaluation of Remedial Alternatives). PB90-155581. Office of Emergency and Remedial Response, U.S. Environmental Protection Agency, Washington, D.C.

USEPA. 1991c. Conducting Remedial Investigations/Feasibility Studies for CERCLA Municipal Landfill Sites. Office of Emergency and Remedial Response. EPA/540/ P-91/001 (OSWER Directive 9355.3-11). U.S. Environmental Protection Agency, Washington, D.C.

Walton, W.C. 1984. *Practical Aspects of Ground Water Modeling.* National Water Well Association. Dublin, OH.

Williams, J.R. 1975. Sediment-yield prediction with the universal equation using runoff energy factor. In, "Present and Prospective Technology for Predicting Sediment Yields and Sources. U.S. Department of Agriculture. ARS-S-40. Washington, D.C.

Wonnacott, T.H. and R.J. Wonnacott. 1972. *Introductory Statistics.* 2nd ed. John Wiley & Sons, New York.

WPCF. 1988. *Hazardous Waste Site Remediation: Assessment and Characterization.* Spec. Publ. Water Pollution Control Federation, Technical Practice Committee, Alexandria, VA.

Zirschy, J. H. and D. J. Harris. 1986. Geostatistical Analysis of Hazardous Waste Site Data. *ASCE J. Environ. Eng.* 112 (4).

5   RISK ASSESSMENT APPLICATIONS TO CONTAMINATED SITE PROBLEMS

There are several health, environmental, political, and socioeconomic implications associated with contaminated site problems. It is therefore important to use systematic and technically sound methods of approach in corrective action response programs for such sites. Risk assessment provides one of the best mechanisms for completing the tasks involved. In fact, a systematic and accurate assessment of current and future risks associated with a given contaminated site problem is crucial to the development and implementation of a cost-effective corrective action program for the site. Consequently, risk assessment is generally considered as an integral part of the corrective action assessment processes used in the investigation of potentially contaminated site problems.

Oftentimes, risk assessment is used as a management tool to facilitate effective decision making on the control of environmental pollution problems. In most applications, it is used to provide a baseline estimate of existing risks that are attributable to a specific agent or hazard; the baseline risk assessment consists of an evaluation of the potential threat to human health and the environment in the absence of any remedial action. Risk assessment can also be used to determine the potential reduction in exposure and risk under various corrective action scenarios. In particular, risk assessment can effectively be used to support remedy selection in site restoration programs. Indeed, decisions about remediating contaminated sites are made primarily on the basis of human health and ecological risks.

5.1 RISK ASSESSMENT DEFINED

Several definitions of risk assessment have been published in the literature by various authors (e.g., Asante-Duah, 1990, 1993; Bowles et al., 1987; Hallenbeck and Cunningham, 1988; NRC, 1982, 1983; OTA, 1983; Rowe, 1977; USEPA, 1984) to describe a variety of risk assessment methods and/or protocols. In a generic sense, risk assessment is defined as a systematic process for making estimates of all the significant risk factors that exist over an entire range of failure modes and/or exposure scenarios associated with some hazard situation(s). In its application to the investigation of potentially contaminated site problems, the process encompasses an evaluation of all the significant risk factors associated with all feasible and identifiable exposure scenarios that are the result of contaminant releases into the environment. It involves the characterization of potential adverse consequences or impacts to human and ecological receptors that are potentially at risk from exposure to site contaminants.

Risk assessment is a powerful tool for developing insights into the relative importance of the various types of exposure scenarios associated with potentially hazardous situations. The risk assessment process seeks to estimate the likelihood of occurrence of adverse effects resulting from exposures of humans and ecological receptors to chemical, physical, and/or biological agents present in the environment.

The type and degree of detail of any risk assessment depends on its intended use. Its purpose will shape the data needs, the protocol, the rigor, and related efforts. Current regulatory requirements are particularly important considerations in the application of risk assessment to contaminated site problems. The processes involved in a risk assessment generally require a multidisciplinary approach, covering several areas of expertise in most cases.

5.2 PURPOSE AND ATTRIBUTES OF RISK ASSESSMENT

The conventional paradigm for risk assessment is predictive, which deals with localized effects of a particular action that could result in adverse effects. However, there also is increasing emphasis on assessments of the effects of environmental contaminants associated with existing contaminated site problems. This assessment of past pollutions, with possible ongoing consequences, generally falls under the umbrella of what has been referred to as "retrospective risk assessment" (Suter, 1993). The impetus for a retrospective assessment may be a source, observed effects, or evidence of exposure. Source-driven retrospective assessments result from observed pollution that requires elucidation of possible effects (e.g., hazardous waste sites, spills/accidental releases, etc.); effects-driven retrospective assessments result from the observation of apparent effects in the field that requires explanation (e.g., fish or bird kills, declining populations of a species, etc.); exposure-driven retrospective assessments are prompted by evidence of exposure without prior evidence of a source or effects (e.g., the scare over mercury in swordfish). In all cases, however, the principal objective of risk assessment is to provide a basis for actions that will minimize the impairment of the environment and of public health, welfare, and safety.

The overall goal in a risk assessment is to identify potential system failure modes and exposure scenarios intended to facilitate the design of methods to reduce the probability of failure and the attending human, socioeconomic, and environmental consequences of any failure and/or exposure events. In general, risk assessment, which seems to be one of the fastest evolving tools for developing appropriate strategies relating to contaminated site management decisions, seeks to answer three basic questions:

- What could potentially go wrong?
- What are the chances for this to happen?
- What are the anticipated consequences if this should happen?

A complete analysis of risks associated with a given situation or activity will generate answers to these questions. Subsequently, appropriate mitigative activities can be initiated by implementing the necessary corrective action and risk management decisions. As Whyte and Burton (1980) succinctly indicate, a major objective of risk assessment is to help develop risk management decisions that are more systematic, more comprehensive, more accountable, and more aware of appropriate programs than has often been the case in the past. Ultimately, tasks performed during the risk assessment will help answer the infamous "how safe is safe enough?" and/or "how clean is clean enough?" questions.

5.2.1   The Purpose

The overall purpose of risk assessments is to provide, insofar as possible, a complete information set to risk managers, so that the best possible decision can be made concerning a potentially hazardous situation. Information developed in the risk assessment will typically facilitate decisions about the allocation of resources for safety improvements and hazard/risk reduction, by directing attention and efforts to the features and exposure pathways that dominate the risks. The results of the analysis will generally provide decision makers with a more justifiable basis for determining risk acceptability, and also aid in choosing between possible corrective measures developed for risk mitigation programs. Indeed, it is imperative to make site-specific risk assessment an integral part of all corrective action assessment programs that are designed for potentially contaminated site problems.

The use of risk assessment techniques in site cleanup plans in particular, and corrective action programs in general, is becoming increasingly important in several places. Risk assessment serves as a useful tool for evaluating the effectiveness of remedies at contaminated sites and also for establishing cleanup objectives (including the determination of cleanup levels) that will produce efficient, feasible, and cost-effective remedial solutions. The goal is to gather sufficient information to adequately and accurately characterize potential risks from the site. The use of a conceptual site model developed and refined previously in a site assessment will help focus the investigation efforts and, therefore, streamline this process. In general, a risk assessment process is utilized to determine whether the level of risk at a contaminated site warrants remediation, and to further project the amount of risk reduction necessary to protect public health and the environment. Subsequently, an appropriate corrective action plan can be developed and implemented for the case site and/or the impacted area.

5.2.2   The Attributes

The risk assessment process will generally utilize the best available scientific knowledge and data to establish case-specific responses to contaminated site management problems. In particular, the assessment of health and environmental risks associated with potentially contaminated sites may contribute in a significant way to the processes involved in corrective action planning, in risk mitigation and risk management strategies, and in the overall management of potentially contaminated site problems. Depending on the scope of the analysis, methods used in estimating risks may be either qualitative or quantitative. Thus, the process may be one of data analysis, or modeling, or a combination of the two. In fact, the process of quantifying risks does, by its very nature, give a better understanding of the strengths and weaknesses of the potential hazards being examined. It shows where a given effort can do the most good in modifying a system in order to improve safety and efficiency. The major attributes of risk assessment that are relevant to the management of contaminated site problems include:

- Identification and ranking of all existing and anticipated potential hazards.
- Explicit consideration of all current and possible future exposure scenarios.
- Qualification and/or quantification of risks associated with the full range of hazard situations, system responses, and exposure scenarios.
- Identification of all contributors to the critical pathways, exposure scenarios, and/or total risks.
- Using risk estimates to determine cost-effective risk reduction policies via the evaluation of risk-based remedial alternatives and/or the adoption of efficient risk management and prevention programs.
- Identification and analysis of sources of uncertainties associated with corrective action assessments.

Each attribute will ultimately play an important role in the overall corrective action assessment program. In general, data generated in a risk assessment are used to determine the need for, and the degree of remediation required for potentially contaminated sites. It is noteworthy, however, that there are inherent uncertainties associated with risk assessments due to the fact that the risk assessor's knowledge of the causative events and controlling factors usually is limited, and also because the results obtained depend, to a reasonable extent, on the methodology and assumptions used. Furthermore, risk assessment can impose potential delays in the implementation of corrective action programs; however, the overall gain in program efficiency is likely to more than compensate for the delays.

5.3   RISK ASSESSMENT IN THE U.S. REGULATORY SYSTEM

The U.S. Environmental Protection Agency (EPA) and several state regulatory agencies recognize the use of risk assessment as a facilitator of remedial action decisions, and also in the enforcement of regulatory standards. Risk assessment techniques have been used, and continue to be used, in various regulatory programs employed by federal, state, and local agencies. For instance, both the feasibility study process under CERCLA (or "Superfund") and the alternate concentration limit (ACL) demonstrations under RCRA involve the use of risk assessment to establish cleanup standards for contaminated sites.

CERCLA (or "Superfund") establishes a national program for responding to releases of hazardous substances into the environment, with the overarching mandate to protect human health and the environment from current and potential threats posed by uncontrolled hazardous substance releases. The primary application of quantitative risk assessment in the Superfund program is to evaluate potential risks posed at each National Priorities List (NPL) facility in order to assist in the identification of appropriate remedial action alternatives (Paustenbach, 1988); NPL is the list of uncontrolled or abandoned hazardous waste sites identified for possible long-term remedial actions under Superfund. The federal EPA also uses a risk-based evaluation method, the Hazard Ranking System (HRS), to identify uncontrolled and abandoned hazardous waste sites falling under Superfund programs. The HRS allows the selection or rejection of a site for placement on the NPL; it is used for prioritizing sites so that those apparently posing the greatest risks receive quicker response.

ACLs (which can be considered as surrogate cleanup criteria) can be established, when hazardous constituents are identified in groundwater at RCRA facilities, by applying risk assessment procedures in the analytical processes involved. In fact, nearly every process for developing cleanup criteria incorporates some concept that can be classified as a risk assessment. Thus, all decisions on setting cleanup standards for potentially contaminated sites include, implicitly or explicitly, some risk assessment concepts.

It is noteworthy that the basic steps involved in performing risk assessment at both Superfund and non-Superfund sites are fundamentally the same, except for the degree of detail necessary for the various steps involved.

5.3.1   Legislative-Regulatory Perspectives

The National Contingency Plan (NCP) (40 CFR 300) requires that applicable or relevant and appropriate requirements (ARARs), standards, or criteria be considered when developing remedial actions as part of a remedial investigation/feasibility study (RI/FS) process for contaminated site problems. *Applicable requirements* are those cleanup standards, standards of control, and other substantive environmental protection requirements, criteria, or limitations promulgated under federal or state laws that specifically address a hazardous substance, pollutant, contaminant, remedial action, location, or other circumstances at a CERCLA site (USEPA, 1988a). *Relevant and appropriate requirements* are those cleanup standards, standards of control, and other substantive environmental protection requirements, criteria, or limitations promulgated under federal or state laws that, while not "applicable" to a hazardous substance, pollutant, contaminant, remedial action, location, or other circumstances at a CERCLA site, address problems or situations sufficiently similar to those encountered at the CERCLA site, that their use is well suited to the particular site (USEPA, 1988a). The requirement can be either one of these categories, but not both. Potential ARARs are identified through a review of current local, state, and federal guidelines available in several guidance documents and databases.

Also, to-be-considered materials (TBCs) may be used, where appropriate, in the RI/FS process. TBCs are nonpromulgated advisories or guidances issued by federal or state governments that are not legally binding and do not have the status of potential ARARs. However, in many circumstances TBCs will be considered along with ARARs as part of the risk assessment, and may be used in determining the necessary level of cleanup required for the protection of public health and the environment. In fact, the cleanup requirements necessary to meet cleanup goals will generally be based not only on ARARs, but also on risk-based criteria, TBCs, health advisories, and other guidance for the state or region in which a site is located. In general, the identification of ARARs must be done on a site-specific basis.

ARARs (and TBCs to some extent) may be chemical specific (i.e., based on the nature and toxicity of the chemicals of potential concern), location-specific (e.g., based on sensitive habitats or ecosystems, wetlands, flood plains, etc.), or action-specific (i.e., based on remedial actions). A comprehensive tabulation of the requirements, administrative agencies, and summaries of potential federal and state ARARs, together with other federal, state, and local criteria, advisories, and TBC guidance will generally provide important input to contaminated site risk assessments. In fact, requirements under several federal, state, and local laws and regulations (such as

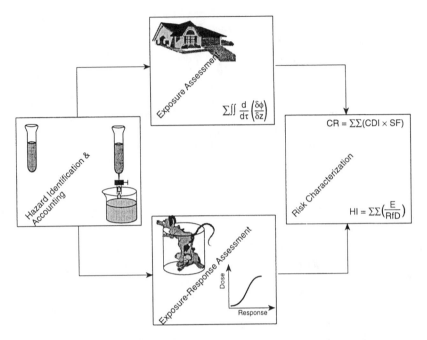

Figure 5.1   Illustrative elements of a risk assessment process.

CAA, RCRA, SDWA, CWA, and TSCA) serve as typical sources of such information for contaminated site management programs. The details of procedures and criteria for selecting ARARs, TBCs, and other guidelines are discussed elsewhere in the literature (e.g., USEPA, 1988a, 1989a).

5.4   THE RISK ASSESSMENT PROCESS

Specific forms of risk assessment generally differ considerably in their levels of detail. Most risk assessments, however, share the same general logic consisting of four basic elements (Figure 5.1). A discussion of these fundamental elements follows, with more detailed elaboration given elsewhere in the risk analysis literature (e.g., Asante-Duah, 1990, 1993; Bowles et al., 1987; CAPCOA, 1990; Hallenbeck and Cunningham, 1988; Huckle, 1991; NRC, 1983; Paustenbach, 1988; Rowe, 1977; USEPA, 1984, 1989b, 1989c).

Hazard identification and accounting involves a qualitative assessment of the presence of, and the degree of hazard that a contaminant could have on potential receptors. In the context of corrective action assessment for potentially contaminated site problems this may consist of the identification of contaminant sources, a compilation of the lists of all contaminants present at the site, the identification and selection of the specific chemicals of potential concern (that should become the focus of the risk assessment) based on their specific hazardous properties (such as persistence, bioaccumulative properties, toxicity, and general fate and transport properties), and a compilation of summary statistics for the key constituents selected for further investigation and evaluation. In identifying the chemicals of potential concern, an attempt is generally made to select all chemicals that could represent the major part (≥95%) of the risks associated with site-related exposures.

The *exposure-response* component of the risk assessment undertaken for contaminated site problems will generally include a toxicity assessment and/or a dose-response evaluation. It considers the types of adverse effects associated with chemical exposures, the relationship between the magnitude of exposure and adverse effects, and related uncertainties (such as the weight-of-evidence of a particular chemical's carcinogenicity in humans). The toxicity assessment consists of compiling toxicological profiles for the chemicals of potential concern. Dose-response relationships are then used to quantitatively evaluate the toxicity information and to characterize the relationship between dose of the contaminant administered or received and the incidence of adverse effects on the exposed population. From the quantitative dose-response relationship, appropriate toxicity values can be derived and subsequently used to estimate the incidence of adverse effects occurring in populations at risk for different exposure levels.

An *exposure assessment* is conducted to estimate the magnitude of actual and/or potential receptor exposures to environmental contaminants, the frequency and duration of these exposures, the nature and size of the populations potentially at risk (i.e., the risk group), and the pathways by which the risk group may be exposed. Several physical and chemical characteristics of the chemicals of concern will provide an indication of the critical exposure features. These characteristics can also provide information necessary to determine the chemical's distribution, intake, metabolism, residence time, excretion, magnification, and half-life or breakdown to new chemical compounds. To complete a typical exposure assessment, populations potentially at risk are identified, and concentrations of the chemicals of concern are determined in each medium to which potential receptors may be exposed. Finally, using the appropriate site-specific exposure parameter values, the intakes of the chemicals of concern are estimated. The exposure estimates can then be used to determine if any threats exist based on existing exposure conditions at or near the potentially contaminated site.

Risk characterization is the process of estimating the probable incidence of adverse impacts to potential receptors under various exposure conditions. Typically, the risk characterization summarizes and then integrates outputs of the exposure and toxicity assessments in order to qualitatively and/or quantitatively define risk levels. This usually will include an elaboration of uncertainties associated with the risk estimates. An adequate characterization of risks and hazards at a potentially contaminated site allows corrective action programs to be better focused. Exposures resulting in the greatest risk can be identified; site mitigation measures can then be selected to address the situation in order of priority, according to the levels of imminent risks.

The results of a risk assessment can be used to determine if corrective actions are needed at a potentially contaminated site and/or if compliance with applicable regulatory requirements are violated. In this sense, the risk assessment process integrates the information obtained during a site characterization or remedial investigation activity into a coherent set of goals for a feasibility study or the restoration of potentially contaminated sites. Ultimately, cleanup criteria can be developed based on an acceptable benchmark risk level for the specific situation.

5.5 METHODS OF APPROACH IN CONTAMINATED SITE RISK ASSESSMENT

Risk assessment is a process used to evaluate the collective demographic, geographic, physical, chemical, biological, and related factors at potentially contaminated sites, in order to determine and characterize possible risks to public health and the environment. The overall objective of such an assessment is to determine the magnitude and probability of actual or potential harm that the site poses to human health and the environment. A procedure generally used in contaminated site risk assessments will comprise of the following elements:

- Identification of the sources of contamination
- Definition of the contaminant migration pathways
- Identification of populations potentially at risk
- Determination of the specific chemicals of potential concern
- Determination of frequency of potential receptor exposures to site contaminants
- Evaluation of contaminant exposure levels
- Determination of receptor response to chemical exposures
- Estimation of impacts or damage resulting from receptor exposures to the chemicals of potential concern

Potential risks are estimated by considering the probability or likelihood of occurrence of harm, the intrinsic harmful features or properties of specified hazards, the populations potentially at risk, the exposure scenarios, and the extent of expected harm and potential effects.

A number of techniques are available for conducting risk assessments. Invariably, the methods of approach consist of the selection of site contaminants of significant concern, an exposure assessment (consisting of a pathway analysis and the estimation of chemical intakes), a toxicity assessment (for human and ecological receptors), and a risk characterization (for both human health and ecological effects). The more commonly used risk assessment approaches that are relevant to the management of contaminated site problems are elaborated in the subsequent sections. It is noteworthy that, in general, much of the effort in the development of risk assessment methodologies has been directed at human health risk assessments (as reflected by the differences in the depth of presentation of human health vs. ecological risk assessment found in the literature). However, the fundamental components of the risk assessment process for other biological organisms parallel those for human receptors, and can be described in similar terms.

5.5.1 Baseline Risk Assessments

The baseline risk assessment is an analysis of the potential adverse health and environmental effects (current or future) caused by hazardous substance releases from a contaminated site in the absence of any actions to control or mitigate these releases (i.e., under an assumption of "no-action"). Thus, the baseline risk assessment provides an estimate of the potential risks to human health and the environment resulting from receptor exposure to site contaminants in the absence of remediation. Because this type of assessment identifies the primary health and environmental threats at the site, it also provides valuable input to the development and evaluation of alternative site restoration options. In fact, baseline risk assessments are generally conducted to evaluate the need for, and the extent of remediation required at potentially contaminated sites. That is, they provide the basis and rationale as to whether or not remedial action is necessary.

A baseline risk assessment contributes to the contaminated site characterization process. It further facilitates the development, evaluation, and selection of appropriate corrective action response alternatives; this may include the evaluation and selection of a "no-action" remedial alternative, where appropriate. The results of the baseline risk assessment are generally used to:

- Document the magnitude of risk at a site, and the primary causes of that risk
- Help determine whether additional response action is necessary at the site
- Prioritize the need for remediation
- Provide the basis for quantifying remedial action objectives
- Modify preliminary remediation goals
- Support and justify "no further action" decisions by documenting the threats posed by a site based on expected exposure scenarios

Baseline risk assessments are site specific and therefore may vary in both detail and the extent to which qualitative and quantitative analyses are used. The level of effort required to conduct a baseline risk assessment depends largely on the complexity and particular circumstances (such as regulatory systems, criteria, and guidances) associated with the site.

5.5.2   Planning and Design of Risk Assessment Programs

Human populations and ecological receptors are continuously in contact with varying amounts of environmental contaminants present in air, water, soil, and food. Thus, managing the health and environmental risks associated with environmental contaminants usually requires an integrated model to effectively assess the environmental fate and transport, as well as to determine potential human and ecosystem exposures. Methods for linking contaminant sources in multiple environmental media (such as air, water, and soil) to human and ecological receptor exposures are often necessary to facilitate the development of sound corrective action response programs. Ultimately, the following specific tasks are typically carried out in order to achieve the risk assessment goals established for case-specific contaminated site problems:

- Data Evaluation

    Perform DQO assessment
    Identify, quantify, and group site contaminants
    Screen and select chemicals of potential concern
    Carry out statistical analysis of relevant site data

- Exposure Assessment

    Compile information on the physical setting of the site
    Identify source areas, significant migration pathways, and potentially impacted/receiving media

Determine the important environmental fate and transport processes for the chemicals of potential concern
Identify populations potentially at risk
Determine likely and significant receptor exposure pathways
Construct representative conceptual site model(s)
Develop exposure scenarios (to include current and potential future land uses)
Estimate/model exposure point concentrations for the chemicals of potential concern
Compute potential receptor intakes and resultant doses for the chemicals of potential concern (for all potential receptors and significant pathways of concern)

- Toxicity Assessment

  Compile toxicological profiles (to include the intrinsic toxicological properties of the chemicals of potential concern, such as their acute, subchronic, chronic, carcinogenic, and reproductive effects)
  Determine appropriate toxicity indices (such as the acceptable daily intakes or reference doses, cancer slope or potency factors, lethal doses, lethal concentrations, etc.)

- Risk Characterization

  Estimate carcinogenic risks from carcinogens
  Estimate noncarcinogenic hazard quotients and indices for systemic toxicants
  Estimate ecological quotients for ecological receptors
  Perform sensitivity analyses, evaluate uncertainties associated with the risk estimates, and summarize the risk information

- Development of Cleanup Criteria

  Determine benchmark level of risks that can be tolerated by potential receptors
  "Back-model" to obtain target/acceptable cleanup levels for contaminants

In actuality, the development of cleanup criteria (elaborated in Chapter 6) is not necessarily an integral component of the risk assessment process, since this belongs more in the realm of risk management. Oftentimes, however, this is included in corrective action assessment programs.

The risks and/or hazards associated with residual contaminations to be left at a contaminated site following the implementation of a remedial action can also be evaluated as part of the site restoration program. This is accomplished via the simulation of risk characterization scenarios for future land use and exposure conditions at the restored site.

5.6   HUMAN HEALTH RISK ASSESSMENTS

Human health risk assessment is defined as the characterization of the potential adverse health effects associated with human exposures to environmental hazards

(NRC, 1983). In the health risk assessment process, the extent to which potential human receptors have been, or could be exposed to chemical(s) present at a potentially contaminated site is determined. The extent of exposure is then considered in relation to the type and degree of hazard posed by the chemical(s), thereby permitting an estimate to be made of the present or future health risks to the populations-at-risk (PARs).

Quantitative human health risk assessment is often an integral part of corrective action assessment and also site mitigation programs designed for contaminated sites. Figure 5.2 shows the basic components and steps involved in a comprehensive human health risk assessment. Several important aspects of the health risk assessment methodology are enumerated below; further details can be found elsewhere in the literature (e.g., Asante-Duah, 1993; Huckle, 1991; NRC, 1983; USEPA, 1989b).

5.6.1 Data Evaluation

The data evaluation aspect of a human health risk assessment consists of an identification and analysis of the chemicals present at a potentially contaminated site that should become the focus of the investigation. In this process, an attempt is generally made to select all chemicals that could represent the major part of the risks associated with site-related exposures; typically, this will consist of all carcinogens together with noncarcinogenic constituents contributing ≥95% of the site risks. Chemicals are screened based on such parameters as toxicity, carcinogenic classification, concentration of detected chemicals, frequency of detection in the sampled matrix, etc. Sample blanks should also be evaluated as part of the data screening process.

Evaluation of "Control" Samples
The analysis of blank samples provides a way to determine if contamination has been introduced into a sample set either in the field while the samples were being collected and transported to the laboratory, or in the laboratory during sample preparation and analysis (DTSC, 1994; USEPA, 1989b).

To prevent the inclusion of nonsite-related contaminants in the risk assessment, the concentrations of the chemicals detected in blanks must be compared with concentrations of the same chemicals detected in site samples. In general, blanks containing common laboratory contaminants are evaluated differently from blanks which contain chemicals that are not common laboratory contaminants. For instance, acetone, 2-butanone (methyl ethyl ketone), methylene chloride, toluene, and the phthalate esters are generally considered by the EPA to be common laboratory contaminants (USEPA, 1989b, 1990a). Thus, if the blank contains detectable levels of these common laboratory contaminants, then the environmental sample results are considered as positive only if the concentrations in the sample exceed ten times the maximum amount detected in any blank (DTSC, 1994; USEPA, 1989b, 1990a). For blanks containing detectable levels of one or more organic or inorganic chemicals that are not considered to be common laboratory contaminants, site sample results are considered as positive only if the concentration of the chemical in the site sample exceeds five times the maximum amount detected in any blank (DTSC, 1994; USEPA, 1989b).

Statistical Evaluation of Site Data
Statistical analysis procedures that reflect the character of the underlying data distribution should be used in the data evaluation. Indeed, as discussed in Chapter 4,

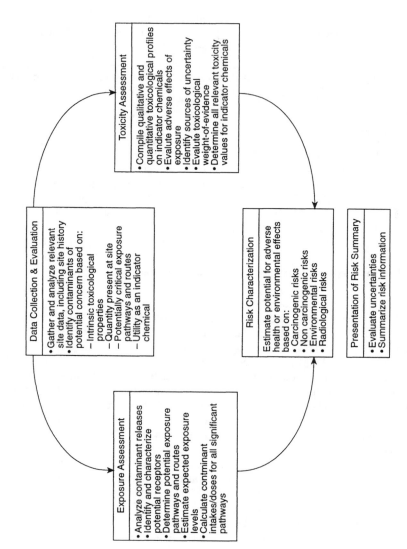

Figure 5.2 Components of a human health risk assessment process for contaminated site problems.

statistical procedures used for data evaluation can significantly affect the final results of the overall assessment. For instance, the use of a normal distribution (whose central tendency is measured by the arithmetic mean) to describe environmental contaminant distribution, rather than lognormal statistics (whose central tendency is defined by the geometric mean) will often result in significant overestimation. Appropriate statistical methods should therefore be utilized in the evaluation of environmental sampling data necessary to support a risk assessment. Furthermore, contamination levels and exposures may have temporal variations and the dynamic nature of such parameters should be incorporated in a full analysis.

For any potentially contaminated site, contaminant concentrations are likely to vary widely across the site and also within the various environmental media. To represent the natural variability and heterogeneous nature of site data, it often is convenient to use continuous probability distributions to describe the media contamination. The probability distribution can also be stated in the form of discrete uncertainty nodes. For instance, consider a representation of three possible values of soil concentrations for which there are 40, 50, and 10% chances that the average soil concentrations are 100 ppm (low), 500 ppm (moderate/nominal), and 1000 ppm (high), respectively. This may equivalently be considered as representing the spatial distribution of contamination for the whole site, in which 40, 50, and 10% of the site are respectively at low, moderate, and high contaminant levels. In most screening analyses, the "nominal" level will be considered as representing an average concentration, and this value will be used in the evaluation. However, for a more comprehensive analysis, the uncertainties in the average and other pertinent variables should be explicitly defined and analyzed. This may include the use of weighted averages as being relatively more representative of the contaminant concentrations. In general, contaminant concentrations used during an exposure assessment should reflect the average concentration to which potential receptors may be exposed (usually represented by the 95% UCL).

Statistical Treatment of Non-Detects in Environmental Samples

All laboratory analytical techniques have detection limits (DLs) below which only "less than" values may be reported; the reporting of such values provides a degree of quantification. In fact, it is customary to assign non-zero values to all environmental sampling data reported as non-detects (NDs). This is important because even at or near their detection limits, certain chemical constituents may be of considerable importance in a risk assessment. However, uncertainty about the actual values below the DL can bias or preclude subsequent statistical analyses.

One approach in the calculation of the applicable statistical values during the evaluation of data containing NDs involves the use of a value of one-half of the sample quantitation limit (SQL) (a.k.a., practical quantitation limit, PQL), which is a quantifiable number used in practice to define the DL. This approach assumes that the samples are equally likely to have any value between the detection limit and zero. However, when the sample values above the ND level are log-normally distributed, it generally may be assumed that the ND values are also log-normally distributed. The best estimate of the ND values for such situations is the reported SQL divided by the square root of two {i.e., $\frac{SQL}{1.414}$} (CDHS, 1990; USEPA, 1989b).

In general, during the analysis of environmental sampling data that contains some NDs, a fraction of the SQL is usually assumed (as a proxy or estimated concentration) for nondetectable levels, instead of assuming a value of zero or neglecting such values.

This procedure is used provided there is at least one detected value from the analytical results and/or if there is reason to believe that the chemical is possibly present in the sample at a concentration below the SQL. The method conservatively assumes that some level of the chemical is present (even when an ND is recorded) and arbitrarily sets that level at the appropriate percentage of the SQL (i.e., $\frac{SQL}{2}$ if the data set is assumed to be normally distributed or $\frac{SQL}{1.414}$ for a log-normally distributed data set). In fact, it has been suggested in some cases (e.g., USEPA, 1989b) that the SQL value itself be used if there is strong reason to believe that the chemical concentration is closer to this value rather than to a fraction of the SQL. Where it is apparent that serious biases could result, more sophisticated analytical and evaluation methods may be warranted.

5.6.2   Exposure Assessment

The exposure assessment phase of the human health risk assessment involves:

- A characterization of the physical and exposure setting of the site
- The identification of significant migration and exposure pathways
- The identification of potential receptors, or the PARs
- The development of conceptual site model(s) and exposure scenarios under both current and future site conditions
- An analysis of contaminant fate and transport, including intermedia transfers
- The estimation of exposure point concentrations for the critical pathways and environmental media
- The calculation of chemical intakes for all potential receptors and significant pathways of concern

The process is used to estimate the rates at which chemicals are absorbed by potential receptors. Since humans (and indeed ecological receptors also) tend to be exposed to chemicals from a variety of sources and/or in different environmental media, an evaluation of the relative contributions of each medium or source to total chemical intake could be critical in a multipathway exposure scenario. In fact, the accuracy with which exposures are characterized could be a major determinant of the ultimate validity of a risk assessment.

Several techniques may be used for the exposure assessment, including the modeling of anticipated future exposures, environmental monitoring of current exposures, and biological monitoring to determine past exposures. The physicochemical properties of the contaminants of potential concern and the impacted media are important considerations in the exposure modeling. In general, a variety of exposure models and conservative but realistic assumptions regarding contaminant migration and equilibrium partitioning are used to facilitate the exposure quantification. Site-related exposure point concentrations can be determined once the potentially affected populations are identified and the exposure scenarios are defined. If the transport of compounds associated with the site is under steady-state conditions, monitoring data are generally adequate to determine potential exposure concentrations. If there are no data available, or if conditions are transient (such as pertains to a migrating plume in groundwater), models are better used to predict exposure concentrations. Many factors — including the fate and transport properties of the chemicals of concern — must be considered in the model selection process. In any case, in lieu of an established trend in historical data indicating the contrary, a potentially contaminated site may be considered to be in steady state with its surroundings.

Exposure Pathways

Exposure pathways are one of the most important elements of the exposure assessment, consisting of the routes that contaminants follow to reach potential receptors. An exposure pathway is considered complete only if all of the following elements are present:

- Contaminant source(s)
- Mechanism(s) of contaminant release into the environment
- Contaminant migration pathway(s) and exposure route(s)
- Receptor exposure in the affected media

Exposure pathways are determined by integrating information from an initial site characterization with knowledge about potentially exposed populations and their likely behavior. The significance of migration pathways is evaluated on the basis of whether the contaminant migration could cause significant adverse human exposures and impacts. The route and duration of exposure will significantly influence the impacts on the affected receptors. Exposure routes may consist of inhalation, ingestion, and/or dermal contacts; the exposure duration may be short term (acute) or long term (chronic). For receptor exposure to occur, a complete pathway must be present, which is used to define the exposure scenarios that will realistically exist for the site. Failure to identify and address any significant exposure pathway may seriously detract from the validity of any risk assessment.

Chemical Intake Vs. Dose

Once exposure point concentrations in all media of concern have been estimated, the intakes and/or doses to potentially exposed populations can be determined. Intake is defined as the amount of chemical coming into contact with the receptor's body or exchange boundaries (such as the skin, lungs, or gastrointestinal tract); dose is the amount of chemical absorbed by the body into the bloodstream. The absorbed dose differs significantly from the externally applied dose (called exposure or intake).

For each exposure pathway considered in the exposure assessment, an intake per event is developed. This value quantifies the amount of a chemical contacted during each exposure event, where "event" may have different meanings depending on the nature of exposure scenario being considered (e.g., each day's inhalation of an air contaminant constitutes one inhalation exposure event). The quantity of a chemical absorbed into the bloodstream per event, represented by the dose, is calculated by further considering pertinent physiological parameters (such as gastrointestinal absorption rates). When the systemic absorption from an intake is unknown, or cannot be estimated by a defensible scientific argument, intake and dose are considered to be the same (i.e., 100% absorption into the bloodstream from contact is assumed). This approach provides a conservative estimate of the actual exposures. In general, conservative estimates assume that the potential receptor is always in the same location, exposed to the same ambient concentration, and that there is 100% absorption on exposure. These assumptions hardly represent any real-life situation. In fact, lower exposures will be expected due to the fact that potential receptors will generally be exposed to lower or even near-zero levels of the chemicals of concern for the period of time spent outside the impacted areas.

Event-based intake values are converted to final intake values by multiplying the intake per event by the frequency of exposure events over the time frame being considered. Chronic daily intake (CDI), which measures long-term (chronic) exposures, are based on the number of events that are assumed to occur within an assumed 70-year lifetime for human receptors; subchronic daily intake (SDI), which represents projected human exposures over a short-term period, consider only a portion of a lifetime (USEPA, 1989b). These are calculated by multiplying the average or the reasonable maximum exposure (RME) media concentrations — usually represented by the 95% UCL — by the appropriate human exposure and body weight factors. SDIs are generally used to evaluate subchronic noncarcinogenic effects, whereas CDIs are used to evaluate both carcinogenic risks and chronic noncarcinogenic effects.

Intakes and doses are normally calculated in the same step of the exposure assessment, whereby the former, multiplied by an absorption factor, yields the latter value. In general, determination of the carcinogenic effects (and sometimes the chronic noncarcinogenic effects) from a contaminated site involve estimating the lifetime average daily dose (LADD). For noncarcinogenic effects, the average daily dose (ADD) is usually used. The ADD differs from the LADD in that the former is not averaged over a lifetime; rather, it is the average daily dose pertaining to the days of exposure. The maximum daily dose (MDD) will typically be used in estimating acute or subchronic exposures.

The Exposure Estimation Model

The methods by which each type of exposure is estimated are well documented in the literature of exposure assessment (e.g., CAPCOA, 1990; DOE, 1987; USEPA, 1988b, 1989b, 1989d). Typically, potential receptor exposures to environmental contaminants is conservatively estimated according to the generic equation shown in Box 5.1. The exposure estimation equations for specific major routes of exposure are presented in Appendix F, together with an illustration of the computational steps involved in the calculation of receptor intakes and doses.

The concentration in the various media may be obtained from field measurements, estimated by simple mass balance analyses or other appropriate contaminant transport models, or may be determined from equilibrium and partitioning relations. The various exposure parameters may be derived on a case-specific basis, or they may be compiled from regulatory guidance manuals and documents (e.g., DTSC, 1994; OSA, 1993; USEPA, 1989b, 1989d, 1991, 1992). These parameters are usually based on information relating to the maximum exposure level resulting from specified categories of human activity and/or exposures.

Receptor Age-Adjustment Factors

In a refined and comprehensive evaluation, it is generally recommended to incorporate age adjustment factors in the exposure assessment, where appropriate. Age adjustments, necessary when receptor exposures occur from childhood through adult life, are meant to account for the transitioning of a potential receptor from childhood (requiring one set of intake assumptions and exposure parameters) to adulthood (that requires a different set of intake assumptions and exposure parameters). Further details on the development of age-adjusted factors is provided elsewhere

---

BOX 5.1

General Equation for Estimating Receptor Exposures to Environmental Contaminants

$$I = \frac{(CONC \times CR \times CF \times FI \times ABS_s \times EF \times ED)}{(BW \times AT)}$$

where:

I	=	intake (i.e., the amount of chemical at the exchange boundary), adjusted for absorption (mg/kg-day)
CONC	=	average chemical concentration contacted over the exposure period from the media of concern (e.g., $\mu g/m^3$ [air], or mg/L [water], or mg/kg [soil])
CR	=	contact rate, i.e., the amount of contaminated medium contacted per unit time or event (e.g., inhalation rate in m^3/day, ingestion rate in mg/day [soil], or L/day [water])
CF	=	conversion factor
FI	=	fraction of intake from contaminated source (unitless)
ABS_s	=	bioavailability/absorption factor (%)
EF	=	exposure frequency (days/years)
ED	=	exposure duration (years)
BW	=	body weight, i.e., the average body weight over the exposure period (kg)
AT	=	averaging time (period over which exposure is averaged — days).

---

in the literature (e.g., DTSC, 1994; OSA, 1993; USEPA, 1989b). For simplicity, such age adjustments are usually not made part of most screening-level computational processes.

Incorporating Contaminant Degradation into Exposure Calculations

When certain chemical compounds undergo degradation, potentially more toxic daughter products result (such as is the case when trichloroethylene, TCE, biodegrades to produce vinyl chloride). On the other hand, there are situations where the end products of degradation are less toxic than the parent compounds. Since receptor exposures could be occurring over long time periods, a more valid approach in exposure modeling will be to take contaminant degradation (or indeed other transformation processes) into consideration during an exposure assessment. Under such circumstances, if significant degradation is likely to occur, then exposure calculations become much more complicated. In that case, contaminant concentrations at release sources are calculated at frequent and short time intervals, and then summed over the exposure period.

To illustrate the concept of incorporating contaminant degradation into exposure assessment, assume first-order kinetics for a hypothetical problem. An approximation of the degradation effects for this scenario can be obtained by multiplying contaminant concentrations by a degradation factor, DGF, that is defined by:

$$DGF = \frac{\left(1 - e^{-kt}\right)}{kt}$$

where k [days^{-1}] is a chemical-specific degradation rate constant and t [days] is the time period over which exposure occurs. For a first-order decaying substance, k is estimated from the following relationship:

$$T_{1/2} \,[days] = \frac{0.693}{k}$$

or

$$k\left[days^{-1}\right] = \frac{0.693}{T_{1/2}}$$

where $T_{1/2}$ is the chemical half-life, which is the time after which the mass of a given substance will be one-half its initial value.

The degradation factor is generally ignored in most exposure calculations. This is especially justifiable if the degradation product is of potentially equal toxicity and of comparable amounts. Although it cannot always be proven that the daughter products result in receptor exposures of comparable levels to the parent compound, the DGF is nevertheless ignored in most screening-level exposure assessments.

5.6.3 Toxicity Assessment

A toxicity assessment is conducted as part of the human health risk assessment, to qualitatively and quantitatively determine the potential for adverse health effects to occur as a result of human exposure to environmental contaminants. The toxicity assessment evaluates the types of adverse health effects associated with chemical exposures, the relationship between the magnitude of exposure and adverse effects, and related uncertainties such as the weight-of-evidence of a particular chemical's carcinogenicity in humans.

A detailed toxicity assessment for chemicals found at contaminated sites is generally accomplished in two steps: hazard assessment and dose-response assessment. Hazard assessment is the process used to determine whether exposure to an agent can cause an increase in the incidence of an adverse health effect (e.g., cancer, birth defects, etc.); it involves a characterization of the nature and strength of the evidence of causation. Dose-response assessment is the process of quantitatively evaluating the toxicity information and characterizing the relationship between the dose of the contaminant administered or received (i.e., exposure to an agent) and the incidence of adverse health effects in the exposed populations; it is the process by which the potency of the compounds is estimated by use of dose-response relationships. These steps are discussed in more detail elsewhere in the literature (e.g., Klaassen et al., 1986; USEPA, 1989b).

Carcinogenic Vs. Noncarcinogenic Chemicals

For the purpose of human health risk assessment, chemicals are usually categorized into carcinogenic and noncarcinogenic groups. Chemicals that give rise to toxic endpoints other than cancer and gene mutations are often referred to as "systemic toxicants" because of their effects on the function of various organ systems; the toxic endpoints are referred to as "noncancer or systemic toxicity". Most chemicals that produce noncancer toxicity do not cause a similar degree of toxicity in all organs, but usually demonstrate major toxicity to one or two organs. These are referred to as the target organs of toxicity for the chemicals (Klaassen et al., 1986; USEPA, 1989b). In addition, chemicals that cause cancer and gene mutations also commonly evoke other toxic effects (i.e., systemic toxicity). Carcinogenic chemicals are generally classified into several categories, depending on the weight-of-evidence or the strength-of-evidence available on the particular chemical's carcinogenicity (Appendix G).

Threshold Vs. Nonthreshold Chemicals

Noncarcinogens operate by threshold mechanisms, i.e., the manifestation of systemic effects requires a threshold level of exposure or dose to be exceeded during a continuous exposure episode. Thus, noncancer or systemic toxicity is generally treated as if there is an identifiable exposure threshold below which there are no observable adverse effects. This characteristic distinguishes systemic endpoints from carcinogenic and mutagenic endpoints, which are often treated as nonthreshold processes. That is, the threshold concept and principle is not applicable for carcinogens, since it is believed that no thresholds exist for this group. It is noteworthy, however, that there is a belief among some toxicologists that certain carcinogens require a threshold exposure level to be exceeded in order to provoke carcinogenic effects.

5.6.3.1 *Determination of Toxicological Parameters*

Often, it becomes necessary to compare receptor intakes of chemicals with doses shown to cause adverse effects in humans or experimental animals. The dose at which no effects are observed in human populations or experimental animals is referred to as the "no-observed-effect-level" (NOEL). Where data identifying a NOEL are lacking, a "lowest-observed-effect-level" (LOEL) may be used as the basis for determining safe threshold doses. For acute effects, short-term exposures/doses shown to produce no adverse effects are involved; this is called the "no-observed-adverse-effect-level" (NOAEL). A NOAEL is an experimentally determined dose at which there has been no statistically or biologically significant indication of the toxic effect of concern. In cases where a NOAEL has not been demonstrated experimentally, the term "lowest-observed-adverse-effect level" (LOAEL) is used. For chemicals possessing carcinogenic potentials, the LADD is compared with the NOEL identified in long-term bioassay experimental tests; for chemicals with acute effects, the MDD is compared with the NOEL observed in short-term animal studies.

Appendix H presents a number of relevant procedures that may be used in the derivation and interconversion of toxicity parameters required for human health risk assessments.

Toxicity Parameters for Noncarcinogenic Effects

Traditionally, risk decisions on systemic toxicity have been made using the concept of "acceptable daily intake" (ADI) derived from the experimentally determined NOAEL, or by using the reference dose (RfD). The ADI is the amount of a chemical (in mg/kg body weight/day) to which a receptor can be exposed to on a daily basis over an extended period of time — usually a lifetime — without suffering a deleterious effect. The RfD is defined as the maximum amount of a chemical (in mg/kg body weight/day) that the human body can absorb without experiencing chronic health effects. For exposure of humans to the noncarcinogenic effects of chemicals, the ADI or RfD is used as a measure of exposure considered to be without adverse effects. Although often used interchangeably, the RfD is based on a more rigorously defined methodology and is therefore a preferred parameter to use, rather than the ADI.

The RfD provides an estimate of the continuous daily exposure of a noncarcinogenic substance for the general human population (including sensitive subgroups) which appears to be without an appreciable risk of deleterious effects. RfDs have been established as thresholds of exposure to toxic substances below which there should be no adverse health impact. These thresholds have been established on a substance-specific basis for oral and inhalation exposures, taking into account evidence from both human epidemiologic and laboratory toxicologic studies.

Typically, subchronic RfD is used to refer to cases involving only a portion of the lifetime, whereas chronic RfD refers to lifetime exposures. In assessing the chronic and subchronic effects of noncarcinogens and also noncarcinogenic effects associated with carcinogens, the experimental dose value (e.g., NOEL) is typically divided by a safety (or uncertainty) factor to yield the RfD, as illustrated in Appendix H.

Toxicity Parameters for Carcinogenic Effects

Assuming a no-threshold situation for carcinogenic effects, an estimate of the excess cancer per unit dose called the unit cancer risk (UCR) or the cancer slope factor (SF), is used to develop risk decisions for contaminated site problems. Under the nonthreshold assumption, exposure to any level of a carcinogen is considered to have a finite risk of inducing cancer. The SF is a measure of the carcinogenic toxicity or potency of a chemical. It is the cancer risk (proportion affected) per unit of dose (i.e., risk per mg/kg/day).

In evaluating risks from chemicals found in certain environmental sources, dose-response measures may be expressed as risk per concentration unit. These measures may include the unit risk (UR) for air (i.e., inhalation UR) and the unit risk for drinking water (i.e., oral UR). The continuous lifetime exposure concentration units for air and drinking water are usually expressed in micrograms per cubic meter ($\mu g/m^3$) and micrograms per liter ($\mu g/L$), respectively.

Use of Surrogate Toxicity Parameters

In general, the toxicity parameters are dependent on the route of exposure. However, oral RfDs and SFs will normally be used for both ingestion and dermal exposures to some chemicals that affect receptors through a systemic action; this will be inappropriate if the chemical affects the receptor contacts through direct local action at the point of application. Thus, in several (but certainly not all) situations, it is appropriate to use oral SFs and RfDs as surrogate values to estimate systemic toxicity as a result of dermal absorption of a chemical (DTSC, 1994; USEPA, 1989b).

It is noteworthy, however, that use of the oral SF or oral RfD directly does not correct for differences in absorption and metabolism between the oral and dermal routes. Typically, absorption fractions corresponding to 10% and 1% are applied to organic and inorganic chemicals, respectively. Also, direct toxic effects on the skin are not accounted for. Thus, the use of an oral SF or oral RfD for the dermal route may result in an over- or underestimation of the risk or hazard depending on the chemical involved. Consequently, the use of the oral toxicity value as a surrogate for a dermal value will tend to increase the uncertainty in the estimation of risks and hazards; however, this is not generally expected to significantly underestimate the risk or hazard relative to the other routes of exposure that are evaluated in the risk assessment (DTSC, 1994). Furthermore, in the evaluation of the inhalation pathways, when an inhalation SF or RfD is not available for a compound, the oral SF or RfD may be used in its place for screening analyses. Similarly, inhalation SFs and RfDs may be used as surrogates for both ingestion and dermal exposures for those chemicals lacking oral toxicity values.

In other situations, toxicity values to be used in characterizing risks are available only for certain chemicals within a chemical class. In such cases, rather than eliminate those chemicals without toxicity values from the quantitative evaluation, it usually is prudent to group data for such class of chemicals (e.g., according to structure-activity relationships or other similarities) for consideration in the risk assessment; such grouping should not be based solely on toxicity class or carcinogenic classifications. Significant uncertainties will likely result by using this type of approach. Hence, if and when this type of grouping is carried out it should be acknowledged and documented in the risk assessment summary, indicating the fact that the action may have produced over- or underestimates of the true risk.

The introduction of additional uncertainties in an approach that relies on surrogate toxicity parameters cannot be overemphasized, and such uncertainties should be documented as part of the risk evaluation process. Further elaboration on the derivation of RfDs, ADIs, SFs, and UCRs is included in Appendix H of this book; an in-depth discussion can be found in the literature elsewhere (e.g., Dourson and Stara, 1983; USEPA, 1986, 1989d, 1989i).

5.6.4   Risk Characterization

Risk characterization consists of estimating the probable incidence of adverse impacts to potential receptors under various exposure conditions that are associated with a hazard situation. It involves an integration of the toxicity and exposure assessments in order to arrive at an estimate of risk to the exposed population. During risk characterization, chemical-specific toxicity information is compared against both field-measured and estimated contaminant exposure levels (and in some cases, those levels predicted through fate and transport modeling) in order to determine whether contaminant concentrations at or near a potentially contaminated site are of significant concern.

Risk characterization is the final step in the risk assessment process, and becomes the first input into risk management programs. It does indeed serve as a bridge between risk assessment and risk management, and is therefore a key step in the decision-making process developed for contaminated site problems. The risks to potentially exposed populations as a result of exposure to the COCs are characterized through a calculation of noncarcinogenic hazard quotients and indices and/or carcinogenic risks

(Appendix I). These parameters can then be compared with benchmark standards in order to arrive at risk decisions about the site.

Adjustments for Absorption

Absorption adjustments may be necessary to ensure that the site exposure estimate and the toxicity value being compared are both expressed as absorbed doses or both expressed as administered doses (i.e., intakes). Adjustments may also be necessary for different vehicles of exposure (e.g., water, food, or soil). Furthermore, adjustments may be needed for different absorption efficiencies, depending on the medium of exposure. In particular, correction for fractional absorption is generally appropriate when interaction with environmental media or other contaminants may alter absorption from that expected for the pure compound, and/or when assessment of exposure was via a different route of contact than what was utilized in the experimental studies used to establish the toxicity parameters (i.e., SFs and RfDs).

Absorption factors should not be used to modify exposure estimates in those cases where absorption is inherently factored into the toxicity/risk parameters used for the risk characterization. Thus, "correction" for fractional absorption is appropriate only for those values derived from experimental studies based on absorbed dose. That is, absorbed dose should be used in risk characterization only if the applicable toxicity parameter (e.g., SF or RfD) has been adjusted for absorption; otherwise, intake data (unadjusted for absorption) are used for the calculation of risk levels. Absorption efficiency adjustment procedures are discussed elsewhere in the literature (e.g., USEPA, 1989b). In the absence of reliable information, 100% absorption is normally used for most chemicals; for metals, approximately a 10% absorption may be considered as a reasonable upper bound for other than the inhalation exposure route.

Aggregate Effects of Chemical Mixtures

The contaminants found at contaminated sites tend to be heterogeneous and variable mixtures that may contain several distinct compounds distributed over wide spatial regions and across several environmental compartments. The toxicology of complex mixtures is not well understood, complicating the problem in regard to the potential for these compounds to cause various health and environmental effects. Nonetheless, there is the need to assess the cumulative health risks for the chemical mixtures despite potential large uncertainties that may exist. The risk assessment process must address the multiple endpoints or effects and also the uncertainties in the dose-response functions for each effect. Risk characterization involves the quantitative estimation of the actual and potential risks and/or hazards due to exposure to each key chemical constituent, and also the possible additive effects of exposure to mixtures of the COCs.

The common method of approach assumes additivity of effects for carcinogens when evaluating chemical mixtures or multiple carcinogens. Prior to the summation of aggregate risks, however, estimated cancer risks should preferably be segregated by weight-of-evidence category for the contaminants at the site, the goal being to provide a clear understanding of the risk contribution of each category of carcinogen.

For multiple pollutant exposures to noncarcinogens and noncarcinogenic effects of carcinogens, constituents should be grouped by the same mode of toxicological action (i.e., those which induce the same toxicological endpoint, such as liver toxicity). Toxicological endpoints that will normally be considered with respect to chronic toxicity include the cardiovascular system (CVS), central nervous system (CNS), immune system, reproductive system (including teratogenic and developmental effects), kidney, liver, and respiratory system. Cumulative noncarcinogenic risk is evaluated through the use of a hazard index that is generated for each health "endpoint". In fact, in a strict sense, constituents should not be grouped together unless they have the same toxicological endpoint. Thus, it becomes necessary to segregate chemicals by organ-specific toxicity, since strict additivity without consideration for target-organ toxicities could overestimate potential hazards (USEPA, 1986, 1989b). Consequently, the hazard index is preferably calculated only after putting chemicals into groups with same physiologic endpoints.

5.6.4.1 *Estimation of Carcinogenic Risks*

For potential carcinogens, risk is defined by the incremental probability of an individual developing cancer over a lifetime as a result of exposure to a carcinogen. The risk of contracting cancer can be estimated by combining information about the carcinogenic potency of a chemical and exposure to the substance. Specifically, carcinogenic risks are estimated by multiplying the route-specific cancer slope factor (which is the upper 95% confidence limit of the probability of a carcinogenic response per unit intake over a lifetime of exposure) by the estimated intakes; this yields the excess or incremental individual lifetime cancer risk.

The carcinogenic effects of the COCs at potentially contaminated sites typically are calculated using the linear low-dose and one-hit models, represented by the following relationships (USEPA, 1989b):

$$CR = CDI \times SF \text{ for the linear low-dose model}$$

$$CR = 1 - \exp(-CDI \times SF) \text{ for the one-hit model}$$

where:

CR = probability of an individual developing cancer (dimensionless)
CDI = chronic daily intake for long-term exposure (i.e., averaged over 70-year lifetime) (mg/kg-day)
SF = slope factor ($[\text{mg/kg-day}]^{-1}$).

In reality, and for all practical purposes, the linear low-dose cancer risk model is valid only at low risk levels (i.e., estimated risks <0.01). For sites where chemical intakes may be high (i.e., potential risks >0.01), the one-hit model represents the more appropriate tool to use.

The method of approach for assessing the cumulative health risks from chemical mixtures generally assumes additivity of effects for carcinogens when evaluating chemical mixtures or multiple carcinogens. For multiple carcinogenic chemicals and multiple exposure routes/pathways, the aggregate cancer risk for all exposure pathways and all contaminants associated with a potentially contaminated site can be estimated by the equations represented in Box 5.2. The combination of risks across exposure pathways is based on the assumption that the same receptors would consistently experience the reasonable maximum exposure via the multiple pathways. Hence, if specific pathways do not affect the same individual or receptor group, risks should not be combined under those circumstances.

BOX 5.2
General Equation for Calculating Carcinogenic Risks

For the linear low-dose model at low risk levels:

$$\text{Total Cancer Risk} = \sum_{j=1}^{p} \sum_{i=1}^{n} \left(CDI_{ij} \times SF_{ij} \right)$$

For the one-hit model used at high carcinogenic risk levels:

$$\text{Total Cancer Risk} = \sum_{j=1}^{p} \sum_{i=1}^{n} \left[1 - \exp\left(-CDI_{ij} \times SF_{ij} \right) \right]$$

where:

CDI_{ij} = chronic daily intake for the ith contaminant and j^{th} pathway
SF_{ij} = slope factor for the ith contaminant and j^{th} pathway/exposure route
n = total number of carcinogens
p = total number of pathways or exposure routes

As a rule of thumb, incremental risks of between 10^{-4} and 10^{-7} are generally perceived as reasonably acceptable levels for the protection of human health and the environment, with 10^{-6} used as point of departure. In reality, however, populations may be exposed to the same constituents from sources unrelated to a specific site. Consequently, it is preferred that the estimated carcinogenic risk be well below the 10^{-6} benchmark level, to allow for a reasonable margin of protectiveness.

Population Excess Cancer Burden

The two important parameters or measures for describing carcinogenic effects are the individual cancer risk and the estimated number of cancer cases — the cancer burden. The individual cancer risk from simultaneous exposure to several carcinogens is assumed to be the sum of the individual cancer risks from each individual chemical. The risk experienced by the individual receiving the greatest exposure is referred to as the "maximum individual risk".

To assess the population cancer burden associated with contaminated sites, the number of cancer cases due to an emission source or the presence of a potentially contaminated site within a given community can be estimated by multiplying the individual risk experienced by a group of people by the number of people in that group. Thus, if 10 million people experience an estimated cancer risk of 10^{-6} over their lifetimes, it would be estimated that 10 (i.e., 10 million $\times$ 10^{-6}) additional cancer cases could occur. The number of cancer incidents in each receptor area can be added to estimate the number of cancer incidents over an entire region. Thus, the excess cancer burden, B_{gi}, is given by:

$$B_{gi} = R_{gi} \times P_g$$

where:

B_{gi} =   population excess cancer burden for i^{th} chemical for exposed group, G
R_{gi} =   excess lifetime cancer risk for i^{th} chemical for the exposed population
                 group, G
P_g =   number of persons in exposed population group, G

Assuming the cancer burden from each carcinogen is additive, then the total population group excess cancer burden is

$$B_g = \sum_{i=1}^{N} B_{gi} = \sum_{i=1}^{N} \left(R_{gi} \times P_g \right)$$

and

total population burden, $B = \sum_{g=1}^{G} B_g = \sum_{g=1}^{G} \left\{ \sum_{i=1}^{N} B_{gi} \right\} = \sum_{g=1}^{G} \left\{ \sum_{i=1}^{N} \left(R_{gi} \times P_g \right) \right\}$

Generally, the risk assessment is site-specific and the calculated risks should be combined for pollutants originating from a given site or group of sites in the case-study affecting same receptor groups. Where possible, cancer risk estimates are expressed in terms of both individual and population risk. For the population risk, the individual upper bound estimate of excess lifetime cancer risk for an average exposure scenario is simply multiplied by the size of the potentially exposed population.

5.6.4.2 Estimation of Noncarcinogenic Hazards

The total potential noncancer effect associated with contaminants present at a potentially contaminated site is usually expressed by the hazard quotient (HQ) and/or the hazard index (HI). The HQ is calculated as the ratio of the estimated chemical exposure level to the route-specific reference dose, represented as follows (USEPA, 1989b):

$$\text{Hazard Quotient, HQ} = \frac{E}{RfD}$$

where

E   =   chemical exposure level or intake (mg/kg-day)
RfD   =   reference dose (mg/kg-day)

For multiple pollutant exposures to noncarcinogens and noncarcinogenic effects of carcinogens, constituents are normally grouped by the same mode of toxicological action (i.e., those which induce the same physiologic endpoint, such as liver or kidney toxicity). Cumulative risk is evaluated through the use of a hazard index that is generated for each health "endpoint". Chemicals with the same endpoint are generally included in a hazard index calculation. For multiple noncarcinogenic effects of several chemical compounds and multiple exposure routes/pathways, the aggregate noncancer

---

BOX 5.3
General Equation for Calculating Noncarcinogenic Risks

$$\text{Total Hazard Index} = \sum_{j=1}^{p} \sum_{i=1}^{n} \frac{E_{ij}}{RfD_{ij}}$$

$$= \sum_{j=1}^{p} \sum_{i=1}^{n} [HQ]_{ij}$$

where:

E_{ij} = exposure level (or intake) for the ith contaminant and j^{th} pathway

RfD_{ij} = acceptable intake level (or reference dose) for i^{th} contaminant and j^{th} pathway/exposure route

n = total number of chemicals presenting noncarcinogenic effects

p = total number of pathways or exposure routes

---

risk for all exposure pathways and all contaminants associated with a potentially contaminated site can be estimated by the equation shown in Box 5.3.

The combination of hazard quotients across exposure pathways is based on the assumption that the same receptors would consistently experience the reasonable maximum exposure via the multiple pathways. Hence, if specific pathways do not affect the same individual or receptor group, hazard quotients should not be combined under those circumstances. Furthermore, in the strictest sense, constituents should not be grouped together unless the toxicological endpoint is known to be the same, otherwise the process will likely overestimate and overstate potential effects from the site.

In accordance with guidelines on the interpretation of hazard indices, for any given chemical there may be potential for adverse health effects if the hazard index exceeds unity (1). It is noteworthy that since the RfD incorporates a large margin of safety, it is possible that no toxic effects may occur even if the benchmark level is exceeded. However, in interpreting the results, a reference value of HI less than or equal to 1 should be taken as the acceptable benchmark. For HI values greater than unity (i.e., HI >1), the higher the value the greater is the likelihood of adverse noncarcinogenic health impacts. In fact, since populations may be exposed to the same constituents from sources unrelated to a specific site, it is preferred that the estimated noncarcinogenic hazard index be well below the benchmark level of unity, to allow for an additional margin of protectiveness. Indeed, if any calculated hazard index exceeds unity, then the health-based criterion for the chemical mixture has been exceeded and the need for interim corrective measures must be considered.

Distinction Between Chronic and Subchronic Noncarcinogenic Effects

Human receptor exposures to environmental contaminants can occur over long-term periods (i.e., chronic exposures), or over short-term periods (i.e., subchronic exposures). The chronic noncancer hazard index is represented by the following modification to the general equation presented above:

$$\text{Total Chronic Hazard Index} = \sum_{j=1}^{p} \sum_{i=1}^{n} \frac{CDI_{ij}}{RfD_{ij}}$$

where:

CDI_{ij} = chronic daily intake for the ith contaminant and j^{th} pathway
RfD_{ij} = chronic reference dose for i^{th} contaminant and j^{th} pathway/exposure route

The subchronic noncancer hazard index is represented by the following modification to the general equation presented above:

$$\text{Total Subchronic Hazard Index} = \sum_{j=1}^{p} \sum_{i=1}^{n} \frac{SDI_{ij}}{RfD_{sij}}$$

where:

SDI_{ij} = subchronic daily intake for the ith contaminant and j^{th} pathway
RfD_{sij} = subchronic reference dose for i^{th} contaminant and j^{th} pathway/exposure route

Appropriate chronic and subchronic toxicity parameters and intakes are used in the estimation of noncarcinogenic effects associated with the different exposure durations.

5.6.5 Human Health Risk Assessment for a Contaminated Site Problem

This section consists of a baseline risk assessment for an inactive site that previously housed the A2Z facility. This hypothetical facility, that operated for about two decades before being permanently closed, was used as a machine components manufacturing plant and also as an automotive repair workshop. Under a current decommissioning program, it is expected that the land parcel at the A2Z facility could be zoned for residential or light industrial developments in the near future.

Soils and groundwater present at the A2Z site appear to have been significantly impacted as a result of releases of chemical materials that were used in the past manufacturing processes and related activities. An extensive site characterization program has been udertaken to define the nature and extent of the soil and groundwater contamination within the site boundary. Chemicals found in soils and groundwater at the site consist of both organic and inorganic constituents (Table 5.1).

The objective of this illustrative example is to complete a baseline health risk assessment for the A2Z site, consisting of an evaluation of the potential risks posed by the site contaminants under a variety of exposure conditions.

Table 5.1 Summary of the Chemicals of Potential Concern at the A2Z Site

Chemicals of Potential Concern in Soils	Important Synonyms or Trade Names, or Chemical Formula	Chemical Abstracts Service number (CAS No.)	95% UCL Soil Concentration (mg/kg)
Inorganic chemicals			
Antimony	Sb	7440-36-0	4.0
Arsenic	As	7440-38-2	17.2
Beryllium	Be	7440-41-7	1.4
Cadmium	Cd	7440-43-9	4.5
Chromium (total)	Cr	16065-83-1	92.2
Chromium (hexavalent)	Cr(VI)	7440-47-3	4.6
Cobalt	Co	7440-48-4	7.8
Manganese	Mn	7439-96-5	512
Mercury	Hg	7439-97-6	0.25
Molybdenum	Mo	7439-98-7	16.6
Nickel	Ni	7440-02-0	90.9
Selenium	Se	7782-49-2	1.4
Vanadium	V	7440-62-2	62.9
Zinc	Zn	7440-66-6	413
Organic compounds			
Chloroform	Trichloromethane	67-66-3	0.008
Trichloroethene	TCE	79-01-6	32
1,1,2-Trichloroethane	1,1,2-TCA	79-00-5	0.020
Tetrachloroethene	PCE	127-18-4	0.033
Ethylbenzene	EB; Phenylethane	100-41-4	0.009
Xylenes (mixed)	Dimethylbenzene	1330-20-7	0.044

Chemicals of Potential Concern in Groundwater	Important Synonyms or Trade Names, or Chemical Formula	Chemical Abstracts Service number (CAS No.)	95% UCL Groundwater Concentration (µg/L)
Inorganic chemicals			
Antimony	Sb	7440-36-0	3.4
Arsenic	As	7440-38-2	7.6
Cadmium	Cd	7440-43-9	4.6
Chromium (total)	Cr	16065-83-1	8,690
Chromium (hexavalent)	Cr(VI)	7440-47-3	435
Manganese	Mn	7439-96-5	299
Molybdenum	Mo	7439-98-7	164
Nickel	Ni	7440-02-0	82.2
Vanadium	V	7440-62-2	54
Zinc	Zn	7440-66-6	64.6
Organic compounds			
Vinyl chloride	VC	75-01-4	0.729
1,1-Dichloroethene	1,1-DCE	75-35-4	3.31
trans-1,2-Dichloroethene	1,2-trans-DCE	156-60-5	11.8
cis-1,2-Dichloroethene	cis-1,2-DCE	540-59-0	1,220
Trichloroethene	TCE	79-01-6	548

The Baseline Risk Assessment

The following tasks typically will be carried out in order to accomplish the overall goal of the risk assessment:

1. Compile and characterize the list of contaminants present at the site.
2. Compile the toxicological profiles of the chemicals of potential concern (COCs).
3. Investigate all possible contaminant migration pathways, and determine the pathways of concern.
4. Identify targets in the vicinity of the site, and all other populations potentially at risk (including possible sensitive receptors).
5. Develop a representative conceptual model for the site.
6. Develop exposure scenarios, by integrating information on the populations potentially at risk with the likely and significant migration/exposure pathways.
7. Calculate carcinogenic risks and noncarcinogenic hazard indices for the various receptor groups that have been determined to be potentially at risk.

Figure 5.3 provides a graphical summary of the anticipated migration and exposure pathways at the A2Z site. An exposure evaluation for the site considers potential receptor exposures to the COCs under "no-action" (i.e., baseline) conditions.

Exposure scenarios involving three different population groups are evaluated using the methods of approach discussed in the preceding sections, with the results shown in Tables 5.2 through 5.4. The 95% upper confidence level (UCL) concentrations of the COCs in the environmental samples were used as the exposure point concentrations in this evaluation. Also, case-specific exposure parameters obtained from the literature (viz., DTSC, 1994; OSA, 1993; USEPA, 1989b, 1989d, 1991, 1992) were used in the modeling effort. Toxicity values used pertain to those promulgated into federal regulations and found in toxicological databases, in this case the Integrated Risk Information System (Appendix E). Calculation of potential carcinogenic risks and noncarcinogenic hazards under the existing conditions at the A2Z site were performed for the three different population groups identified in the conceptual site model (Figure 5.3). For the noncarcinogenic effects, it is assumed for the sake of simplicity that all the COCs have the same physiologic endpoint.

Hypothetical site resident — Table 5.2A consists of an evaluation of the potential risks asssociated with a hypothetical residential development at A2Z, assuming the contaminated soils present at the site remain in place. It is assumed that potential receptors may be exposed via inhalation of airborne contamination (consisting of particulates and/or volatile emissions), through the ingestion of contaminated soils, and by dermal contact with the contaminated soils at the site. Default exposure parameters indicated in the literature (viz., DTSC, 1994; OSA, 1993; USEPA, 1989b, 1989d, 1991, 1992) were used for the calculations shown in this spreadsheet. Based on this scenario, it is determined that risks exceed the generally accepted benchmark of 10^{-6}. The noncarcinogenic hazard index also exceeds the reference level of unity. The major "risk drivers" here are arsenic, beryllium, and hexavalent chromium.

Table 5.2B consists of an evaluation of the potential risks associated with a hypothetical population exposure to impacted groundwater originating from the A2Z site, assuming the contaminated water is not treated before going into a public water supply system. It is assumed that potential receptors may be exposed through the inhalation of volatiles during domestic usage of contaminated water, from the ingestion of contaminated water, and by dermal contact with contaminated waters. Default

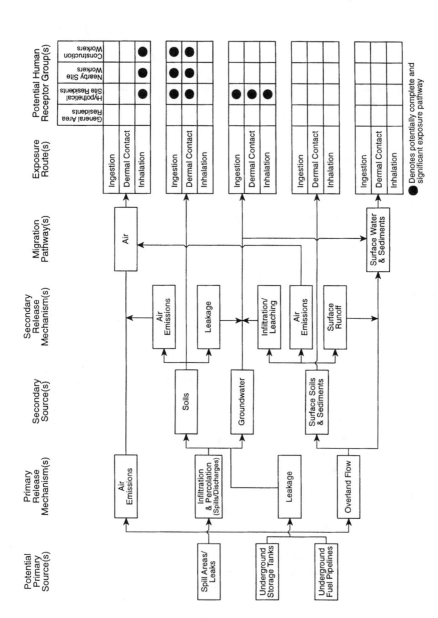

Figure 5.3 Conceptual site model representation for the baseline risk assessment at the A2Z site.

Table 5.2A Risk Screening for a Hypothetical Residential Population Exposure to Soils at the A2Z Site

Chemical of Potential Concern	95% UCL Soil Concentration (mg/kg)	Chemical-Specific Dermal Absorption (ABS$_a$)	Toxicity Criteria				Risk for Air	Hazard for Air	Risk for Soil	Hazard for Soil	Total Risk (Air + Soil)	Total Hazard (Air + Soil)
			Oral RfD (mg/kg-day)	Oral SF (1/mg/kg-day)	Inhalation RfD (mg/kg-day)	Inhalation SF (1/mg/kg-day)						
Inorganic chemicals												
Antimony	4.0	0.01	4.00E − 04				0.00E + 00	3.20E − 04	0.00E + 00	1.41E − 01		
Arsenic	17.2	0.03	3.00E − 04	1.75E + 00		1.20E + 01	1.54E − 06	1.83E − 03	6.41E − 05	9.54E − 01		
Beryllium	1.4	0.01	5.00E − 03	4.30E + 00		8.40E + 00	8.76E − 08	8.95E − 05	1.06E − 05	3.94E − 03		
Cadmium	4.5	0.001	5.00E − 04			1.50E + 01	5.03E − 07	2.88E − 04	0.00E + 00	1.16E − 01		
Chromium (total)	92.2	0.01	1.00E + 00				0.00E + 00	2.95E − 06	0.00E + 00	1.30E − 03		
Chromium (VI)	4.6	0.00	5.00E − 03	4.20E − 01		5.10E + 02	1.75E − 05	2.94E − 05	3.03E − 06	1.18E − 02		
Cobalt	7.8	0.01	2.90E − 04				0.00E + 00	8.59E − 04	0.00E + 00	3.79E − 01		
Manganese	512	0.01	1.40E − 01		2.90E − 04		0.00E + 00	1.49E − 01	0.00E + 00	5.15E − 02		
Mercury	0.25	0.01	3.00E − 04		1.10E − 04		0.00E + 00	9.29E − 05	0.00E + 00	1.17E − 02		
Molybdenum	16.6	0.01	5.00E − 03		8.60E − 05		0.00E + 00	1.06E − 04	0.00E + 00	4.67E − 02		
Nickel	90.9	0.01	2.00E − 02		5.00E − 03	9.10E − 01	6.16E − 07	1.45E − 04	0.00E + 00	6.40E − 02		
Selenium	1.4	0.01	5.00E − 03				0.00E + 00	8.95E − 06	0.00E + 00	3.94E − 03		
Vanadium	62.9	0.01	7.00E − 03				0.00E + 00	2.87E − 04	0.00E + 00	1.27E − 01		
Zinc	413	0.01	3.00E − 01				0.00E + 00	4.40E − 05	0.00E + 00	1.94E − 02		
Organic compounds												
Chloroform	0.008	0.10	1.00E − 02	3.10E − 02	1.00E − 02	1.90E − 02	1.32E − 09	2.98E − 05	8.53E − 10	2.05E − 05		
Trichloroethene	32	0.10	6.00E − 03	1.50E − 02	6.00E − 03	1.00E − 02	2.01E − 06	1.44E − 06	1.65E − 06	1.37E − 01		
1,1,2-Trichloroethane	0.020	0.10	4.00E − 03	5.70E − 02	4.00E − 03	5.60E − 02	3.54E − 09	6.77E − 05	3.92E − 09	1.28E − 04		
Tetrachloroethene	0.033	0.10	1.00E − 02	5.10E − 02	1.00E − 02	5.10E − 02	5.18E − 09	4.36E − 05	5.79E − 09	8.45E − 05		
Ethylbenzene	0.009	0.10	1.00E − 01		2.90E − 01		0.00E + 00	5.47E − 07	0.00E + 00	2.30E − 06		
Xylenes	0.044	0.10	2.00E + 00		2.00E − 01		0.00E + 00	3.16E − 06	0.00E + 00	5.63E − 07		
							2.22E − 05	0.30	7.94E − 05	2.1	1.02E − 04	2.4

Notes: The computational formulas and models used in this evaluation are discussed in this chapter and further elaborated in Appendices F and I. Risk/hazard for air accounts for the volatilization effects (for volatile organic compounds) and airborne emissions of contaminated particulates (for nonvolatile chemicals) present at the site (see Section 5.8.1 and DTSC, 1994).

Case-specific exposure parameters used in the calculations were obtained from the following sources — DTSC (1994), OSA (1993), and USEPA (1989b, 1989d, 1992).

Table 5.2B Risk Screening for a Hypothetical Residential Population Exposure to Groundwater at the A2Z Site

Chemical of Potential Concern	95% UCL Water Concentration (µg/L)	Chemical-Specific K_p (cm/hr)	Toxicity Criteria				Risk for Water	Hazard for Water
			Oral RfD (mg/kg-day)	Oral SF (1/mg/kg-day)	Inhalation RfD (mg/kg-day)	Inhalation SF (1/mg/kg-day)		
Inorganic chemicals								
Antimony	3.4	1.60E − 04	4.00E − 04				0.00E + 00	5.43E − 01
Arsenic	7.6	1.60E − 04	3.00E − 04	1.75E + 00		1.20E + 01	1.98E − 04	1.62E + 00
Cadmium	4.6	1.60E − 04	5.00E − 04			1.50E + 01	0.00E + 00	5.88E − 01
Chromium (total)	8,690	1.60E − 04	1.00E + 00				0.00E + 00	5.55E − 01
Chromium (hexavalent)	435	1.60E − 04	5.00E − 03	4.20E − 01		5.10E + 02	2.72E − 03	5.56E + 00
Manganese	299	1.60E − 04	5.00E − 03		1.10E − 04		0.00E + 00	3.82E + 00
Molybdenum	164	1.60E − 04	5.00E − 03		5.00E − 03		0.00E + 00	2.10E + 00
Nickel	82.2	1.60E − 04	2.00E − 02			9.10E − 01	0.00E + 00	2.63E − 01
Vanadium	54	1.60E − 04	7.00E − 03				0.00E + 00	4.93E − 01
Zinc	64.6	1.60E − 04	3.00E − 01				0.00E + 00	1.38E − 02
Organic compounds								
Vinyl Chloride	0.729	7.30E − 03		2.70E − 01		2.70E − 01	5.91E − 06	0.00E + 00
1,1-Dichloroethene	3.31	1.60E − 02	9.00E − 03	6.00E − 01	9.00E − 03	1.20E + 00	8.98E − 05	4.74E − 02
trans-1,2-Dichloroethene	11.8	1.00E − 02	2.00E − 02		2.00E − 02		0.00E + 00	7.58E − 02
cis-1,2-Dichloroethene	1,220	1.00E − 02	1.00E − 02		1.00E − 02		0.00E + 00	1.57E + 01
Trichloroethene	548	1.60E − 02	6.00E − 03	1.50E − 02	1.00E − 02	1.50E − 02	2.49E − 04	9.43E + 00
							3.27E − 03	40.8

Notes: The computational formulas and models used in this evaluation are discussed in this chapter and further elaborated in Appendices F and I.

Risk/hazard for water account for both volatile chemical emissions and nonvolatile chemical contributors present at the site (see Section 5.8.1 and DTSC, 1994).

Case-specific exposure parameters used in the calculations were obtained from the following sources — DTSC (1994), OSA (1993), and USEPA (1989b, 1989d, 1992).

K_p = chemical-specific dermal permeability coefficient for water.

Table 5.3   Risk Screening for a Nearby Worker Exposure to Soils at the A2Z Site

Chemical of Potential Concern	95% UCL Soil Concentration (mg/kg)	Chemical-Specific Dermal Absorption (ABS_s)	Toxicity Criteria				Risk for Air	Hazard for Air	Risk for Soil	Hazard for Soil	Total Risk (Air + Soil)	Total Hazard (Air + Soil)
			Oral RfD (mg/kg-day)	Oral SF (1/mg/kg-day)	Inhalation RfD (mg/kg-day)	Inhalation SF (1/mg/kg-day)						
Inorganic chemicals												
Antimony	4.0	0.01	4.00E − 04				0.00E + 00	9.78E − 05	0.00E + 00	1.06E − 02		
Arsenic	17.2	0.03	3.00E − 04	1.75E + 00		1.20E + 01	7.21E − 07	5.61E − 04	2.36E − 05	1.26E − 01		
Beryllium	1.4	0.01	5.00E − 03	4.30E + 00		8.40E + 00	4.11E − 08	2.74E − 06	2.27E − 06	2.96E − 04		
Cadmium	4.5	0.001	5.00E − 04			1.50E + 01	2.36E − 07	8.81E − 05	0.00E + 00	4.91E − 03		
Chromium (total)	92.2	0.01	1.00E + 00				0.00E + 00	9.02E − 07	0.00E + 00	9.74E − 05		
Chromium (VI)	4.6	0.00	5.00E − 03	4.20E − 01	2.90E − 04	5.10E + 02	8.20E − 06	9.00E − 06	3.38E − 07	4.50E − 04		
Cobalt	7.8	0.01	2.90E − 04				0.00E + 00	2.63E − 04	0.00E + 00	2.84E − 02		
Manganese	512	0.01	1.40E − 01		1.10E − 04		0.00E + 00	4.55E − 02	0.00E + 00	3.86E − 03		
Mercury	0.25	0.01	3.00E − 04		8.60E − 05		0.00E + 00	2.84E − 05	0.00E + 00	8.81E − 04		
Molybdenum	16.6	0.01	5.00E − 03		5.00E − 03		2.89E − 07	3.25E − 05	0.00E + 00	3.51E − 03		
Nickel	90.9	0.01	2.00E − 02			9.10E − 01	0.00E + 00	4.45E − 05	0.00E + 00	4.80E − 03		
Selenium	1.4	0.01	5.00E − 03				0.00E + 00	2.74E − 06	0.00E + 00	2.96E − 04		
Vanadium	62.9	0.01	7.00E − 03				0.00E + 00	8.79E − 05	0.00E + 00	9.50E − 03		
Zinc	413	0.01	3.00E − 01				0.00E + 00	1.35E − 05	0.00E + 00	1.45E − 03		
Organic compounds												
Chloroform	0.008	0.10	1.00E − 02	3.10E − 02	1.00E − 02	1.90E − 02	5.31E − 13	7.83E − 09	5.46E − 10	4.93E − 06		
Trichloroethene	32	0.10	6.00E − 03	1.50E − 02	6.00E − 03	1.00E − 02	1.12E − 09	5.22E − 05	1.06E − 06	3.29E − 02		
1,1,2-Trichloroethane	0.020	0.10	4.00E − 03	5.70E − 02	4.00E − 03	5.60E − 02	3.91E − 12	4.89E − 08	2.51E − 09	3.08E − 05		
Tetrachloroethene	0.033	0.10	1.00E − 02	5.10E − 02	1.00E − 02	5.10E − 02	5.88E − 12	3.23E − 08	3.71E − 09	2.03E − 05		
Ethylbenzene	0.009	0.10	1.00E − 01		2.90E − 01		0.00E + 00	8.81E − 10	0.00E + 00	5.55E − 07		
Xylenes	0.044	0.10	2.00E + 00		2.00E − 01		0.00E + 00	2.15E − 10	0.00E + 00	1.36E − 07		
							9.49E − 06	0.05	2.72E − 05	0.23	3.67E − 05	0.3

Notes: The computational formulas and models used in this evaluation are discussed in this chapter and further elaborated in Appendices F and I.

Risk/hazard for air accounts for only the airborne emissions of contaminated particulates for all chemicals present at the site (see Section 5.8.1 and DTSC, 1994); volatilization effects are not included in this screening analysis.

Case-specific exposure parameters used in the calculations were obtained from the following sources — DTSC (1994), OSA (1993), and USEPA (1989b, 1989d, 1992).

Table 5.4   Risk Screening for a Construction Worker Exposure to Soils at the A2Z Site

Chemical of Potential Concern	95% UCL Soil Concentration (mg/kg)	Chemical-Specific Dermal Absorption (ABS_d)	Toxicity Criteria Oral RfD (mg/kg-day)	Oral SF (1/mg/kg-day)	Inhalation RfD (mg/kg-day)	Inhalation SF (1/mg/kg-day)	Risk for Air	Hazard for Air	Risk for Soil	Hazard for Soil	Total Risk (Air + Soil)	Total Hazard (Air + Soil)
Inorganic chemicals												
Antimony	4.0	0.01	4.00E − 04				0.00E + 00	1.96E − 03	0.00E + 00	5.26E − 02		
Arsenic	17.2	0.03	3.00E − 04	1.75E + 00		1.20E + 01	5.77E − 07	1.12E − 02	2.75E − 06	3.67E − 01		
Beryllium	1.4	0.01	5.00E − 03	4.30E + 00		8.40E + 00	3.29E − 08	5.48E − 05	4.53E − 07	1.47E − 03		
Cadmium	4.5	0.001	5.00E − 04			1.50E + 01	1.89E − 07	1.76E − 03	0.00E + 00	4.28E − 02		
Chromium (total)	92.2	0.01	1.00E + 00				0.00E + 00	1.80E − 05	0.00E + 00	4.85E − 04		
Chromium (VI)	4.6	0.00	5.00E − 03	4.20E − 01		5.10E + 02	6.56E − 06	1.80E − 04	1.30E − 07	4.32E − 03		
Cobalt	7.8	0.01	2.90E − 04		2.90E − 04		0.00E + 00	5.26E − 03	0.00E + 00	1.42E − 01		
Manganese	512	0.01	1.40E − 01		1.10E − 04		0.00E + 00	9.11E − 01	0.00E + 00	1.93E − 02		
Mercury	0.25	0.01	3.00E − 04		8.60E − 05		0.00E + 00	5.69E − 04	0.00E + 00	4.39E − 03		
Molybdenum	16.6	0.01	5.00E − 03		5.00E − 03		0.00E + 00	6.50E − 04	0.00E + 00	1.75E − 02		
Nickel	90.9	0.01	2.00E − 02			9.10E − 01	2.31E − 07	8.89E − 04	0.00E + 00	2.39E − 02		
Selenium	1.4	0.01	5.00E − 03				0.00E + 00	5.48E − 05	0.00E + 00	1.47E − 03		
Vanadium	62.9	0.01	7.00E − 03				0.00E + 00	1.76E − 03	0.00E + 00	4.73E − 02		
Zinc	413	0.01	3.00E − 01				0.00E + 00	2.69E − 04	0.00E + 00	7.25E − 03		
Organic compounds												
Chloroform	0.008	0.10	1.00E − 02	3.10E − 02	1.00E − 02	1.90E − 02	4.25E − 13	1.57E − 07	3.67E − 11	8.30E − 06		
Trichloroethene	32	0.10	6.00E − 03	1.50E − 02	6.00E − 03	1.00E − 02	8.95E − 10	1.04E − 03	7.11E − 08	5.53E − 02		
1,1,2-Trichloroethane	0.020	0.10	4.00E − 03	5.70E − 02	1.00E − 03	5.60E − 02	3.13E − 12	9.78E − 07	1.69E − 10	5.19E − 05		
Tetrachloroethene	0.033	0.10	1.00E − 02	5.10E − 02	1.00E − 02	5.10E − 02	4.71E − 12	6.46E − 07	2.49E − 10	3.42E − 05		
Ethylbenzene	0.009	0.10	1.00E − 01		2.90E − 01		0.00E + 00	6.07E − 09	0.00E + 00	9.33E − 07		
Xylenes	0.044	0.10	2.00E + 00		2.00E − 01		0.00E + 00	4.31E − 08	0.00E + 00	2.28E − 07		
							7.59E − 06	0.94	3.41E − 06	0.79	1.10E − 05	1.7

Notes: The computational formulas and models used in this evaluation are discussed in this chapter and further elaborated in Appendices F and I.

Risk/hazard for air accounts for only the airborne emissions of contaminated particulates for all chemicals present at the site (see Section 5.8.1 and DTSC, 1994); volatilization effects are not included in this screening analysis.

Case-specific exposure parameters used in the calculations were obtained from the following sources — DTSC (1994), OSA (1993), and USEPA (1989b, 1989d, 1992).

exposure parameters indicated in the literature (viz., DTSC, 1994; OSA, 1993; USEPA, 1989b, 1989d, 1991) were used for the calculations presented in this spreadsheet. Based on this scenario, it is apparent that risks exceed the generally accepted benchmark risk level of 10^{-6}. The reference hazard index of 1 is also exceeded for receptor exposures to raw/untreated groundwater from the A2Z site. The major contributors to the carcinogenic risk is from the general population exposure to arsenic, hexavalent chromium, and TCE in groundwater. The most significant contributors to noncarcinogenic effects are hexavalent chromium, manganese, *cis*-1,2-DCE, and TCE in groundwater.

Nearby site worker — Table 5.3 consists of an evaluation of the potential risks associated with a nearby site worker being exposed to the COCs at the A2Z site, assuming the contaminated soils remain in place. It is assumed that potential receptors may be exposed via inhalation of airborne contamination (consisting predominantly of particulate emissions from fugitive dust), through the incidental ingestion of contaminated soils, and by dermal contact with the contaminated soils at the site. Case-specific exposure parameters used in this evaluation conservatively assume that the nearby site worker will be exposed at a frequency of 250 days per year over a 25-year period. Additional parameters include using a soil ingestion rate of 50 mg/day and airborne particulate emission rate of 50 µg/m³. Other default exposure parameters indicated in the literature (viz., DTSC, 1994; OSA, 1993; USEPA, 1989b, 1989d, 1991, 1992) were used for the calculations shown in this spreadsheet. Based on this scenario, it is determined that risks exceed the generally accepted benchmark risk level of 10^{-6}. The noncarcinogenic hazard index, however, is within the reference index of 1. The major "risk drivers" in this case are arsenic and hexavalent chromium.

Site construction worker — Table 5.4 consists of an evaluation of the potential risks associated with a construction worker being exposed to the COCs at the A2Z site, assuming the contaminated soils remain in place. It is assumed that potential receptors may be exposed via inhalation of airborne contamination (consisting predominantly of particulate emissions from fugitive dust), through the incidental ingestion of contaminated soils, and by dermal contact with the contaminated soils at the site. Case-specific exposure parameters used in this evaluation conservatively assume that the construction worker will be exposed at a frequency of 250 days per year over a 1-year period. Additional parameters include using a soil ingestion rate of 480 mg/day and airborne particulate emission rate of 1000 µg/m³. Other default exposure parameters indicated in the literature (viz., DTSC, 1994; OSA, 1993; USEPA, 1989b, 1989d, 1991, 1992) were used for the calculations shown in this spreadsheet. Based on this scenario, it is apparent that potential risks to a construction worker at the A2Z site could exceed a benchmark risk level of 10^{-6}. The noncarcinogenic hazard index also marginally exceeds the reference index of 1. The major "risk drivers" are hexavalent chromium and manganese.

A Risk Management Decision

Table 5.5 summarizes the results of the baseline risk assessment for the A2Z site. In general, a cancer risk estimate greater than 10^{-6} or a noncarcinogenic hazard index greater than 1 indicate the presence of contamination that may pose a significant threat to human health. Overall, the levels of both carcinogenic and noncarcinogenic risks associated with the A2Z site should *not* require time-critical removal action for contaminated soils present at this site. However, the use of raw/untreated groundwater from aquifers underlying this site as a potable water supply source could potentially pose significant risks to exposed individuals, or to a community that uses such water for culinary purposes.

Table 5.5   Summary of the Risk Screening for the A2Z Site

Receptor Group	Risk Parameter	Exposure Routes and Pathways				
		Inhalation Exposure to Soils	Oral Exposure to Soils	Total Exposure to Soils	Total Exposure to Groundwater	Overall Total Risk
Hypothetical resident	Cancer Risk	2.2×10^{-5}	7.9×10^{-5}	1.0×10^{-4}	3.3×10^{-3}	3.4×10^{-3}
	Hazard Index	0.3	2.1	2.4	40.8	43.2
Nearby worker	Cancer Risk	9.5×10^{-6}	2.7×10^{-5}	3.7×10^{-5}	—	3.7×10^{-5}
	Hazard Index	0.1	0.2	0.3	—	0.3
Construction worker	Cancer Risk	7.6×10^{-6}	3.4×10^{-6}	1.1×10^{-5}	—	1.1×10^{-5}
	Hazard Index	0.9	0.8	1.7	—	1.7

Note: Risk due to oral exposure is the total contribution from ingestion and dermal absorption of chemicals present in the contaminated medium.

The risk evaluation presented above indicates that the COCs present in soils at A2Z may pose significant risks to human receptors potentially exposed via the soil and groundwater media. Consequently, if this site is to be used for future residential developments, or if raw/untreated groundwater from this site is to be used as a potable water supply source, then a comprehensive site restoration program is required to abate the imminent risks the site poses. That is, general response actions (comprised of an integrated soil and groundwater remediation program) should be developed for each of the potentially impacted environmental medium associated with the A2Z site.

Ultimately, limited site control measures will probably be adequate to protect nearby site workers and construction workers at the A2Z site, whereas a more extensive corrective action program will be needed for a residential development.

5.7   ECOLOGICAL RISK ASSESSMENTS

Traditionally, contaminated site endangerment assessments have focused almost exclusively on risks to human health, often ignoring potential ecological effects. This bias has resulted in part from anthropocentrism and in part from the common but mistaken belief that protection of human health automatically protects nonhuman organisms (Suter, 1993).

In fact, human health risks in most situations are more substantial than ecological risks; considering that the mitigative actions taken to alleviate risks to human health are often sufficient to mitigate potential ecological risks at the same time, extensive ecological investigation is usually not required for most contaminated site problems. However, it should be recognized that in some situations nonhuman organisms, populations, or ecosystems may be more sensitive than human receptors. Consequently, ecological risk assessment programs have become an important part of the management of contaminated site problems, to facilitate the attainment of an adequate level of protectiveness for both human and ecological populations potentially at risk.

Purpose and Scope

An ecological (or environmental) risk assessment (ERA) involves the qualitative and/or quantitative appraisal of the actual or potential effects of contaminated sites on plants and animals other than humans and domesticated species (USEPA, 1989c, 1990b). The objectives of an ERA include identifying and estimating the potential ecological impacts associated with site contaminants, with the specific focus being to determine:

- Biological and ecological characteristics of the study area
- Types, forms, amounts, distribution, and concentration of the contaminants of concern
- Migration pathways to, and exposure of ecological receptors to pollutants
- Habitats potentially affected and populations potentially exposed to contaminants
- Actual and/or potential ecological effects/impacts and overall nature of risks

ERAs can be divided into two general types based on habitats: terrestrial (land), and aquatic (freshwater and marine), with avifauna belonging to either or both groups. Although the assessment focuses on the impacts of contaminants on the terrestrial and aquatic flora and fauna that inhabit the site and vicinity, ecological assessments may also identify new or unexpected exposure pathways which could potentially affect human populations through the food chain or through changes in the ecosystem.

The ERA allows for the identification of habitats and organisms that may be affected by the chemicals of potential concern present at a contaminated site. Typical ecological effects include changes in the aquatic and terrestrial natural resources brought about by exposure to environmental contaminants. In general, a contaminant entering the environment may cause adverse effects only if the contaminant exists in a form and concentration sufficient to cause harm, the contaminant comes in contact with organisms or environmental media with which it can interact, and the interaction that takes place is detrimental to life functions.

The ecological assessment seeks to determine the nature, magnitude, and transience or permanence of observed or expected effects of contaminants introduced into an ecosystem. This type of assessment is usually directed at investigating the loss of habitat, reduction in population size, changes in community structure, and changes in ecosystem structure and function. In fact, knowledge of environmental effects is important in analyzing both nonhuman and potential human risks associated with chemical releases into the environment.

5.7.1 General Considerations in the Ecological Risk Assessment Process

The process used to evaluate environmental or ecological risks parallels that used in the evaluation of human health risks (Section 5.6). In both cases, potential risks are determined by the integration of information on chemical exposures with toxicological data for the contaminants of potential concern. Ideally, ERA would estimate the potential for occurrence of adverse effects that are manifested as changes in the diversity, health, and behavior of the constellation of organisms that share a given environment over time. Ecological areas included in an ERA should therefore not be limited by property boundaries of a study area if affected environments or habitats are located beyond the property boundaries.

ERAs will normally be conducted in phases. The phased approach facilitates the utilization of cost-effective programs by eliminating unnecessary and expensive sampling and analysis efforts. For instance, a 'Phase I' level of effort for an ERA will generally consist of the following tasks:

- Compilation of relevant site data for the study area
- Identification of the contaminants of potential ecological concern
- Ecotoxicity assessment and/or bioassay of indicator species

- Determination of background (ambient) concentrations of the ecological contaminants in the study area
- Development of an appropriate conceptual site model (CSM) and/or a food chain diagram
- Identification and location of habitats and environments at the study area and in its vicinity
- Selection of indicator species or habitats
- Establishment of appropriate assessment endpoints for all chemicals of potential ecological concern (COECs)
- Characterization of exposure based on environmental fate and transport of the COECs as predicted based on the CSM and ecological food chain considerations
- The development of ecological risk characterization parameters for the indicator species/habitats

Where deemed necessary, a 'Phase II' assessment that comprises a biological diversity analysis, and then also a 'Phase III' assessment consisting of population studies, may become part of the overall ERA process.

Design Elements of an ERA Program

In the design of an ERA program for a potentially contaminated site the common elements of populations, communities, and ecosystems should be defined. This becomes the basis for developing a logical framework to characterize risks at the site. Typically, the following elements are given in-depth consideration during the ecological investigation at potentially contaminated sites:

- Definition and role of the ERA within the context of the contaminated site assessment and restoration
- Establishment of a concept of acceptable ecological risk
- Evaluation and selection of appropriate ecological endpoints at the population, community, and ecosystem levels
- Evaluation of exposure and biomarkers of exposure
- Validation of strategy adopted for the ERA (including the basis for its acceptability and appropriateness for the case-specific problem)
- Design of field sampling programs, data analysis and evaluation plans, and ecological monitoring programs
- Determination of acute and chronic risks and secondary hazards
- Application of ERA results to site restoration plans (i.e., for derivation of site-specific remediation objectives)

The environmental transport media of greatest interest in ERAs are surface water and soil because these are the media that are most frequently contacted by the organisms of interest. Whereas surface water is of primary interest to aquatic ecosystems, terrestrial ecosystems involve both soil and surface water. The reason that both media are of concern in terrestrial assessments is because many terrestrial receptors contact surface water bodies for such reasons as drinking, development through some of the life stages (e.g., tadpole stage of frogs and toads, and larval stage of dragonflies), and living in or near the water (e.g., beaver, muskrat, and some snakes). Consequently, areas of contaminated soil, and territories near contaminated surface water bodies as well as near contaminated soils, are likely to require ecological risk assessments.

Unlike endangerment assessment for human populations, ERAs often lack a significant amount of critical and credible data necessary for a comprehensive quantitative evaluation. Nonetheless, the pertinent data requirements should be identified and categorized as far as practical.

Background Information Needs

The collection and review of the existing information on terrestrial and aquatic ecosystems, wetlands and flood plains, threatened and endangered species, soils, and other topics relevant to the study should form a prime basis for identifying any data gaps. For instance, survey information on soil types, vegetation cover, and residential migratory wildlife may be required for terrestrial habitats, whereas comparative information needed for freshwater and marine habitats will most likely include survey data on abundance, distribution, and kinds of populations of plants and animals living in the water column and in or on the bottom. The biological and ecological information collected should include a general identification of the flora and fauna associated in and around the site, with particular emphasis placed on sensitive environments, especially endangered species and their habitats and those species consumed by humans or found in human food chains. Furthermore, the biological, chemical, and environmental factors perceived to influence the ecological effects of contaminants should be identified and described.

Types of Ecosystems

The types of ecosystems vary with climatic, topographical, geological, chemical, and biotic factors. Each ecosystem type has unique combinations of physical, chemical, and biological characteristics, and thus may respond to contamination in its own unique way. The physical and chemical structure of an ecosystem may determine how contaminants affect its resident species, and the biological interactions may determine where and how the contaminants move in the environment and which species are exposed to particular concentrations. The following general ecosystems will normally be investigated in an ERA:

- Terrestrial ecosystems (to be categorized according to the vegetation types that dominate the plant community and terrestrial animals)
- Wetlands (which are areas in which topography and hydrology create a zone of transition between terrestrial and aquatic environments)
- Freshwater ecosystems (in which environment, the dynamics of water temperature, and movement of water can significantly affect the availability and toxicity of contaminants)
- Marine ecosystems (which are of primary importance because of their vast size and critical ecological functions)
- Estuaries (which support a multitude of diverse communities, are more productive than their marine or freshwater sources, and are important breeding grounds for numerous fish, shellfish, and bird species)

The ecosystem types pertaining to a case-specific study should be defined and integrated into the overall ERA. In fact, a wide variety of other possible measures of community structure can be employed in ERA programs.

Evaluation of Ecological Habitats

Different evaluation strategies are generally employed in ERAs, depending on the level of refinement required to define the conditions within an ecological community. For instance, an evaluation of the condition of aquatic communities may proceed from two directions, as discussed below.

The first direction will consist of examining the structure of the lower trophic levels as an indication of the overall health of the aquatic ecosystem. This approach emphasizes the base of the aquatic food chain, and may involve studies of plankton (microscopic flora and fauna), periphyton (including bacteria, yeast, molds, algae, and protozoa), macrophyton (aquatic plants), and benthic macroinvertebrates (e.g., insects, annelid worms, mollusks, flatworms, roundworms, and crustaceans). Benthic macroinvertebrates are commonly used in studies of aquatic communities. These organisms usually occupy a position near the base of the food chain. Just as importantly, however, their range within the aquatic environment is restricted, so that their community structure may be referenced to a particular stream reach or portion of lake substrate. By comparison, fish are generally mobile within the aquatic environment, and evidence of stress or contaminant load may not be amenable to interpretation with reference to specific releases. The presence or absence of particular benthic macroinvertebrate species, sometimes referred to as "indicator species", may provide evidence of a response to environmental stress. A "species diversity index" provides a quantitative measure of the degree of stress within the aquatic community; this is an example of the common basis for interpreting the results of studies pertaining to aquatic biological communities. Measures of species diversity are most useful for comparison of streams with similar hydrologic characteristics or for the analysis of trends over time within a single stream (USEPA, 1989c).

The second approach to evaluating the condition of an aquatic community focuses on a particular group or species, possibly because of its commercial or recreational importance or because a substantial historic database already exists. This is done through selective sampling of specific organisms, most commonly fish, and evaluation of standard "condition factors" (e.g., length, weight, girth). In many cases, receiving water bodies are recreational fisheries monitored by state or federal agencies. In such cases, it is common to find some historical record of the condition of the fish population, and it may be possible to correlate contaminant release records with alterations in the status of the fish population.

In general, the different levels of an ecological community are studied to determine if they exhibit any evidence of stress. If the community appears to have been disturbed, the goal will be to characterize the source(s) of the stress and, specifically, to focus on the degree to which the release of chemical constituents has caused the disturbance or possibly exacerbated an existing problem.

Nature and Ecological Effects of Contaminants

Although a contaminant may cause illness and/or death to individual organisms, its effects on the structure and function of ecological assemblages or interlinkages may be measured in terms quite different from those used to describe individual effects. Consequently, a discussion that includes a wider spectrum of ecological effects on individual organisms as well as the ecological interlinkages is an important part of a comprehensive ERA. Furthermore, the biological, chemical, and environmental factors perceived to influence the ecological effects of contaminants should be identified and described. Typically, a variety of environmental variables (such as temperature, pH, salinity, water hardness, and soil composition) can indeed influence the nature and extent of the effects of a contaminant on ecological receptors. Biological factors such as species susceptibility to contaminants, characteristics affecting population abundance and distribution, and the movement of chemicals in food chains may all have significant influence on the degree of impacts that contaminants exert on ecological systems.

Selection of Target Species

Generally, it is not feasible to evaluate each species that may be present at a potentially contaminated site and vicinity. Consequently, selected target or indicator species will normally be chosen for further evaluation in an ERA. By using reasonably conservative assumptions in the overall assessment, it is rationalized that adequate protection of selected indicator species will provide protection for all other significant environmental species as well. In fact, not every organism may be suitable for use as target species in the evaluation of contaminant impacts on ecological systems. Thus, specific general considerations and assumptions should be applied in selecting target species. The following criteria are used to guide the selection of target species for an ERA (USEPA, 1989c, 1990b):

- Species that are threatened, endangered, rare, or of special concern
- Species that are valuable for several purposes of interest to human populations (i.e., of economic and societal values)
- Species critical to the structure and function of the particular ecosystem they inhabit
- Species that serve as indicators of important changes in the ecosystem
- Relevance of species at the site and its vicinity

The presence of threatened or endangered species and/or habitats critical to their survival should be documented, and the location of such species determined. Similarly, sensitive sport or commercial species and habitats essential for their reproduction and survival should be identified. Information on these may be obtained from appropriate national, federal, provincial, state, regional, local, and/or private institutions and other organizations. It is important to consider both the effects of chemical pollutants on the endangered population as well as on the habitats critical to their survival.

Identification of Ecological Assessment Endpoints

The development of an ERA requires the identification of one or several ecological assessment endpoints. These endpoints define the environmental resources which are to be protected, and which, if found to be impacted, determine the need for corrective action. The selection of appropriate site-specific assessment endpoints is therefore crucial to the development of a cost-effective corrective action program that will be protective of potential ecological receptors.

5.7.2 Exposure Assessment

The objectives of the exposure assessment are to define contaminant behaviors, identify potential ecological receptors, determine exposure routes to receptors, and estimate the degree of contact and/or intakes of the chemicals of concern by the potential receptors.

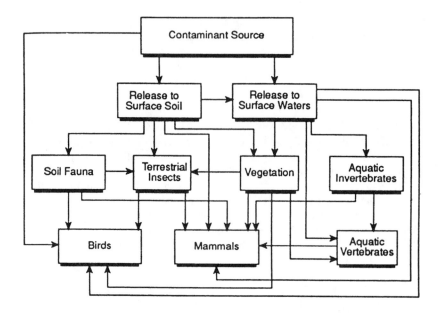

Figure 5.4 A simplified schematic of a food chain diagram showing contaminant migration pathways within an ecosystem.

In a representative situation, a chemical may be released into the environment which is then subject to physical dispersal into the air, water, soils, and/or sediments. The chemical may then be transported spatially and into the biota and perhaps be chemically or otherwise modified or transformed and degraded by abiotic processes (such as photolysis, hydrolysis, etc.) and/or by microorganisms present in the environment. The resulting transformation products may have different environmental behavior patterns and toxicological properties from the parent chemical. Nonetheless, it is the nature of exposure scenarios that determines the potential for any adverse impacts.

A food chain (also called food web) is normally constructed for the target species, to facilitate the development of appropriate exposure scenarios. Figure 5.4 is a simplified and generic conceptual representation of typical interlinkages resulting from the consumption, uptake, and absorption processes associated with an ecological community. The exposure routes are selected based on the behavior patterns and/or ecological niches of the target species and communities. This means that the nature of the target organisms (e.g., birds, fish, etc.) must be identified together with the nature of exposure (such as acute, chronic, or intermittent), as part of the ERA process.

The amount of a target species exposure to contamination from a potentially contaminated site is based on the maximum plausible exposure concentrations of the chemicals in the affected environmental matrices. The total daily exposure (in mg/kg-day) of target species can be calculated by summing the amounts of constituents ingested and absorbed from all sources (e.g., soil, vegetation, surface water, fish tissue, and other target species), and also that absorbed through inhalation and dermal contacts. Analytical procedures used to estimate receptor exposures to chemicals in various contaminated media (such as the daily chemical exposure to wildlife and game and the resulting body burden) are presented elsewhere in the literature of endangerment assessment (e.g., CAPCOA, 1990; Paustenbach, 1988).

5.7.3   Ecotoxicity Assessment

Similar to the human health endangerment assessment (discussed in Section 5.6), the scientific literature is reviewed to obtain ecotoxicity information for the COECs associated with the ecological receptors that were identified during the exposure assessment. Additional data may have to be developed via field sampling and analysis and/or bioassays. Subsequently, critical toxicity values for the contaminants of concern are derived for the target receptor species and ecological communities of concern, to be used in characterizing risks associated with the site.

5.7.4   Risk Characterization

Ecological risk characterization entails both temporal and spatial components, requiring an evaluation of the probability or likelihood of an adverse effect occurring, the degree of permanence and/or reversibility of each effect, the magnitude of each effect, and receptor populations or habitats that will be affected.

Risk characterization steps similar to those discussed under the human endangerment assessment (Section 5.6) are adopted in the process of characterizing ecological risks. That is, the doses determined for the ecological receptors and community during the exposure assessment are integrated with the appropriate toxicity values and information derived in the ecotoxicity assessment, in order to arrive at plausible ecological risk estimates. This process generally includes a discussion of uncertainties.

Typically, a quantitative ecological risk characterization is accomplished by using the ecological quotient (EQ) method, similar to the hazard quotient used in human health risk characterizations. The EQ estimates the risk of a COEC to indicator species, independent of the interactions among species or between different COECs. It integrates the data from the exposure and ecological effects assessment to produce a 'yes/no' determination of risk (USEPA, 1989c). In the EQ approach, the exposure point concentration or estimated daily dose is compared to a benchmark critical toxicity parameter (e.g., the National Ambient Water Quality Criteria) as follows:

$$\text{Ecological Quotient (EQ)} = \frac{\text{exposure point concentration or estimated daily dose}}{\text{benchmark critical ecotoxicity parameter or surrogate}}$$

The denominator represents the concentration that produces an assessment endpoint (e.g., toxic effects) in target species. Where data for specific species or endpoints are unavailable, other toxicity data (e.g., LC_{50} or LOEL) may be used to derive a surrogate parameter. When NOEL-type values are used as surrogate, however, the EQ will have a significantly more conservative meaning as an indicator of risk. In general, if EQ <1, then an acceptable risk is indicated. Conversely, an EQ >1 calls for action or further refined investigations, due to potentially unacceptable levels of risk.

5.8   RISK ASSESSMENT OF AIR CONTAMINANT RELEASES

Contaminants released into the atmosphere are subjected to a variety of physical and chemical influences — including transformation, deposition, and depletion processes — which are but secondary to transport and diffusion processes, albeit important in several situations. Although deposition depletes concentrations of the contaminants in air, it increases the concentrations on vegetation and in soils and surface water bodies. Furthermore, certain types of deposited contaminants are subject to some degree of resuspension, especially through wind erosion for wind speeds exceeding 15 to 25 km/h (≈10 to 15 mi/h).

Air emissions from contaminated sites often are a significant source of human exposure to toxic or hazardous substances. Contaminated sites can therefore pose significant risks to public health as a result of possible airborne release of soil particulate matter laden with toxic chemicals, and/or volatile emissions. In fact, even very low-level air emissions could pose significant threats to exposed individuals, especially if toxic or carcinogenic contaminants are involved. Furthermore, remedial actions — especially those involving excavation — may create much higher emissions than baseline or undisturbed conditions. Consequently, there is increased attention on the assessment of risks associated with air toxic releases.

Classification of Air Emission Sources

Air emissions from contaminated sites may be classified as either point or area sources. Point sources include vents (e.g., landfill gas vents) and stacks (e.g., air stripper releases); area sources are generally associated with ground-level emissions (e.g., from landfills, lagoons, and contaminated surface areas). Area sources are released at ground level and disperse there, with less influence of winds and turbulence; point sources, generally, come from a stack, and are emitted with an upward velocity, often at a height significantly above ground level. Thus, point sources are more readily diluted by mixing and diffusion, further to being at greater heights, so that ground-level concentrations are reduced. This means that area source emissions may be up to 100 times as hazardous on a mass per time basis compared to point sources and stacks. This scenario demonstrates the importance of adequately describing the source type in the assessment of potential air impacts associated with contaminated site problems.

Categories of Air Emissions

Air contaminant emissions from contaminated sites fall into two major categories — gas phase emissions and particulate matter emissions. The emission mechanisms associated with gas phase and particulate matter releases are quite different. Gas phase emissions primarily involve organic compounds but may also include certain metals (such as mercury); these emissions may be released through several mechanisms such as volatilization, biodegradation, photodecomposition, hydrolysis, and combustion (USEPA, 1989e). Particulate matter emissions at contaminated sites can be released through wind erosion, mechanical disturbances, and combustion. For airborne particulates, the particle size distribution plays an important role in inhalation exposures. Large particles tend to settle out of the air more rapidly than small particles, but may be important in terms of noninhalation exposures. Very small particles (2.5 to 10 μm in diameter) are considered to be respirable and thus pose a greater health hazard than the larger particles.

Determination of Air Contaminant Emissions

Once released to the ambient air, a contaminant is subject to simultaneous transport and diffusion processes in the atmosphere; these conditions are significantly affected by meteorological, topographical, and source factors. Additional fundamental atmospheric processes (other than atmospheric transport and diffusion) that affect airborne contaminants include transformation, deposition, and depletion. The extent to which all these atmospheric processes act on the contaminant of concern determines the magnitude, composition, and duration of the release, the route of human exposure, and the impact of the release on the environment.

Methods for estimating air emissions at contaminated sites may consist of emissions measurement (direct and indirect), air monitoring (supplemented by modeling, as appropriate), and predictive emissions modeling. Protocols for estimating emission levels for contaminants from several sources are available in the literature (e.g., CAPCOA, 1990; CDHS, 1986; Mackay and Leinonen, 1975; Mackay and Yeun, 1983; Thibodeaux and Hwang, 1982; USEPA, 1989e, 1989f, 1989g, 1990c, 1990d). In all cases, site-specific data should be used whenever possible, in order to increase the accuracy of the emission rate estimates. In fact, the combined approach of environmental fate analysis and field monitoring should provide an efficient and cost-effective strategy for investigating the impacts of air pathways on potential receptors, given a variety of meteorological conditions. Once the emission rates are determined, and the exposure scenarios are defined, decisions can be made regarding potential air impacts for specified activities.

5.8.1   Air Pathway Exposure Assessment

An air pathway exposure assessment (APEA) is a systematic approach involving the application of modeling and monitoring methods to estimate emission rates and concentrations of air contaminants. It can be used to assess actual or potential receptor exposures to atmospheric contamination. An APEA is basically an exposure assessment for the air pathway, consisting of the following main components:

- Characterization of air emission sources
- Determination of the effects of atmospheric processes on contaminant fate and transport
- Evaluation of populations potentially at risk (for various exposure periods)
- Estimation of receptor intakes and doses

The purpose of the exposure assessment is to estimate the extent of potential receptor exposures to the identified COCs. This involves emission quantification, modeling of environmental transport, evaluation of environmental fate, identification of potential exposure routes and populations potentially at risk from exposures, and the estimation of both short- and long-term exposure levels.

The following information, at a minimum, needs to be collected and reviewed to support the design of an APEA program:

- Source data (to include contaminant toxicity factors, off-site sources, etc.)
- Receptor data (including identification of sensitive receptors, local land use, etc.)
- Environmental data (such as dispersion data, meteorological information, topographic maps, soil and vegetation data, etc.)
- Previous APEA data (to include ARARs summary, air monitoring, emission rate monitoring and modeling, dispersion modeling, etc.)

A first step in assessing air impacts associated with a contaminated site is to evaluate site-specific characteristics and the chemical contaminants present at the site, and to determine if transport of hazardous chemicals to the air is of potential concern. Atmospheric dispersion and emission source modeling can be used with appropriate air sampling data as input to an atmospheric exposure assessment. Finally, exposure calculations are carried out using the appropriate relationships previously discussed in Section 5.6.2, and also in Appendix F.

5.8.1.1 Dispersion Modeling

Atmospheric dispersion modeling has become an integral part of the planning and decision-making process in the assessment of public health and environmental impacts from contaminated site problems. It is an approach that can be used to provide contaminant concentrations at potential receptor locations of interest based on emission rate and meteorological data. A number of general assumptions are normally made in the assessment of contaminant releases into the atmosphere, including the following:

- Air dispersion and particulate deposition modeling of emissions adequately represent the fate and transport of chemical emission to ground level
- The composition of emission products found at ground level is identical to the composition found at source, but concentrations are different
- The potential receptors are exposed to the maximum annual average ground-level concentrations from the emission sources for 24 h/day, throughout a 70-year lifetime — a conservative assumption
- There are no losses of chemicals through transformation and other processes (such as biodegradation or photodegradation) — a conservative assumption

Naturally, the accuracy of the model predictions depends on the accuracy and representativeness of the input data. In general, model input data will include emissions and release parameters, meteorological data, and receptor locations. Existing air monitoring data (if any) for the site area can be used in designing a receptor grid and in selecting indicator chemicals to be modeled. This can also provide insight to background concentrations. Table E.1 (Appendix E) contains a listing of selected models that may be utilized in an APEA; this list is by no means complete and exhaustive, but indicates the range and variations in the available models. There are many levels of complexity implicit in the models used in air pathway assessments; a brief description of the mathematical structure for such models is presented elsewhere (e.g., Asante-Duah, 1993; USEPA, 1992).

Some selected screening-level air emission models are discussed below for illustrative purposes; these include typical computational procedures for both volatile and nonvolatile emissions. All nonvolatile compounds are generally considered to be bound onto particulates. For the purposes of a screening evaluation, a volatile substance may be defined as any chemical having a vapor pressure greater than 1×10^{-3} mmHg or a Henry's law constant greater than 1×10^{-5} atm-m^3/mol (DTSC, 1994). Thus, a chemical with values less than or equal to these are generally considered as nonvolatile compounds.

Screening-Level Estimation of Airborne Dust Concentrations
Particulate emissions from potentially contaminated sites can cause human exposure to contaminants in a variety of ways, including:

- Direct inhalation of respirable particulates
- Deposition on soils, leading to human exposure via dermal absorption or ingestion
- Ingestion of food products that have been contaminated as a result of deposition on crops or pasture lands and introduction into the human food chain
- Ingestion of contaminated dairy and meat products from animals eating contaminated crops
- Deposition on waterways, uptake through aquatic organisms, and eventual human consumption

In the estimation of potential risks from fugitive dust inhalation, an estimate of the respirable (<10 μm aerodynamic diameter, denoted by the symbol PM-10 or PM_{10}) fraction and concentrations are required. The amount of nonrespirable (>10 μm aerodynamic diameter) concentrations may also be needed to estimate deposition of windblown emissions which will eventually reach potential receptors via other routes such as ingestion and dermal exposures.

Air models for fugitive dust emission and dispersion can be used to estimate the applicable exposure point concentrations of respirable particulates from contaminated sites (e.g., CAPCOA, 1989; CDHS, 1986; DOE, 1987; USEPA, 1989h). In such models, evaluated fugitive dust dispersion concentrations are typically represented by a three-dimensional Gaussian distribution of particulate emissions from the source.

For the screening analysis of particulate inhalation, the chemical concentration in air may be calculated according to the following equation:

$$C_a = C_s \times PM_{10} \times CF$$

where:

C_a	=	concentration in air (μg/m^3)
C_s	=	contaminant concentration in soil (mg/kg)
PM_{10}	=	airborne concentration of respirable dust (less than 10 μm in diameter), usually assumed to be 50 μg/m^3
CF	=	conversion factor, equal to 10^{-6} kg/mg

Typically, a screening-level assumption is made that for non-VOCs, particulate contamination levels are directly proportional to the maximum soil concentrations. The particulate concentration used in this type of characterization may be set at 50 μg/m^3. This value is the National Ambient Air Quality Standard for the annual average respirable portion (PM_{10}) of suspended particulate matter (USEPA, 1993). By using the above assumptions, the non-VOC concentration in air is given as:

$$C_a = C_s \times (5 \times 10^{-5})$$

It is noteworthy that, this estimation procedure is not applicable to a site that is particularly dusty — as would be expected in situations where the air quality standard for suspended particulate matter is routinely exceeded (i.e., $PM_{10} > 50$ μg/m^3).

Screening-Level Estimation of Airborne Vapor Concentrations

The most important chemical parameters to consider in the evaluation of volatile air emissions are the vapor pressure and the Henry's law constant. Vapor pressure is a useful screening indicator of the potential for a chemical to volatilize from the media in which it exists. The Henry's law constant is particularly important in estimating the tendency of a chemical to volatilize from a surface impoundment or water; it also indicates the tendency of a chemical to partition between the soil and gas phase from soil water in the vadose zone or groundwater.

A vaporization model may be used to calculate flux from volatiles present in soils into the overlying air zone. Typically, the following general equation will be used to estimate the average emission over a residential lot (USEPA, 1990c, 1992):

$$E_i = \frac{2AD_e P_a K_{as} C_i \times CF}{(3.14\alpha T)^{1/2}}$$

where:

E_i	=	average emission rate of contaminant i over a residential lot during the exposure interval (mg/s)
A	=	area of contamination (cm²), with a typical default value of 4.84×10^6 cm²
D_e	=	effective diffusivity of compound (cm²/s)
	=	$D_i (P_a^{3.33}/P_t^2)$
D_i	=	diffusivity in air for compound i (cm²/s)
P_t	=	total soil porosity (dimensionless)
	=	$1 - (\beta/\rho)$
β	=	soil bulk density (g/cm³), with typical default value of 1.5 g/cm³
ρ	=	particle density (g/cm³), with typical default value of 2.65 g/cm³
P_a	=	air-filled soil porosity (dimensionless)
	=	$P_t - \theta_m \beta$
θ_m	=	soil moisture content (cm³/g), with typical default value of 0.1 cm³/g
K_{as}	=	soil/air partition coefficient (g/cm³)
	=	$(H_c/K_d) \times CF2 = (H_c/K_d) \times 41$
H_c	=	Henry's law constant (atm-m³/mol)
K_d	=	soil-water partition coefficient (cm³ water/g soil, or L/kg)
$CF2$	=	41, a conversion factor that changes H_c into a dimensionless form
C_i	=	bulk soil concentration of contaminant i (i.e., chemical concentration in soil, mg/kg × 10⁻⁶ kg/mg) (g/g soil)
CF	=	10^3 mg/g, a conversion factor
α	=	conversion factor composed of several quantities defined above

$$\alpha = \frac{D_e \times P_a}{\left\{ P_a + \left[\rho(1 - P_a)/K_{as} \right] \right\}}$$

T = exposure interval (s), with a typical default value of 30 years = 9.5×10^8 s

The equation for estimating emission rates of VOCs can be reduced to the following form:

$$E_i = \frac{1.6 \times 10^5 \times D_i \times \dfrac{H_c}{K_d} \times C_i}{\left(D_i \times \dfrac{0.023}{\left\{ 0.284 + \left[0.046 \times \dfrac{K_d}{H_c} \right] \right\}} \right)^{1/2}}$$

It is noteworthy that, this equation is valid only in situations when no "free product" is expected to exist in the soil vadose zone. Analytical procedures for checking the saturation concentrations, intended to help determine if free products exist, are available in the literature (e.g., DTSC, 1994; USEPA, 1990c, 1992). Under circumstances when the soil contaminant concentration is greater than the calculated saturation concentration for the contaminant (implying the presence of free product), a more sophisticated evaluation scheme should be adopted.

The potential air concentration of chemicals in the breathing zone as a result of volatilization of chemicals through the soil surface is calculated above each discrete area of concern. A simple box model (Hwang and Falco, 1986; USEPA, 1990c, 1992) can be used to provide an estimate of ambient air concentrations using the total emission rate calculated above. The length dimensions of the hypothetical box within which mixing will occur is usually based on the minimum dimensions of a residential lot in the applicable locality/region (Hadley and Sedman, 1990). Consequently, a screening-level estimate of the ambient air concentration is given by:

$$C_a = \frac{E}{(LS \times V \times MH)}$$

where:

C_a = ambient air concentration (mg/m^3)

E = total emission rate (mg/s)

LS = length dimension perpendicular to wind (m), with a typical default value of 22 m based on the length dimension of a square residential lot with area 484 m^2 (Hadley and Sedman, 1990; USEPA, 1992)

V = average wind speed within mixing zone (m/s), with typical default value of 2.25 m/s (Hadley and Sedman, 1990; USEPA, 1992)

MH = mixing height (m), with typical default value of 2 m, or the height of the average breathing zone for an adult (Hadley and Sedman, 1990; USEPA, 1992)

By using the above-indicated default values, the ambient air concentration can be estimated by:

$$C_a = \frac{E_i}{99}$$

where E_i would already have been estimated as shown above.

5.8.2   Risk Characterization

The process for characterizing risks from air emissions follows the same steps as previously described under Section 5.6. Once the contaminant inhalation concentrations are determined by appropriate field measurements and/or modeling practices, exposure calculations are performed. The exposure estimates are subsequently integrated with the appropriate toxicity information to generate the corresponding risk and hazard estimates.

5.9   RISK ASSESSMENT APPLICATIONS

Risk assessment techniques and principles can generally be employed to facilitate the development of effectual site characterization and corrective action programs. In addition to providing information about the nature and magnitude of potential health and environmental risks associated with a potentially contaminated site problem, risk assessment also provides a basis for judging the need for remediation. Furthermore, it can be used to compare the risk reductions afforded by different remedial or risk control strategies.

Several advantages may indeed accrue from the use of risk assessment procedures to arrive at consistent and cost-effective corrective action decisions for potentially contaminated site problems. Specific applications include the following:

- Preliminary screening for potential environmental impact problems. This typically will include an identification of factors contributing the most to overall risks of exposures from chemicals originating from a potentially contaminated site. It incorporates an analysis of baseline risks, and a consistent process to document potential public health and environmental threats from potentially contaminated sites.
- Field sampling design and identification of data needs and/or data gaps. Risk assessment can aid in the development of cost-effective field sampling programs. This will then allow an adequate number of samples to be collected and analyzed, with an overall objective of determining the presence or absence of contamination, its extent and distribution, or of verifying the attainment of site mitigation criteria.
- Evaluation and ranking of potential liabilities from potentially contaminated properties. Risk assessment can aid in the prioritization of potentially contaminated sites for remedial action. It helps prioritize cleanup actions by providing consistent data for the rank-ordering of potentially contaminated sites.
- Corrective measures evaluation leading to the selection of remedial alternatives. The risk assessment process allows performance goals of remedial alternatives to be established prior to their implementation. Risk assessment facilitates the determination of cost-effective risk reduction policies through the selection of feasible remedial alternatives protective of public health and the environment. Also, the remedial alternatives can be evaluated to determine whether the proposed remedial action itself will have any deleterious environmental and human health effects.
- Development of target cleanup criteria for potentially contaminated sites. Risk assessment provides the basis for determining the levels of chemicals that can remain at a site, or in environmental matrices, without impacting public health and the environment.

In fact, the use of health and environmental risk assessments in site cleanup decisions in particular, and corrective action programs in general, is becoming ever important. For instance a number of environmental regulations and laws in various jurisdictions (e.g., regulations promulgated under CERCLA, RCRA, CAA, and CWA) increasingly require risk-based approaches in determining cleanup goals and related decision parameters.

Overall, a risk assessment will generally provide the decision maker with scientifically defensible procedures for determining whether or not a potentially contaminated site could represent a significant adverse health and environmental risk that should therefore be considered a candidate for mitigative actions. By using site-specific characteristics rather than default values, risk assessments can be conducted that result in realistic determinations of site risks and cleanup goals. It also helps identify uncertainties associated with the decision-making process.

5.9.1   Risk Management and Corrective Action Decisions

Risk management and corrective action decisions generally are complex processes that involve a variety of technical, political, and socioeconomic considerations. To ensure public safety in all situations, contaminant migration beyond a compliance boundary into the public exposure domain must be below some stipulated risk-based maximum exposure level established through a risk assessment process.

Risk assessments are generally conducted to aid risk management and corrective action decisions. The chief purpose of risk assessment is to aid decision making and this focus should be maintained throughout a given program. Notwithstanding the complexity and the fuzziness of issues involved, the ultimate goal of corrective actions at contaminated sites is to protect public health and the environment. The application of risk assessment can remove some of the ambiguity in the decision-making process. It can also aid in the selection of prudent, technically feasible, and scientifically justifiable corrective actions that will help protect public health and the environment in a cost-effective manner.

As part of a risk management program, data generated in a risk assessment may help determine the need for, and the degree of remediation at a contaminated site. However, to successfully apply the risk assessment process to a potentially contaminated site problem, the process must be tailored to the site-specific conditions and relevant regulatory constraints. Irrespective of the required level of detail, the continuum of acute to chronic hazards and exposures should be fully investigated in a comprehensive assessment; this will then allow the complete spectrum of risks to be defined for subsequent risk management decisions. The decision on the effort involved in the analysis (i.e., qualitative, quantitative, or combinations thereof) will usually be dependent on the complexity of the situation and the level of risk involved, anticipated, or predicted.

Based on the results of a risk assessment, decisions can be made relating to the types of risk management actions needed for a given contaminated site problem. If unacceptable risk levels are identified, the risk assessment process can further be employed in the evaluation of remedial alternatives. This will ensure that net risks to human health and the environment are truly reduced to acceptable levels via the remedial action of choice.

The Role of Risk Assessment

The assessment of health and environmental risks plays an important role in site characterization activities, in corrective action planning, and also in risk mitigation and risk management strategies for potentially contaminated site problems. A major objective of any site-specific risk assessment is to provide an estimate of the baseline risks posed by the existing conditions at a contaminated site, and to further assist in the evaluation of site restoration options. Thus, appropriately applied, risk assessment techniques can be used to estimate the risks posed by site contaminants under various exposure scenarios, and to further estimate the degree of risk reduction achievable by implementing various engineering remedies.

Risk assessment is particularly useful in determining the level of cleanup most appropriate for potentially contaminated sites. By utilizing methodologies that establish cleanup criteria based on risk assessment principles, corrective action programs can be conducted in a cost-effective and efficient manner. Once realistic risk reduction levels potentially achievable by various remedial alternatives are known, the decision maker can use other scientific criteria (such as implementability, reliability, operability, and cost) to select a final design alternative. Subsequently, an appropriate corrective action plan can be developed and implemented for the contaminated site.

It is apparent that some form of risk assessment is inevitable if corrective actions are to be conducted in a sensible and deliberate manner. This is because, the very process of performing risk assessment does lead to a better understanding and appreciation of the nature of the risks inherent in a study, and further helps develop steps that can be taken to reduce such risks. The application of the risk assessment process to contaminated site problems will generally serve to document the fact that risks to human health and the environment have been evaluated and incorporated into the appropriate response actions. Almost invariably, every process for developing corrective action strategies should incorporate some concepts or principles of risk assessment. Thus, all decisions on site restoration plans for potentially contaminated sites will include, implicitly or explicitly, some aspect of risk assessment.

Ecological Risk Considerations in Corrective Action Decisions

Oftentimes, and more so in the past, only limited attention is given to the ecosystems associated with contaminated sites, and also to the protection of ecological resources during site remediation activities. Instead, much of the focus has been on the protection of human health and resources directly affecting public health and safety. In recent times, however, the ecological assessment of contaminated sites is gaining considerable attention. This is the result of prevailing knowledge or awareness of the intricate interactions between ecological receptors/systems and contaminated site cleanup processes.

In a number of situations, site remediation can destroy or otherwise affect uncontaminated ecological resources. Soil removal techniques, alteration of site hydrology, and site preparation are examples of remediation activities that can result in inadvertent damage to ecological resources. Thus, the ecological impacts of a site restoration activity must be understood by the decision makers before remediation plans are approved and implemented. In situations where adverse ecological effects are identified, corrective action alternatives with potentially less damaging impacts must be evaluated as preferred methods of choice. This means that the assessment of potential ecological impacts should be performed for each remedial alternative.

In fact, there are several important ecological concerns associated with contaminated site cleanup programs that should be addressed early in site characterization programs. Furthermore, there currently are a number of legislative requirements for incorporating ecological issues as part of site investigation and cleanup efforts for contaminated sites. To achieve adequate ecological protection and/or regulatory compliance, ecological assessments should address the overall site contamination issues and should be coordinated with all aspects of site cleanup, including human health concerns, engineering feasibility, and economic considerations (Maughan, 1993). It is apparent that ecological resources must be appropriately evaluated in order to achieve the mandate of any comprehensive program designed to ensure the effective management of contaminated site problems.

5.9.2   A Framework for Risk-Based Corrective Action Programs

The application of risk assessment to contaminated site problems helps identify critical pathways, exposure routes, and other extraneous factors contributing most to total risks. It also facilitates the determination of cost-effective risk reduction policies. Used in the corrective action planning process, risk assessment generally provides a useful tool for evaluating the effectiveness of remedies at potentially contaminated sites, and also for determining acceptable cleanup levels. Overall, risk-based corrective action programs facilitate the selection of appropriate and cost-effective site characterization and corrective action measures.

To reduce cleanup costs associated with contaminated site problems, it is important that the decision-making process involved be well defined. A systematic decision framework should therefore be used to develop the risk information necessary to support the corrective action decision-making process (Figure 5.5). Such a formulation will generally provide a rational protocol for determining the level of effort required in the evaluation and implementation of site restoration programs. The approach will indeed allow for a comprehensive evaluation of risk mitigation measures needed for the corrective action program.

Although human health is frequently the major concern in corrective action assessments, an ecological assessment may serve to expand the scope of the investigation for a potentially contaminated site problem by enlarging the area under consideration, or redefining remediation criteria, or both. A detailed assessment may be required to determine whether or not the potential ecological effects of the contaminants at a site warrant remedial action. Ecological data gathered before and during remedial investigations are used to determine the appropriate level of detail for an ecological assessment, to decide if remedial action is necessary based on ecological considerations, to evaluate the potential ecological effects of relevant remedial options, to provide information necessary for mitigation of site threats, and to design monitoring strategies used to assess the progress and effectiveness of remediation.

The decision framework shown in Figure 5.5 will generally aid the decision-making process involved in contaminated site management problems. The principles and ideas of the site restoration framework can be used, on a site-specific basis, to help determine the extent of cleanup required at contaminated sites. The process will generally incorporate a consideration of the complex interactions existing between the hydrogeological environment, regulatory policies, and technical feasibility of remedial technologies. Specifically, the decision processes involved should help environmental analysts identify, rank/categorize and monitor the status of potentially contaminated sites, identify field data needs and decide on the best sampling strategy, establish appropriate remediation goals and select an appropriate cleanup level, and choose the remedial action that is most cost effective in managing site-related risks.

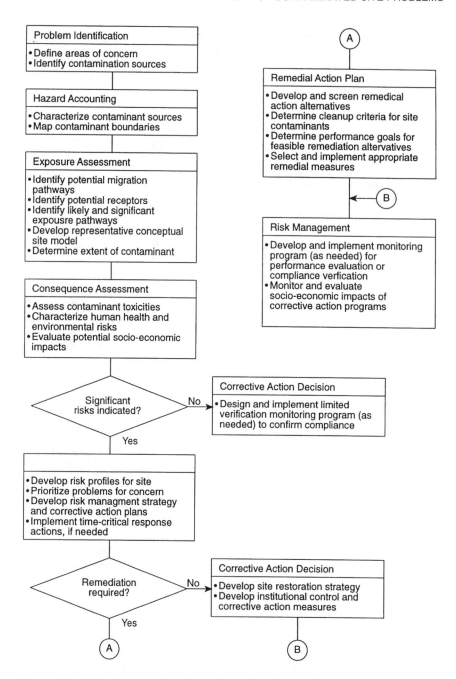

Figure 5.5 Risk management and corrective action decision framework for contaminated site management problems.

REFERENCES

Asante-Duah, D.K. 1990. "Quantitative Risk Assessment as a Decision Tool for Hazardous Waste Management." In, *Proceedings of 44-th Purdue Industrial Waste Conference (May, 1989)*, pp. 111–23. Lewis Publishers, Chelsea, MI.

Asante-Duah, D.K. 1993. *Hazardous Waste Risk Assessment.* CRC Press/Lewis Publishers, Boca Raton, FL.

Bowles, D.S., L.R. Anderson and T.F. Glover. 1987. "Design Level Risk Assessment for Dams", In Proceedings of Struct. Congress, *ASCE.* 210-25, FL.

CAPCOA. 1990. Air Toxics "Hot Spots" Program. Risk Assessment Guidelines. California Air Pollution Control Officers Association, Sacramento.

CDHS. 1986. The California Site Mitigation Decision Tree Manual. California Department of Health Services, Toxic Substances Control Division, Sacramento.

CDHS. 1990. Scientific and Technical Standards for Hazardous Waste Sites. Prepared by the California Department of Health Services, Toxic Substances Control Program, Technical Services Branch, Sacramento.

DOE. 1987. The Remedial Action Priority System (RAPS): Mathematical Formulations. U.S. Department of Energy, Office of Environmental Safety and Health, Washington, D.C.

Dourson, M.L. and J.F. Stara. 1983. Regulatory History and Experimental Support of Uncertainty (Safety) Factors. *Regul. Toxicol. Pharmacol.* 3:224–238.

DTSC. 1994. Preliminary Endangerment Assessment Guidance Manual (A guidance manual for evaluating hazardous substance release sites). California Environmental Protection Agency, Department of Toxic Substances Control, Sacramento.

Hadley, P.W. and R.M. Sedman. 1990. A Health-Based Approach for Sampling Shallow Soils at Hazardous Waste Sites Using the $AAL_{soil\ contact}$ Criterion. In *Environ. Health Perspect.* 18:203–207.

Hallenbeck, W.H. and K.M. Cunningham. 1988. *Quantitative Risk Assessment for Environmental and Occupational Health.* 4th ed. Lewis Publishers, Chelsea, MI.

Huckle, K.R. 1991. Risk Assessment — Regulatory Need or Nightmare. Shell Publications, Shell Center, London.

Hwang, S.T. and J.W. Falco. 1986. Estimation of Multimedia Exposures Related to Hazardous Waste Facilities. In, *Pollutants in a Multimedia Environment.* Y. Cohen (Ed.), Plenum Press, New York.

Klaassen, C.D., M.O. Amdur and J. Doull (Eds.) 1986. *Casarett and Doull's Toxicology: The Basic Science of Poisons.* 3rd ed. Macmillan, New York.

Mackay, D. and P.J. Leinonen. 1975. Rate of Evaporation of Low-Solubility Contaminants from Water Bodies. *Environ. Sci. Technol.* 9:1178–1180.

Mackay, D. and A.T.K. Yeun. 1983. Mass Transfer Coefficient Correlations for Volatilization of Organic Solutes from Water. *Environ. Sci. Technol.* 17:211–217.

Maughan, J.T. 1993. *Ecological Assessment of Hazardous Waste Sites.* Van Nostrand-Reinhold, New York.

NRC, 1982. *Risk and Decision-Making: Perspective and Research.* NRC Committee on Risk and Decision-Making. National Academy Press. Washington, D.C.

NRC. 1983. *Risk Assessment in the Federal Government: Managing the Process*, National Academy Press. Washington, D.C.

OSA. 1993. Supplemental Guidance for Human Health Multimedia Risk Assessments of Hazardous Waste Sites and Permitted Facilities. California Environmental Protection Agency, The Office of Scientific Affairs (OSA), Sacramento.

OTA. 1983. Technologies and Management Strategies for Hazardous Waste Control. Congress of the U.S., Office of Technology Assessment, Washington, D.C.

Paustenbach, D.J. (Ed.). 1988. *The Risk Assessment of Environmental Hazards: A Textbook of Case Studies.* John Wiley & Sons, New York.

Rowe, W.D. 1977. *An Anatomy of Risk.* John Wiley & Sons, New York.

Suter, G.W., II. 1993. *Ecological Risk Assessment.* Lewis Publishers, Chelsea, MI.

Thibodeaux, L.J. and S.T. Hwang. 1982. Landfarming of Petroleum Wastes — Modeling the Air Emission Problem. *Environ. Prog.* February, 1:42–46.

USEPA. 1984. Risk Assessment and Management: Framework for Decision Making. EPA 600/9-85-002, U.S. Environmental Protection Agency, Washington, D.C.

USEPA. 1986. Guidelines for the Health Risk Assessment of Chemical Mixtures. *Federal Register* 51(185): 34014-34025, CFR 2984, September 24, 1986. U.S. Government Printing Office, Washington, D.C.

USEPA. 1988a. CERCLA Compliance with Other Laws Manual. EPA/540/6-89/006, Office of Solid Waste and Emergency Response, U.S. Environmental Protection Agency, Washington, D.C.

USEPA. 1988b. Superfund Exposure Assessment Manual, Report No. EPA/540/1-88/001, OSWER Directive 9285.5-1, U.S. Environmental Protection Agency, Office of Remedial Response, Washington, D.C.

USEPA. 1989a. CERCLA Compliance with Other Laws Manual: Part II — Clean Air Act and Other Environmental Statutes and State Requirements. EPA/540/G-89/009. OSWER Directive 9234.1-02. U.S. Environmental Protection Agency, Washington, D.C.

USEPA. 1989b. Risk Assessment Guidance for Superfund. Vol. I. Human Health Evaluation Manual (Part A). EPA/540/1-89/002. Office of Emergency and Remedial Response, U.S. Environmental Protection Agency, Washington, D.C.

USEPA. 1989c. Risk Assessment Guidance for Superfund. Vol. II. Environmental Evaluation Manual. EPA/540/1-89/001. U.S. Environmental Protection Agency, Office of Emergency and Remedial Response, Washington, D.C.

USEPA. 1989d. Exposure Factors Handbook, EPA/600/8-89/043. Office of Health and Environmental Assessment, U.S. Environmental Protection Agency, Washington, D.C.

USEPA. 1989e. Application of Air Pathway Analyses for Superfund Activities. Air/Superfund National Technical Guidance Study Series. Procedures for Conducting Air Pathway Analyses for Superfund Applications, Vol. I. EPA-450/1-89-001. Interim Final. Office of Air Quality Planning and Standards, U.S. Environmental Protection Agency, Research Triangle Park, NC.

USEPA. 1989f. Estimation of Air Emissions from Cleanup Activities at Superfund Sites. Air/Superfund National Technical Guidance Study Series, Volume III. EPA-450/1-89-003. Interim Final. Office of Air Quality Planning and Standards, U.S. Environmental Protection Agency, Research Triangle Park, NC.

USEPA. 1989g. Procedures for Conducting Air Pathway Analyses for Superfund Applications. Vol. IV. Procedures for Dispersion Modeling and Air Monitoring for Superfund Air Pathway Analysis. Air/Superfund National Technical Guidance Study Series. EPA-450/1-89-004. Interim Final. Office of Air Quality Planning and Standards, U.S. Environmental Protection Agency, Research Triangle Park, NC.

USEPA. 1989h. Review and Evaluation of Area Source Dispersion Algorithms for Emission Sources at Superfund Sites. EPA-450/4-89-020. Office of Air Quality Planning and Standards, U.S. Environmental Protection Agency, Research Triangle Park, NC.

USEPA. 1989i. Interim Methods for Development of Inhalation Reference Doses. EPA/ 600/8-88/066F. Office of Health and Environmental Assessment, U.S. Environmental Protection Agency, Washington, D.C.

USEPA. 1990a. Guidance for Data Useability in Risk Assessment, Interim Final. U.S. Environmental Protection Agency. EPA/540/G-90/008. Office of Emergency and Remedial Response. Washington, D.C.

USEPA. 1990b. State of the Practice of Ecological Risk Assessment Document. USEPA draft report. Office of Pesticides and Toxic Substances, U.S. Environmental Protection Agency, Washington, D.C.

USEPA. 1990c. Estimation of Baseline Air Emissions at Superfund Sites. Air/Superfund National Technical Guidance Study Series. Procedures for Conducting Air Pathway Analyses for Superfund Applications, Vol. II. EPA-450/1-89-002a. Office of Air Quality Planning and Standards, U.S. Environmental Protection Agency, Research Triangle Park, NC.

USEPA. 1990d. Air/Superfund National Technical Guidance Study Series. Development of Example Procedures for Evaluating the Air Impacts of Soil Excavation Associated with Superfund Remedial Actions. EPA-450/4-90-014. Office of Air Quality Planning and Standards, U.S. Environmental Protection Agency, Research Triangle Park, NC.

USEPA. 1991. Risk Assessment Guidance for Superfund, Vol. I. Human Health Evaluation Manual. Supplemental Guidance. "Standard Default Exposure Factors" (Interim Final). March, 1991. OSWER Directive: 9285.6-03. Office of Emergency and Remedial Response, U.S. Environmental Protection Agency, Washington, D.C.

USEPA. 1992. Guideline for Predictive Baseline Emissions Estimation Procedures for Superfund Sites. In, Air/Superfund National Technical Guidance Study Series. Interim Final. EPA 450/I-92-002. Office of Health and Environmental Assessment. U.S. Environmental Protection Agency, Washington, D.C.

USEPA. 1993. National Ambient Air Quality Standard for Particulate Matter. *Federal Register,* 40 CFR, Part 50.6. U.S. Environmental Protection Agency, Washington, D.C.

Whyte, A.V. and I. Burton (Eds.). 1980. *Environmental Risk Assessment.* SCOPE Report 15, John Wiley & Sons, New York.

6

RISK-BASED CLEANUP CRITERIA
FOR CONTAMINATED SITE PROBLEMS

An important consideration in the development of corrective action response programs for contaminated sites is the level of cleanup to be achieved during possible remedial activities. Environmental quality criteria for contaminated sites typically serve as benchmarks in the assessment of the degree of contamination, and also in the determination of the level of cleanup necessary to protect human health and the environment. In general, the benchmarks may be used to:

- Determine the degree of contamination at a site
- Evaluate the need for further investigation
- Provide guidance on the need for further response actions with respect to specified land uses
- Establish remediation goals and strategies
- Verify the adequacy of remedial actions or site cleanup

The scale and urgency of further response actions at contaminated sites depend on the degree to which contaminant levels exceed their respective benchmarks. Where site remediation is not feasible, the environmental quality criteria can be used to guide land-use restrictions or other forms of risk management actions protective of human health and the environment.

Risk assessment principles are typically used to establish the cleanup objectives for contaminated sites that require remediation. In fact, the use of risk assessment to determine the need for, and the extent of remediation required at potentially contaminated sites is becoming increasingly prominent, and has found particularly extensive applications and approvals by several environmental regulatory agencies.

A number of analytical relationships and numerical models that can be used to estimate contaminated site cleanup levels required during site restoration activities are elaborated in this chapter. The contaminant-specific target levels are generally established for all affected environmental media. However, much of the emphasis in this chapter relates to the development of cleanup criteria for soils and sediments. This is because these matrices often serve as important long-term reservoirs for contaminant releases into other environmental media. Furthermore, there generally are no preestablished generic criteria for soils and sediments that are comparable to established water quality criteria (such as MCLs). The basic principles discussed here can, however, be

175

applied to establish cleanup criteria for environmental matrices other than soils/sediments (Appendix J).

6.1 HOW CLEAN IS CLEAN ENOUGH FOR A CONTAMINATED SITE?

An important and yet controversial issue that comes up when developing remediation plans for contaminated sites concerns the "how clean is clean?" question, relating to the level of cleanup to be achieved during site restoration. The cleanup level is a site-specific criterion that a remedial action would have to satisfy in order to keep exposure to potential receptors at or below an "action level" (AL); the AL is considered as the concentration of a chemical in a particular medium that, when exceeded, presents significant risk of adverse impact to potential receptors (CDHS, 1986). Indeed, in a number of situations the ALs tend to drive the cleanup process for a contaminated site. However, the ALs may not always result in "acceptable" risk levels due to the nature of the critical exposure scenarios, receptors, and other conditions that are specific to the candidate site. Consequently, in such a situation it becomes necessary to develop more stringent and health-protective levels that will meet the "acceptable" risk level criteria.

The type of exposure scenarios envisioned for a contaminated site and its vicinity usually will significantly affect the selection and acceptance of appropriate site cleanup criteria. Thus, entirely different cleanup levels may be needed for similar pieces of contaminated sites, based on the differences in the exposure scenarios. That is, the specifics of two contaminated sites may be very similar, and yet two radically different site restoration or cleanup philosophies may have to be implemented for these sites. In fact, the same amount of contamination at similar sites does not necessarily call for the same level of cleanup. In general, however, the cleanup must attain contaminant levels that are protective of both current and future land uses.

It is important that the determination of the extent of cleanup required at a contaminated site be based on an assessment of the potential risks to both human health and the environment. Several factors that are important to the process of establishing contaminated site cleanup criteria relate to the following:

- Nature and level of risks involved
- Populations potentially at risk
- Migration and exposure pathways (from source to receptors)
- Degree to which human health, safety, or welfare may be affected by exposure to site contaminants
- Effect of contamination on the environment
- Individual site characteristics affecting exposure
- Current and future beneficial uses of the affected land and subsurface resources
- Variability in exposure scenarios
- Regulatory requirements and/or guidelines

Media cleanup goals should generally be established at contaminant levels protective of both human health and the environment. Oftentimes, however, cleanup levels established for the protection of human health will also be protective of the environment at the same time, but there may be instances where adverse environmental effects may occur at or below contaminant levels that adequately protect human health. Consequently, sensitive ecosystems (e.g., wetlands) as well as threatened and endangered species or habitats that may be affected by releases of hazardous contaminants or constituents should, if possible, be evaluated separately as part of the process used to establish media cleanup criteria needed for site restoration initiatives.

6.2 GENERAL PROTOCOL FOR ESTABLISHING CLEANUP OBJECTIVES

Due to the possibility for different cleanup levels to be imposed on similar sites potentially contaminated to the same degree, the final remediation costs for such sites may be found to be significantly different. It is imperative, therefore, that a systematic approach is adopted in developing site-specific cleanup criteria for contaminated sites. Figure 6.1 illustrates a general protocol that can be adapted in developing appropriate cleanup criteria as part of the contaminated site restoration activities.

A number of technical and exposure-related factors will generally affect the development of remediation objectives and cleanup goals, including the following:

- The probability of occurrence of an exposure
- The possibilities of human exposures to elevated levels of contamination not related to site activities
- Sensitivity and vulnerability of the populations potentially at risk
- Potential effects of site contamination on ecological receptors
- The extent of cross-media contamination problems that could result from implementing certain types of remedial action alternatives
- The effectiveness and reliability of feasible remedial alternatives identified for the site
- The reliability of scientific data relating to exposure assessment, toxicity data, and risk models
- The availability of reliable and representative media-specific background threshold concentrations for the contaminants of concern — especially for nonanthropogenic chemicals (mostly inorganics)
- The confidence level of laboratory analytical protocols, and the detection or quantitation limits used for the contaminants of significant concern

Typically, cleanup decisions are developed by "back-modeling" from a benchmark risk level in order to obtain acceptable risk-based concentrations (RBCs), or the maximum acceptable concentrations (MACs). In general, once the degree of risk or hazard due to existing levels of contamination has been determined, then by working backwards using information on contaminant dilution-attenuation concepts, degradation properties, partitioning coefficients, and/or mass balance analyses, media cleanup criteria may be established for a potentially contaminated site. The type of exposure scenarios envisioned and the exposure assumptions used may drive the level of cleanup warranted, and will ensure that public health and/or the environment are not jeopardized by any residual contamination. The use of such an approach should aid in the selection of appropriate remediation options capable of achieving a set of performance goals.

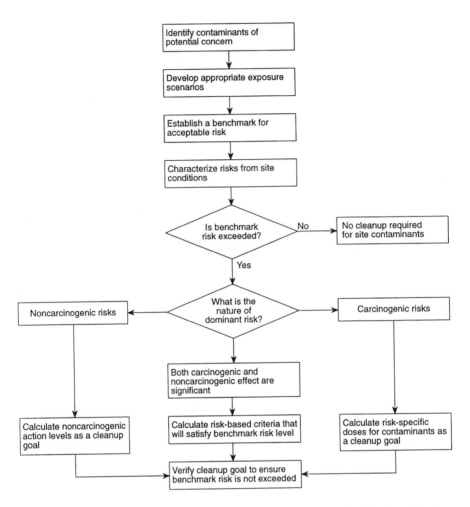

Figure 6.1 General protocol for developing cleanup goals for contaminated site problems.

6.3   DEVELOPMENT OF SITE-SPECIFIC AND RISK-BASED REMEDIATION GOALS

Soils present at potentially contaminated sites can be a major long-term reservoir for chemical contaminants, with the capacity to release contamination into several other environmental media. Yet, because of the inherent variability of soils (both spatially and temporally), a generic list of "acceptable soil contaminant levels" that would be protective of human health and the environment in all possible situations is generally not available for universal use at potentially contaminated sites.

Consider, for illustrative purposes, a potentially contaminated site that is being considered for remediation so that it may be redeveloped for either residential or industrial purposes. Contaminant levels in residential soils in which children might play (which allows for pica behavior in toddlers and infants) must necessarily be lower than the same contaminant levels in soils present at a site designated for large industrial complexes (which effectively prevent direct exposures to contaminated soils). Also, the release potential of several chemical constituents will usually be different for sandy soils vs. clayey soils, thus affecting the possible exposure scenarios and therefore the "acceptable soil contaminant levels" that are designated for the different types of soils. Consequently, it is generally preferable to establish and use site-specific cleanup criteria for most contaminated site problems, especially where soil exposure is critical to the site restoration decisions.

Health-based criteria for carcinogens, commonly referred to as risk-specific dose (RSD) or virtually safe dose (VSD), may be used as the action levels of carcinogenic constituents present at potentially contaminated sites. An RSD is the daily dose of a carcinogenic chemical that, over a lifetime, will result in an incidence of cancer equal to a specified risk level. This yields environmental concentrations that, under site-specific intake assumptions, correspond to a specified cumulative lifetime cancer risk (of, e.g., 10^{-6} for Class A and B carcinogens, 10^{-5} for Class C carcinogens, 10^{-7} for situations when sensitive populations such as nursing homes or schools are potentially exposed, 10^{-4} when the potential for exposures is rather remote, etc.). The RSD (or VSD) concentrations are calculated based on the appropriate risk level, and this becomes the action level for the specific contaminated site problem.

Health-based criteria for noncarcinogenic effects of both carcinogens and noncarcinogens, commonly referred to as reference doses (RfD) or allowable daily intakes (ADI), are estimated from the threshold exposure limit below which no adverse effects are anticipated.

6.3.1   The Media Action Level as a Site Restoration Goal

Media action levels may be established to serve as remediation goals for contaminated site problems. Having determined the media of interest for a contaminated site, it is possible to calculate the media action levels for the chemicals of potential concern (COCs) present at the site. To determine the ALs, algebraic manipulations of the hazard index or carcinogenic risk equations and the exposure estimation equations discussed in Chapter 5 can be used to arrive at the appropriate analytical relationships. This involves a back-calculation process that results in a medium AL that is based on health-protective exposure parameters (i.e., yields a noncancer hazard index ≤ 1 and/or a carcinogenic risk $\leq 10^{-6}$).

Media ALs are usually established for both carcinogenic and noncarcinogenic effects of the site contaminants, with the more stringent criteria usually being selected as a site restoration goal. In addition, the following criteria, assuming dose additivity, must be met by the cleanup action:

$$\sum_{j=1}^{p} \sum_{i=1}^{n} \frac{CMAX_{ij}}{AL_{ij}} < 1$$

where $CMAX_{ij}$ is the maximum concentration of contaminant i in environmental matrix j, and AL_{ij} is the action level for contaminant i in medium j. When these conditions are satisfied, then the AL represents the maximum acceptable contaminant level for site cleanup that will be protective of public health and the environment. Since ALs are considered health-based and environmentally-based levels, exceeding ALs will normally trigger a corrective measures study.

BOX 6.1
General Equation for Calculating Media Action Levels
for Carcinogenic Chemical Contaminants

$$AL_c = \frac{(R \times BW \times LT \times CF)}{(SF \times I \times A \times ED)}$$

where:

AL_c = action level in medium of concern (e.g., mg/kg in soil; mg/L in water)

R = specified benchmark risk level, usually set at 10^{-6} (dimensionless)

BW = body weight (kg)

LT = assumed lifetime (years)

CF = conversion factor (equals 10^6 for soil ingestion exposure; 1.00 for water ingestion)

SF = cancer slope factor ($[mg/kg\text{-}day]^{-1}$)

I = intake assumption (mg/day for soil ingestion rate; L/day for water ingestion)

A = absorption factor (dimensionless)

ED = exposure duration (years)

6.3.1.1 Action Levels for Carcinogenic Contaminants

The governing equation for calculating action levels for carcinogenic constituents present at a potentially contaminated site is shown in Box 6.1; this model assumes that there is only one COC at the site. In the case of a site contaminated by several chemicals, it is assumed (for simplification purposes) that each carcinogen has a different mode of biological action and target organs.

As an example in the application of the equation for calculating media ALs for carcinogenic chemicals, consider a hypothetical site located within a residential setting where children might be exposed. Soils at a nearby playground for children in the neighborhood is contaminated with methylene chloride. Children aged 1 to 6 years could be ingesting up to 200 mg of contaminated soils per day during outdoor activities at the impacted playground. The soil AL associated with the ingestion of 200 mg of soil (contaminated with methylene chloride which has an oral SF of 7.5×10^{-3} $[mg/kg\text{-}day]^{-1}$), by a 16-kg child over a 5-year exposure period, is conservatively estimated to be

$$C_{mc} = \frac{\left[10^{-6} \times 16 \times 70 \times 10^6\right]}{\left[0.0075 \times 200 \times 1 \times 5\right]} = 149.3 \text{ mg/kg}$$

The allowable exposure concentration (represented by the soil AL) for methylene chloride in soils within this residential setting, assuming a benchmark excess lifetime cancer risk level of 10^{-6}, is estimated to be approximately 149 mg/kg. That is, if environmental sampling and analysis indicates contamination levels in excess of 149 mg/kg at this residential playground, then immediate corrective action (such as restricting access to the playground as an interim measure) should be implemented. It is noteworthy that other potentially significant exposure routes have not been accounted for in this illustrative calculation.

In another situation, consider the case of a contaminated site impacting a nearby creek due to overland flow. This surface water body is used both as a culinary water supply source and for recreational purposes. The water AL associated with the ingestion of 2 L of water containing methylene chloride by a 70-kg adult over a 70-year lifetime is given by:

$$C_{mc} = \frac{\left[10^{-6} \times 70 \times 70 \times 1\right]}{\left[0.0075 \times 2 \times 1 \times 70\right]} = 0.005 \text{ mg/L}$$

That is, the allowable exposure concentration (represented by the water AL) for methylene chloride, assuming a benchmark excess lifetime cancer risk level of 10^{-6}, is estimated to be 5 mg/L. If this same potential receptor is exposed through both water and fish consumption (with the fish intake quantified by using its bioconcentration factor, BCF), then a lower AL may be anticipated. In this case, assuming an average daily consumption of aquatic organisms, DIA, of 6.5 g/day, and a BCF of 0.91 L/kg, human exposure levels for ingestion of both water and fish is determined from the following modified equation:

$$C_{mc} = \frac{\left[R \times BW \times LT \times CF\right]}{\left[SF \times \left(I + (DIA \times BCF)\right) \times A \times ED\right]}$$

$$= \frac{\left[10^{-6} \times 70 \times 70 \times 1\right]}{\left[0.0075 \times \left(2 + (0.0065 \times 0.91)\right) \times 1 \times 70\right]} = 0.005 \text{ mg/L}$$

Thus, the allowable exposure concentration (represented by the water AL) for drinking water and eating aquatic organisms contaminated with methylene chloride is also approximately 5 mg/L in this particular case. Obviously, the inclusion of other pertinent exposure routes such as inhalation of vapors during showering/bathing and washing activities will likely result in a lower AL.

In general, regulatory guidance would probably require reducing the contaminant concentration, C_{mc}, to about 20% of the calculated value in view of the fact that there could be other sources of exposure (e.g., air, food, etc.). This thinking should generally be factored into the overall assessment and, in particular, in the decision-making and risk management processes used for contaminated site management problems.

6.3.1.2 Action Levels for Noncarcinogenic Effects of Site Contaminants

The governing equation for calculating action levels for noncarcinogenic effects of chemical constituents present at a potentially contaminated site is shown in Box 6.2; this model assumes that there is only one COC at the site. In the case of a site

---

BOX 6.2
General Equation for Calculating Media Action Levels
for Noncarcinogenic Effects of Chemical Contaminants

$$AL_{nc} = \frac{(RfD \times BW \times CF)}{(I \times A)}$$

where:

AL_{nc} = action level in medium of concern (e.g., mg/kg in soil; mg/L in water)

RfD = reference dose (mg/kg-day)

BW = body weight (kg)

CF = conversion factor (equals 10^6 for soil ingestion exposure; 1.00 for water ingestion)

I = intake assumption (mg/day for soil ingestion rate; L/day for water ingestion)

A = absorption factor (dimensionless)

---

contaminated by several chemicals, it is assumed (for simplification purposes) that each chemical has a different organ-specific noncarcinogenic effects.

As an example in the application of the equation for calculating soil ALs for the noncarcinogenic effects of chemicals present at a contaminated site, consider a hypothetical site located within a residential setting where children might be exposed. Soils at a nearby playground for children in the neighborhood is contaminated with ethylbenzene. Children aged 1 to 6 years could be ingesting up to 200 mg of contaminated soils per day during outdoor activities at the impacted playground. The soil AL associated with the ingestion of 200 mg of soil (contaminated with ethylbenzene which has an oral RfD of 0.1 mg/kg-day) by a 16-kg child over a 5-year exposure period is conservatively estimated as:

$$C_{ebz} = \frac{\left[0.1 \text{ mg/kg} - \text{day} \times 16 \text{ kg} \times 10^6\right]}{\left[200 \text{ mg/day} \times 1\right]} = 8000 \text{ mg/kg}$$

The allowable exposure concentration (represented by the soil AL) for ethylbenzene in soils within this residential setting is estimated to be approximately 8000 mg/kg. That is, if environmental sampling and analysis indicates contamination levels in excess of 8000 mg/kg at this residential playground, then immediate corrective action (such as restricting access to the playground as an interim measure) should be implemented. Notice that this example computation ignores possible exposures via dermal contact and inhalation — both of which will affect the AL determined for this receptor group.

In another situation, consider the case of a contaminated site impacting a multipurpose surface waterbody due to overland flow. This surface water body is used both as a culinary water supply source and for recreational purposes. The water AL associated with the ingestion of 2 L/day of water containing ethylbenzene by a 70-kg adult is given by:

$$C_{ebz} = \frac{[0.1 \text{ mg/kg} - \text{day} \times 70 \text{ kg}]}{[2 \text{ L/day} \times 1]} = 3500 \text{ } \mu g/L$$

Thus, the allowable exposure concentration (represented by the water AL) for ethylbenzene is estimated to be 3500 mg/L. If this same potential receptor is exposed through both contaminated water and fish consumption (with the fish intake quantified by using its bioconcentration factor, BCF), then a lower AL may be anticipated. In this case, assuming an average adult daily consumption of aquatic organisms, DIA, of 6.5 g/day, and a BCF of 37.5 L/kg, human exposure levels for ingestion of both water and fish is determined from the following modified equation:

$$C_{ebz} \text{ [mg/L]} = \frac{[\text{RfD mg/kg} - \text{day} \times \text{BW kg}]}{[2 \text{ L/day} + (0.0065 \text{ kg} \times \text{BCF L/kg})] \times 1}$$

$$= \frac{[0.1 \text{ mg/kg} - \text{day} \times 70 \text{ kg}]}{[2 \text{ L/day} + (0.0065 \text{ kg} \times 37.5 \text{ L/kg})]} = 3120 \text{ } \mu g/L$$

Thus, the allowable exposure concentration (represented by the water AL) for drinking water and eating aquatic organisms contaminated with ethylbenzene is approximately 3120 μg/L. Of course, additional exposures via inhalation during showering/bathing and washing activities can also be incorporated to yield an even lower AL.

In general, for substances that are both carcinogenic and systemically toxic, media ALs are independently calculated for the carcinogenic and noncarcinogenic effects; the lower of the two criteria is then selected in the corrective action decision. Invariably, the carcinogenic AL tends to be more stringent in most situations where both values exist.

6.3.2   Acceptable Risk-Based Concentration Goals for Corrective Action

Risk-based action goals (RBAGs) may be established for various environmental matrices at a contaminated site by manipulating the risk and exposure relationships discussed in Chapter 5. After defining the critical pathways and exposure scenarios associated with the site, it often is possible to calculate the various media concentrations at or below which potential receptor exposures will pose no significant risks to the exposed population.

To determine the RBAGs for a chemical compound present at a contaminated site, algebraic manipulations of the hazard index or carcinogenic risk equations and the exposure estimation equations discussed in Chapter 5 can be used to arrive at the appropriate analytical relationships. This step-wise computational process involves a back-calculation process that results in an acceptable media concentration based on health-protective exposure parameters (i.e., it yields a noncancer hazard index ≤ 1 and/or a carcinogenic risk $\leq 10^{-6}$). For chemicals with carcinogenic effects, a target risk of

1×10^{-6} is generally used in the back-calculation; a target hazard index of 1.0 is generally used for noncarcinogenic effects.

The importance of soil remediation levels for contaminated sites cannot be overemphasized, considering the importance of contaminated soils in the corrective action investigation process. The computational procedures for developing RBAGs for soils are therefore elaborated below. The same principles can be extended in the formulation and development of RBAGs for water and other environmental matrices.

6.3.2.1 Soil RBAGs for Carcinogenic Constituents

As discussed in Chapter 5, the cancer risk (CR) for the significant human exposure routes (comprised of inhalation, ingestion, and dermal exposures) is given by:

$$CR = \Sigma \left\{ \sum_{i=1}^{p} CDI_p \times SF_p \right\}$$

$$= \left[CDI_i \times SF_i \right]_{inhalation} + \left[CDI_o \times SF_o \right]_{ingestion} + \left[CDI_d \times SF_o \right]_{dermal\ contact}$$

where the CDIs represent the chronic daily intakes, adjusted for absorption (mg/kg-day), and the SFs are the route-specific cancer slope factors; the subscripts i, o, and d refer to the inhalation, oral ingestion, and dermal contact exposures, respectively. By adopting the simplistic assumption that there is only one chemical constituent present in the environmental matrix of concern, this model can be reformulated to calculate the RBAG for the medium of interest. Box 6.3 contains the general equation for calculating the soil RBAG for a carcinogenic chemical constituent. This has been derived by back-calculating from the chemical intake equations presented in Appendix F for inhalation of soil emissions, ingestion of soils, and dermal contact with soils.

As an example, assume that there is only one chemical carcinogenic constituent present in soils at a hypothetical site, and that only exposures via the dermal and ingestion routes contribute to, or dominate the total benchmark carcinogenic risk (of $CR = 10^{-6}$). Therefore,

$$\Sigma\ CDI = \frac{CR}{SF_{oral}} = RSD$$

or,

$$\left(CDI_{ing} + CDI_{der} \right) = \frac{CR}{SF_o}$$

i.e.,

$$\frac{\left(RBAG_c \times SIR \times CF \times FI \times ABS_{si} \times EF \times ED \right)}{\left(BW \times AT \times 365 \right)} +$$

$$\frac{\left(RBAG_c \times CF \times SA \times AF \times ABS_{sd} \times SM \times EF \times ED \right)}{\left(BW \times AT \times 365 \right)} = \frac{CR}{SF_o}$$

BOX 6.3
Equation for Calculating Acceptable Soil Concentrations
for Carcinogenic Chemical Constituents

$RBAG_c$

$$= \frac{CR}{\left(\frac{EF \times ED \times CF}{BW \times AT \times 365}\right) \times \left\{[SF_i \times IR \times RR \times ABS_a \times AEF \times CF_a] + [(SF_o \times SIR \times FI \times ABS_{si}) + (SF_o \times SA \times AF \times ABS_{sd} \times SM)]\right\}}$$

$$= \frac{(\text{Target Risk}) \times (BW \times AT \times 365)}{\left(EF \times ED \times 10^{-6}\right) \times \left\{[SF_i \times IR \times RR \times ABS_a \times AEF \times CF_a] + SF_o [(SIR \times FI \times ABS_{si}) + (SA \times AF \times ABS_{sd} \times SM)]\right\}}$$

where:

$RBAG_c$	=	acceptable risk-based action goal of contaminant in soil (mg/kg)
CR	=	target risk, usually set at 10^{-6} (unitless)
SF_i	=	inhalation slope factor ($[mg/kg\text{-}day]^{-1}$)
SF_o	=	oral slope factor ($[mg/kg\text{-}day]^{-1}$)
IR	=	inhalation rate (m³/day)
RR	=	retention rate of inhaled air (%)
ABS_a	=	percent chemical absorbed into bloodstream (%)
AEF	=	air emissions factor, i.e., PM_{10} particulate emissions or volatilization (kg/m³)
CF_a	=	conversion factor for air emission term (10^6)
SIR	=	soil ingestion rate (mg/day)
CF	=	conversion factor (10^{-6} kg/mg)
FI	=	fraction ingested from contaminated source (unitless)
ABS_{si}	=	bioavailability absorption factor for ingestion exposure (%)
ABS_{sd}	=	bioavailability absorption factor for dermal exposures (%)
SA	=	skin surface area available for contact, i.e., surface area of exposed skin (cm²/event)
AF	=	soil to skin adherence factor, i.e., soil loading on skin (mg/cm²)
SM	=	factor for soil matrix effects (%)
EF	=	exposure frequency (days/years)
ED	=	exposure duration (years)
BW	=	body weight (kg)
AT	=	averaging time (i.e., period over which exposure is averaged) (years)

Consequently,

$$RBAG_c = \frac{(BW \times AT \times 365) \times (RSD)}{(CF \times EF \times ED)\left\{(SIR \times FI \times ABS_{si}) + (SA \times AF \times ABS_{sd} \times SM)\right\}}$$

where RSD represents the risk-specific dose, defined by the ratio of the target risk to the slope factor; CF is a conversion factor equal to 10^{-6}.

6.3.2.2 *Soil RBAGs for Noncarcinogenic Effects of Chemical Constituents*

As discussed in Chapter 5, the hazard index (HI) for the significant human exposure routes (comprised of inhalation, ingestion, and dermal exposures) is given by:

$$HI = \Sigma \left\{ \sum_{i=1}^{p} \frac{CDI_p}{RfD_p} \right\}$$

$$= \left[\frac{CDI_i}{RfD_i} \right]_{\text{inhalation}} + \left[\frac{CDI_o}{RfD_o} \right]_{\text{ingestion}} + \left[\frac{CDI_d}{RfD_o} \right]_{\text{dermal contact}}$$

where the CDIs represent the chronic daily intakes, adjusted for absorption (mg/kg-day), and the RfDs are the route-specific reference doses; the subscripts i, o, and d refer to the inhalation, oral ingestion, and dermal contact exposures, respectively. By adopting the simplistic assumption that there is only one chemical constituent present in the environmental matrix of concern, this model can be reformulated to calculate the RBAG for the medium of interest. Box 6.4 contains the general equation for calculating the soil RBAG for the noncarcinogenic effects of chemical constituents. This has been derived by back-calculating from the chemical intake equations presented in Appendix F for inhalation of soil emissions, ingestion of soils, and dermal contact with soils.

As an example, assume that there is only one chemical constitiuent present in soils at a hypothetical site, and that only exposures via the dermal and ingestion routes contribute to, or dominate the total benchmark hazard index (of HI =1). Therefore,

$$\Sigma\, CDI = RfD$$

or,

$$(CDI_{ing} + CDI_{der}) = RfD_o$$

i.e.,

$$\frac{\left(RBAG_{nc} \times SIR \times CF \times FI \times ABS_{si} \times EF \times ED\right)}{\left(BW \times AT \times 365\right)} +$$

$$\frac{\left(RBAG_{nc} \times CF \times SA \times AF \times ABS_{sd} \times SM \times EF \times ED\right)}{\left(BW \times AT \times 365\right)} = RfD_o$$

Consequently,

$$RBAG_{nc} = \frac{\left(BW \times AT \times 365\right) \times \left(RfD_o\right)}{\left(CF \times EF \times ED\right)\left\{\left(SIR \times FI \times ABS_{si}\right) + \left(SA \times AF \times ABS_{sd} \times SM\right)\right\}}$$

assuming a benchmark hazard index of unity; CF is a conversion factor equal to 10^{-6}.

BOX 6.4

Equation for Calculating Acceptable Soil Concentrations for the Noncarcinogenic Effects of Chemical Constituents

$$RBAG_{nc}$$

$$= \frac{\text{Target Hazard}}{\left(\dfrac{EF \times ED \times 10^{-6}}{BW \times AT \times 365}\right) \times \left\{\left[\dfrac{IR \times RR \times ABS_a}{RfD_i} \times AEF \times CF_a\right] + \left[\left(\dfrac{SIR}{RfD_o} \times FI \times ABS_{si}\right)\right] + \left[\dfrac{SA \times AF \times ABS_{sd} \times SM}{RfD_o}\right]\right\}}$$

$$= \frac{(BW \times AT \times 365)}{\left(EF \times ED \times 10^{-6}\right) \times \left\{\left[\dfrac{IR \times RR \times ABS_a}{RfD_i} \times AEF \times CF_a\right] + \dfrac{1}{RfD_o}\left[(SIR \times FI \times ABS_{si}) + (SA \times AF \times ABS_{sd} \times SM)\right]\right\}}$$

where:

$RBAG_{nc}$ = acceptable risk-based action goal of contaminant in soil (mg/kg)

RfD_i = inhalation reference dose (mg/kg-day)

RfD_o = oral reference dose (mg/kg-day)

IR = inhalation rate (m^3/day)

RR = retention rate of inhaled air (%)

ABS_a = percent chemical absorbed into bloodstream (%)

AEF = air emission factor, i.e., PM$_{10}$ particulate emissions or volatilization (kg/m^3)

CF_a = conversion factor for air emission term (10^6)

SIR = soil ingestion rate (mg/day)

CF = conversion factor (10^{-6} kg/mg)

FI = fraction ingested from contaminated source (unitless)

ABS_{si} = bioavailability absorption factor for ingestion exposure (%)

ABS_{sd} = bioavailability absorption factor for dermal exposures (%)

SA = skin surface area available for contact, i.e., surface area of exposed skin (cm^2/event)

AF = soil to skin adherence factor, i.e., soil loading on skin (mg/cm^2)

SM = factor for soil matrix effects (%)

EF = exposure frequency (days/years)

ED = exposure duration (years)

BW = body weight (kg)

AT = averaging time (i.e., period over which exposure is averaged) (years)

6.3.3 Miscellaneous Methods for Establishing Site Cleanup Goals

Several possibilities exist to use various analytical tools in the development of alternative and media-specific contaminant concentration limits and remediation goals. For instance, based on the assumption that the distribution of contaminants among various compartments in sediment is controlled by continuous equilibrium

exchanges, chemical-specific partition coefficients can be used to predict contaminant concentrations in sediment, biota, and/or water. That is, sediment-water equilibrium partitioning concepts can be used to predict contaminant concentrations in
sediment and/or water. Similarly, a sediment-biota partitioning coefficient can be
determined and used to predict distribution of the contaminant between sediment
and a benthic organism and/or interstitial water and a benthic organism (assuming
bioaccumulation factors are constant and independent of the organism or sediment).
Other equilibrium relationships can be employed for the estimation of various
environmental quality criteria.

Some of the procedures relevant to establishing site cleanup goals are briefly
annotated below. All the approaches represent reasonably conservative ways of setting
cleanup criteria. The use of such conservative methods is tantamount to being protective of public health by ensuring that risks are not underestimated. In general, alternate
cleanup goals based simply on attenuation mechanisms may not be acceptable unless
it is accompanied by an in-depth evaluation and discussion of all pertinent processes
involved.

Use of Attenuation-Dilution Factors
Several authors (e.g., Brown, 1986; Dawson and Sanning, 1982; Santos and
Sullivan, 1988; USEPA, 1987b) have given various methods for computing cleanup
limits that either account for attenuation alone or dilution only. To take account of both
contaminant attenuation and dilution effects, a relationship that integrates both total
attenuation and net dilution concurrently seems appropriate, if this is to become the
basis for a realistic site cleanup goal. In the application of attenuation-dilution factors
as the basis for developing site cleanup limits for potentially contaminated sites, the
following relationship that accounts for both total attenuation and net dilution concurrently, may be adopted:

$$SCL = \{Std\} \times \Pi_m \{AF\} \times \Pi_m \{DF\}$$

where:

SCL	=	source/soil cleanup level (mg/kg)
Std	=	receiving media criteria or regulatory standard to be attained (e.g., drinking water standard)
$\Pi_m^{\{AF\}}$	=	cumulative attenuation factor, defining the loss of contaminant during transport, i.e., product of the intermedia distribution constants
$\Pi_m^{\{DF\}}$	=	cumulative dilution factor during transport, i.e., product of the ratios of receiving media to source concentrations

For a single intermedia transfer, the soil cleanup level may be estimated by a more
simplified relationship. For example, if leachate migrates from soil into groundwater
that is a source of a drinking water supply, the SCL is estimated by:

$$SCL = DWS \times AF \times DF$$

where:

DWS = drinking water standard
AF = attenuation of contaminant in soil (typical values in the range 1 to 1000)
DF = dilution of contaminant by groundwater (typical values in the range 1 to 100)

Indeed, soil cleanup levels that ensure that groundwater quality standards are met may be established by using such simplistic relationships that are also reasonably conservative (i.e., adequately protective of public health).

Mass Balance Analyses

Simple mass balance analytical relationships can be applied between various environmental compartments in order to derive cleanup limits. For intermedia transfer of contaminants, current contaminant loadings may be coupled with allowable loadings (as represented by available media standards) and a back-modeling procedure can be used to obtain the required cleanup criteria, according to the following simple relationship:

$$C_{max} = \frac{C_{std}}{C_r} \times C_s$$

where:

C_{max} = source cleanup limit (i.e., the maximum acceptable source concentration)
C_{std} = receiving media criteria for receptors (i.e., regulatory standard)
C_r = receiving media concentration
C_s = source concentration prior to cleanup

Based on the appropriate maximum acceptable media concentration at the source, C_{max}, a site may be cleaned up to such levels as not to impact the receiving media.

For example, consider a situation where water quality standards exist that should be met for a creek adjoining a contaminated site. By performing back-calculations, based on contaminant concentrations in the creek as a result of the current constituents loading from the site, a conservative estimate can be made for the C_{max}. The use of such a concentration limit in a corrective action process ensures that the creek is not adversely impacted based on the exposure scenario defined for the site. Consequently, if the site is cleaned up to such levels, then the surface water quality is not expected to be impacted.

Alternate Concentration Limit Applications

Rather than consistently rely on background levels or other regulatory limits (such as MCLs), standards for the protection of groundwater from potentially contaminated sites may be established by the use of alternate concentration limits (ACLs). The ACL process is used in the U.S. regulatory system to allow owners and operators of

RCRA hazardous waste facilities to demonstrate that the hazardous constituents detected in a groundwater system will not pose any substantial present or potential hazard to human health or the environment at the specified ACL values (USEPA, 1987a). ACL demonstrations tend to be site-specific, and may be a simple manipulation of such parameters as the MCLs, RfDs, RSDs, or other risk-specific levels. A detailed health risk assessment will typically be required of ACL demonstrations intended to determine allowable chemical concentrations at the point of exposure.

6.3.4   A Recommended Site Cleanup Limit

Oftentimes, the preliminary remediation goal that has been established based on an acceptable risk level or hazard index are for single contaminants in one environmental matrix. Therefore, the risk and hazard associated with multiple contaminants in a multimedia setting are not fully accounted for during the back-modeling process used to establish the acceptable media concentrations. In contrast, the evaluation of risks associated with a given contaminated site problem involves a set of equations designed to estimate hazard and risk for all compounds present at the site, and for all exposure pathways. Consequently, if the computed "residual" risks that could remain after site remediation is likely to exceed acceptable health-protective limits, a recommended site cleanup limit (RSCL) should be established for the remedial selection process. To obtain the RSCL, the "acceptable contaminant level" is estimated in the same way as the media action level, the RBAGs, or by using other similar methods suitable for establishing cleanup goals; however, the the cumulative effects of multiple contaminants is taken into account through a process of apportioning target risks and hazards.

The assumption used for apportioning the excess carcinogenic risk is that all carcinogens have the same mode of biological actions and target organs; otherwise, excess carcinogenic risk is not apportioned among carcinogens, but rather each assumes the same value in the computational efforts. A more comprehensive approach to partitioning or combining risks would involve more complicated mathematical manipulations, such as by the use of linear programming algorithms. For carcinogens, such an algorithm will ensure that the sum of all risks from the chemicals involved over all pathways is less than or equal to the set target risk (i.e., $\leq 10^{-6}$ as an example), and for noncarcinogenic effects it will ensure that the sum of all hazard quotients over all pathways for chemicals with the same toxicological endpoints is less than or equal to the hazard index criterion of 1.0.

6.3.4.1   Health-Protective Cleanup Limit
for Carcinogenic Chemicals

The acceptable risk level may be apportioned between the chemical constituents contributing to the overall target risk, assuming each constituent contributes equally or proportionately to the total acceptable risk. The "risk fraction" obtained for each constituent can then be used to derive the RSCL by working from the relationships established previously for the computation of the media action levels or their equivalents.

The RSCL may be estimated by proportionately aggregating — or rather, disaggregating — the target cancer risk between the chemicals of potential concern. By using the approach to estimating media ALs, this is carried out according to the following approximate relationship:

BOX 6.5
Computational Steps for Calculating Media RSCLs
for Carcinogenic Chemical Constituents

(I) Apportion total acceptable carcinogenic risk (TCR) among all N carcinogen groups, as for example in the following:

$$\text{Target specified risk for chemical i, } R_i = \frac{TCR}{N}$$

(II) Calculate target CDI (mg/kg-day) for each chemical

$$\text{Target CDI for chemical i, } TCDI_i = \frac{R_i}{SF_i} = \frac{(I)}{SF_i}$$

(III) Calculate intake factor, IF, for all chemicals as follows:

$$\text{Intake Factor, } IF_i = \frac{(I \times A_i \times ED)}{(BW \times LT \times CF)}$$

(IV) Calculate RSCL (mg/kg) for each chemical

$$\text{RSCL for chemical i, } RSCL_i = \frac{TCDI_i}{IF_i} = \frac{(II)}{(III)}$$

$$RSCL = \frac{(\% \times R \times BW \times LT \times CF)}{(SF \times I \times A \times ED)}$$

All the terms are the same as defined previously in Section 6.3.1, and % represents the proportionate contribution from a specific chemical constituent to the overall target risk level. One may also choose to use weighting factors (based, for instance, on carcinogenic classes such that class A carcinogens are given twice as much weight as class B, etc.) in apportioning the individual chemical contributions to the target risk levels.

Box 6.5 summarizes the steps for the computational process involved in calculating the carcinogenic RSCL.

6.3.4.2 *Health-Protective Cleanup Limit*
for Noncarcinogenic Effects

The acceptable hazard level may be apportioned between the chemical constituents contributing to the overall hazard index, assuming each constituent contributes equally or proportionately to the total acceptable hazard index. The "hazard fraction"

BOX 6.6
Computational Steps for Calculating RSCLs
From the Noncarcinogenic Effects of Chemical Constituents

(I) Apportion total acceptable hazard index (THI) among all N chemical groups as follows:

$$\text{Target specified risk for chemical i, } HI_i = \frac{THI}{N} = \frac{1}{N}$$

(II) Calculate target CDI (mg/kg-day) for each chemical

$$\text{Target CDI for chemical i, } TCDI_i = HI_i \times RfD_i = (I) \times RfD_i$$

(III) Calculate intake factor, IF, for all chemicals as follows:

$$\text{Intake Factor, } IF_i = \frac{I \times A_i}{BW \times CF}$$

(IV) Calculate RSCL (mg/kg) for each chemical

$$\text{RSCL for chemical i, } RSCL_i = \frac{TCDI_i}{IF_i} = \frac{(II)}{(III)}$$

---

obtained for each constituent can then be used to derive the RSCL by working from the relationships established previously for the computation of the media action levels or their equivalents.

The RSCL may be estimated by proportionately aggregating — or rather, disaggregating — the noncancer hazard index among the chemicals of potential concern. By using the approach to estimating media ALs, this is carried out according to the following approximate relationship for noncarcinogenic effects of chemicals having the same toxicological endpoints:

$$RSCL = \frac{(\% \times RfD \times BW \times CF)}{I \times A}$$

All the terms are the same as defined previously in Section 6.3.1, and % represents the proportionate contribution from a specific chemical constituent to the overall target hazard index for noncarcinogenic effects of chemicals with the same physiologic endpoint.

Box 6.6 summarizes the steps for the computational process involved in calculating the RSCL for noncarcinogenic effects.

6.3.4.3 Incorporating Degradation Rates into the Estimation of Cleanup Criteria

Oftentimes, the effects of contaminant decay are not incorporated into the estimation of cleanup criteria for contaminated sites. On the other hand, since exposure scenarios used in calculating the RSCL consider the fact that exposures could be occurring over long time periods (up to a lifetime of 70 years), it is prudent, in a detailed analysis, to consider the fact that degradation or other transformation of the chemical at the source could occur. In such cases, the degradation properties of the contaminants of concern should be closely evaluated. Henceforth, a modified cleanup limit (called the adjusted RSCL) can be estimated that is based on the original RSCL, a degradation rate coefficient, and a specified exposure duration. The new calculated limit would represent the true cleanup limit, given by:

$$RSCL_a = \frac{RSCL}{degradation\ factor\ (DGF)}$$

where $RSCL_a$ is the adjusted RSCL (i.e., remedial goals based on the RSCL and a degradation rate coefficient). Assuming first-order kinetics, an approximation of the degradation effects can be obtained by the following factor:

$$DGF = \frac{\left(1 - e^{-kt}\right)}{kt}$$

where k is a chemical-specific degradation rate constant (days^{-1}), and t is a time period over which exposure occurs (days). For a first-order decaying substance, k is estimated from the following relationship:

$$T_{1/2}[days] = \frac{0.693}{k} \qquad or \qquad k[days^{-1}] = \frac{0.693}{T_{1/2}}$$

where $T_{1/2}$ is the half-life, which is the time after which the mass of a given substance will be one-half its initial value. Consequently,

$$RSCL_a = RSCL \times \frac{kt}{\left(1 - e^{-kt}\right)}$$

This relationship assumes that a first-order degradation/decay is occurring during the complete exposure period; decay/degradation is initiated at time t = 0 years, and the RSCL is the average allowable concentration over the exposure period. In fact, if significant degradation is likely to occur, the $RSCL_a$ calculations become much more complicated. In that case, predicated source contaminant levels must be calculated at frequent intervals and summed over the exposure period.

6.4 THE CLEANUP DECISION IN SITE RESTORATION PROGRAMS

Cleanup of contaminated sites is an important environmental issue in several regions of the world. In general, all cleanup actions target the remediation of previous contamination, using health-based cleanup goals. Consequently, similar cleanup protocols may be used in all site restoration programs, regardless of whether the cleanup action is associated with a RCRA, a Superfund, or indeed any other corrective action program.

The use of site-specific cleanup criteria will normally result in significant cost savings, because this allows a site management team to employ cost-effective corrective action strategies to achieve significant risk reduction for the particular situation. Indeed, the cleanup criteria could become the driving force behind remediation costs. It is therefore prudent to allocate adequate resources to develop contaminated site cleanup criteria. The site-specific cleanup criteria should facilitate the selection and design of cost-effective remedial action alternatives.

To arrive at a responsible decision on the acceptable cleanup criteria to adopt for a contaminated site problem, the corrective action program should, among other things, carefully evaluate the following important factors:

1. The *Level of Risk* indicated by the contaminants of concern.
2. The *Background Level* of the contaminants of concern at upgradient, upstream, and/or upwind locations relative to the source(s) of contamination or release(s).
3. *Natural Attenuation Effects* of the contaminants of concern (via processes such as evaporation, photolysis, dilution, biodegradation, etc.).
4. The *Asymptotic Level* of the contaminants of concern in the impacted media, which represents the cleanup level corresponding to a point of diminishing returns (i.e., the point when monitoring indicates that little additional progress can be made in reducing the contaminant levels). It represents the attainment of contaminant levels below which continued remediation produces negligible reductions in contaminant levels.
5. *Best Available Technologies (BATs)* that can be proven to offer feasible and cost-effective remediation methods and processes in the site restoration program.

Other principal considerations relate to the cost of cleanup, time required to complete site remediation, and the possibility of a cleanup activity creating potential liability problems.

In general, the use of risk-based cleanup levels are likely to result in timely, cost-effective, and adequate site restoration programs. As a rule of thumb, remedies whose cumulative effects fall within the risk range of approximately 10^{-4} to 10^{-7} for carcinogens, or meet acceptable levels for noncarcinogenic effects, are generally considered protective of human health. Where necessary, however, the potential ecological impacts should also be determined before a final corrective action decision is made.

Choice of Media Cleanup Standards

Oftentimes, ARARs (such as MCLs) are used to define cleanup goals if determined to represent an acceptable level with respect to site-specific factors (including the characteristics of the remedial action, the hazardous chemicals present at the site, or the physical setting of the site). However, ARARs may not always be available or they may not be adequate if multiple contaminants, multiple pathways, or other factors result in an "unacceptable" aggregate risk for the site-specific circumstances. In such situations, the appropriate level of protection is determined using alternative cleanup decision strategies. The RSCLs derived for the various pathways from defined exposure scenarios will aid in developing cleanup options, such that public health and/or the environment are not jeopardized by any residual contamination.

In attempts to establish environmental quality or cleanup criteria, it usually is necessary to establish a benchmark level of risk for the contaminants of concern. Media cleanup standards are generally established within the risk range of 10^{-7} to 10^{-4}, with a lifetime excess cancer risk of 10^{-6} normally used as a point-of-departure. The following guidelines may be used in the process of establishing media cleanup standards:

- The cumulative risk posed by multiple contaminants should not exceed a 1×10^{-4} cancer risk.
- Sensitive ecosystems and habitats or threatened or endangered species may require more stringent standards for protection.
- If nearby populations are exposed to hazardous constituents from other sources (e.g., a lead smelter), lower cleanup levels may be required than would ordinarily be necessary.
- If exposures to certain hazardous constituents occur through multiple pathways (e.g, air and water), lower cleanup levels may be prescribed.
- A well-proven technology that can achieve a 1×10^{-5} risk level may be considered and selected over a remedy that can theoretically treat constituents to concentrations equivalent to a 1×10^{-6} risk level, but which latter technology has not been proven under similar conditions for the site.

If site-specific cleanup criteria are to be developed for a site, a substantial wealth of information must also be collected on site soil and groundwater characteristics. Typical soil characteristics required to determine site-specific cleanup criteria include porosity, particle size, moisture content, organic carbon content, partition coefficients, soil pH, and depth of contamination; general aquifer characteristics required to determine site-specific cleanup criteria include effective porosity, hydraulic conductivity, bulk density, longitudinal and transverse dispersivities, aquifer saturated thickness, hydraulic gradient, depth to water table, and average groundwater velocity (Lesage and Jackson, 1992). These parameters are also useful in a general sense for interpreting the results of a site investigation, which are used to characterize the site and subsequently to develop the corrective action objectives.

REFERENCES

Asante-Duah, D.K. 1993. *Hazardous Waste Risk Assessment.* CRC Press/Lewis Publishers, Boca Raton, FL.

Brown, H.S. 1986. A Critical Review of Current Approaches to Determining "How Clean is Clean" at Hazardous Waste Sites. In *Hazardous Wastes and Hazardous Materials,* Vol. 3, No. 3, 233–260. Mary Ann Liebert, New York.

CDHS. 1986. The California Site Mitigation Decision Tree Manual. California Department of Health Services, Toxic Substances Control Division, Sacramento.

Dawson, G.W. and D. Sanning. 1982. Exposure-Response Analysis for Setting Site Res-
   toration Criteria. In Proc. Natl. Conf. Manage. Uncontrolled Hazardous Waste Sites,
   Washington, D.C.
Lesage, S. and R.E. Jackson (Eds.). 1992. *Groundwater Contamination and Analysis at
   Hazardous Waste Sites.* Marcel Dekker, New York.
Santos, S.L. and J. Sullivan. 1988. The Use of Risk Assessment for Establishing Corrective
   Action Levels at RCRA Sites. In *Hazardous Wastes and Hazardous Materials,* Mary
   Ann Liebert, New York.
USEPA. 1987a. Alternate Concentration Limit Guidance. Rep. No. EPA/530-SW-87-017,
   OSWER Directive 9481-00-6C, Office of Solid Waste, U.S. Environmental Protec-
   tion Agency, Washington, D.C.
USEPA. 1987b. RCRA Facility Investigation (RFI) Guidance, EPA/530/SW-87/001, U.S.
   Environmental Protection Agency, Washington, D.C.

7

A SYNOPSIS OF CONTAMINATED SITE RESTORATION METHODS AND TECHNOLOGIES

The need to clean up contaminated sites is an important and sensitive environmental issue confronting modern society. Several corrective action assessment tools will usually assist in determining whether or not remediation is necessary for a contaminated site problem. When required, the choice of a remediation strategy and technology is driven by site conditions, contaminant types, sources of contamination, source control measures, and potential impacts of possible remedial alternatives.

This chapter offers a general overview of the commonly used technologies and processes that find extensive application in contaminated site cleanup programs. A more detailed description of the processes, equipment, and controls, as well as the detailed design elements of the various technologies, can be found in the literature (e.g., ARB, 1991; Cairney, 1987, 1993; Jolley and Wang, 1993; Nyer, 1993; OBG, 1988; Purdue University, 1990, 1991, 1992, 1993; Sims, 1990; Sims et al., 1986; USEPA, 1984, 1985, 1988, 1989, 1990).

7.1 CONTAMINANT TREATMENT PROCESSES

The most widely used contaminant treatment processes and technologies that are applicable to the frequently encountered environmental contaminants may be classified into the following general and broad categories: physical, chemical, biological, and thermal.

Physical treatment — Some chemical constituents present at contaminated sites can be treated through separation and purification processes, consisting of techniques such as filtration, centrifugation, floatation, distillation, evaporation, solvent extraction, reverse osmosis, ion exchange, activated carbon adsorption, decantation, and constituent immobilization by solidification. These types of techniques generally do not alter the chemical composition of the contaminants. Physical processes are rarely used as the final treatment option for any contaminated material. In general, the objectives of most physical treatment methods are to separate hazardous materials from those that are considered to be less hazardous, to separate different types of hazardous materials into various streams that require different treatment methods, or to pretreat a contaminated material prior to final disposal.

Chemical treatment — Certain types of chemicals or contaminated materials can be either separated or rendered less hazardous through chemical treatment processes.

For instance, certain solvents can be used to remove chlorine atom(s) from chlorinated hazardous materials, with the result that toxic compounds are converted to less toxic, more water-soluble compounds; the reaction products typically are more easily removed from soil and more easily treated. Chemical treatment technologies often applied to contaminated materials include neutralization, coagulation-precipitation, and oxidation-reduction (redox).

Biological treatment — Biological treatment has become a viable and cost-effective alternative technology for treating a wide variety of organic contamination. By using microorganisms (natural or engineered) to degrade chemical constituents, biological processes may transform toxic materials into nontoxic elements such as water, carbon dioxide, and other innocuous products. Biological treatment is highly sensitive to changes in the organic composition and concentrations of the material being treated. For instance, biodegradation is less efficient at low substrate concentrations, because if the concentration is too low the compound may not be metabolized by a microbial population which may favor another substrate that is available in higher concentrations. On the other hand, very high substrate concentrations may be toxic to the microbial community. Also, biological treatment generally has no effect on dissolved inorganic substances, even though significant levels of some inorganic chemicals may inhibit biological activity or may even kill the microbes. Careful monitoring and control of dissolved oxygen, nutritional factors, and potentially toxic substances must be conducted during biological treatments to ensure the general viability and effectiveness of the microorganisms.

Thermal treatment — Thermal treatment technologies employ heat to destroy or change the contaminants of concern. Thermal destruction processes control temperature and oxygen availability, and convert hazardous materials to carbon dioxide, water, and other products of combustion. Thermal degradation is applicable to contaminated materials that contain significant concentrations of organic compounds, and can be implemented through different types of incineration (i.e., controlled, high-temperature burning in the presence of oxygen) or pyrolysis (i.e., chemical degradation due to elevated temperatures, not requiring the presence of oxygen). Depending on the type of contaminated material being treated, a variety of end products may result from thermal treatments. Thermal degradation of organic compounds results primarily in the formation and atmospheric releases of byproducts such as water, nitrogen, oxygen, carbon dioxide, acid gases, and particulates. If metals are present in the incinerated materials or wastes, some portion of these metals may also be emitted into the atmosphere. Furthermore, products of incomplete combustion may be formed and released. In general, if significant quantities of inorganic materials are present in the waste, residual ash and slag will additionally be produced by thermal degradation processes.

7.2 CONTAMINATED SITE REMEDIATION METHODS AND TECHNOLOGIES

Several site restoration technologies are currently available for contaminated site cleanup programs. However, no one particular technology or process is usually appropriate for all contaminant types and/or under the variety of site-specific conditions that exist at different geographical locations. In any case, an in-place or on-site technology will generally be given preference as a remedial alternative of choice in site restoration programs.

Several of the restoration methods employed in contaminated site remediation programs may be classified as variations or combinations of physical, chemical, biological, and/or thermal treatment techniques. Techniques finding relatively widespread applications in the management of contaminated site problems are discussed below, with the depth of discussion somewhat reflecting the degree of sophistication and/or interest in the particular site restoration method or technology.

7.2.1   Activated Carbon Adsorption

Activated carbon adsorption technology is based on the principle that certain organic constituents preferentially adsorb to organic carbon. In the process involved, granular activated carbon (GAC) is packed in vertical columns, and contaminated water flows through it by gravity. GAC has a high surface area to volume ratio, and many compounds readily bond to the carbon surfaces. Contaminants from water are thus adsorbed to the carbon, and the effluent water has a lower contaminant concentration. Water may be passed through several of these columns to complete contaminant removal. Activated carbon systems are indeed capable of efficiently removing very low concentrations of dissolved organics from groundwater.

In GAC adsorption, contaminants are adsorbed to the carbon and spent carbon (i.e., carbon that has reached its maximum adsorption capacity) is typically regenerated by incineration. Oftentimes, the highest cost involved in using a carbon adsorption system is the disposal of used carbon and its replacement with new carbon (Nyer, 1993). In fact, a complete carbon adsorption treatment system design must account for the final disposition of the spent carbon.

7.2.2   Asphalt Batching

Asphalt batching (also referred to as *asphalt incorporation*) is a method for treating hydrocarbon-contaminated soils. It is a relatively new remedial technique that involves the incorporation of petroleum-laden soils into hot asphalt mixes as a partial substitute for stone aggregate. This mixture can then be utilized for pavings.

The process of asphalt batching consists of an initial thermal treatment, followed by incorporation of the treated soil into aggregate for asphalt. During the incorporation process, the mixture (including the impacted soils) is heated. This results in the volatilization of the more volatile hydrocarbon constituents at various temperatures. The remainder of the compounds become incorporated into the asphalt matrix during cooling, thereby limiting constituent migration.

7.2.3   Bioremediation

Bioremediation (also called *biorestoration*) has become a viable and cost-effective alternative technology for treating a wide variety of contaminants (such as petroleum and aromatic hydrocarbons, chlorinated solvents, and pesticides). Biorestoration relies on microorganisms (especially bacteria and fungi) to transform hazardous compounds found in soil and groundwater systems (or even air streams) into innocuous or less toxic metabolic products. An optimized biotransformation condition may be attained by manipulating the physical environment and controlling nutrient supplements. The technique requires careful process control to establish the appropriate microbial population. By using microorganisms (natural or engineered) to

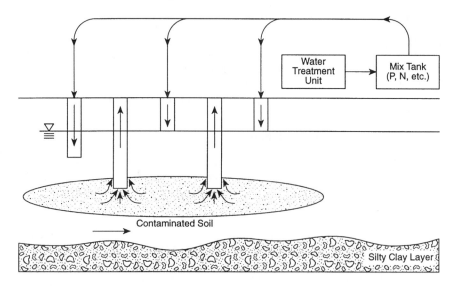

Figure 7.1   A simplified schematic of the bioremediation technology concept.

degrade contaminants in soil, groundwater, or air, bioremediation transforms hazard-ous/toxic materials into nontoxic elements such as water, carbon dioxide, and other innocuous products.

Figure 7.1 is a diagrammatic representation of the general design elements of a bioremediation system. In general, the biodegradation of a compound under field conditions is affected by temperature, type of target compounds, soil dissolved-oxygen levels, soil moisture content, soil permeability, oxidation-reduction potential, soil pH, compound availability and concentrations, availability of nutrients, and the natural microbial community. These factors act together to determine the biodegrad-ability of the contaminants in a particular setting. For instance, the biodegradation rate of most petroleum hydrocarbons increases with increasing temperatures due to in-creased biological activity. Also, aerobic conditions are required for the degradation of hydrocarbons; reduced soil oxygen levels lead to sharply reduced hydrocarbon utilization by microbes. Because oxygen transfer is a key factor in *in situ* biodegrada-tion processes, the soils must be fairly permeable to allow this transfer to occur. Also, because of its limited solubility in water, oxygen often becomes rapidly limiting in the presence of excess biodegradable organic carbon. However, the development of anoxic conditions does not necessarily mean that metabolism of organic contaminants ceases. Furthermore, it is believed that microorganisms can also affect inorganic contaminant mobility by both direct (e.g., oxidation, reduction, methylation, denitri-fication, etc.) and indirect (e.g., biosorption, changes in pH, sulfide production, etc.) processes (Fredrickson et al., 1993). Microbial reduction reactions can indeed influ-ence heavy metal solubility and hence may serve as rather useful remediation strategies.

Oftentimes, biorestoration can be enhanced by using the native microorganisms and injecting nutrients, including oxygen, or by injecting microorganisms to the subsurface environment. Special methods may be necessary to enhance biorestoration of certain compounds. In this process, the toxicity of degradation byproducts should also be taken into consideration; in some cases (such as the degradation of trichloroethylene, TCE, to vinyl chloride) the daughter products are more toxic than the parent compound.

Since most potentially contaminated sites require prompt remedial actions, an acceleration of the natural biodegradation process (i.e., enhanced biodegradation) is generally desirable. Under such circumstances, the soil system may have to be modified to promote the activity of the naturally found organisms. Promotion methods may include the addition of nutrients and aeration of the soil. If the intrinsic microorganism flora do not work on some set of contaminants, selectively adapted inoculants can also be added to the soil.

In general, there are three major schools of thought relating to the approach one takes to biorestoration: natural *in situ* bioremediation, biostimulation, and bioaugmentation (Barnhart, 1992; Fredrickson et al., 1993).

Natural *in situ* bioremediation — Involves the attenuation of contaminants by indigenous (native) microorganisms without any manipulation. In fact, in some situations it is more economical to allow bioremediation to proceed naturally at the sacrifice of higher removal rates and greater effectiveness than are often achieved with the addition of electron acceptors and other nutrients.

Biostimulation — Is a process whereby the addition of selected amounts of nutrient materials stimulate or encourage the growth of the indigenous bacteria in the soil, resulting in the degradation of the target contaminant(s).

Bioaugmentation — Is a process in which specially selected bacteria cultures that are predisposed to metabolize the target compound(s) are added to the impacted media/soil, along with the nutrient materials, to encourage degradation of the contaminants of concern.

The enhanced biodegradation process may be achieved as a conventional waterborne biodegradation in which water is used to uniformly transport nutrients (nitrogen and phosphorus) and oxygen through the unsaturated zone, or as a soil-venting enhanced biodegradation in which a system is engineered to increase the microbial biodegradation in the vadose zone utilizing pumped air as the oxygen source, or as a variation thereof the previously stated.

As an innovative technology, a very important advantage of bioremediation relates to the permanent removal of contaminants by biodegradation, and therefore the reduction of potential liabilities (since the contamination is destroyed rather than removed in the biorestoration process).

7.2.4   Chemical Fixation

Chemical fixation, or *stabilization*, is a technique to chemically fix or modify the chemical structure of contaminants by applying specific reagents. The fixation process consists of immobilizing the contaminants in-place, thereby preventing migration into unaffected enviromental media. Chemical fixation can be applied to many organic and metal-bearing contaminants or waste streams. In its application, it generally is necessary to perform bench-scale treatability studies prior to a field/full-scale treatment.

In the chemical fixation technology, the contaminated soil or material is blended with precise amounts of reagent(s) to stabilize and/or encapsulate chemical constituents, which are then stockpiled and allowed to cure. Oftentimes, the treated material is rendered nonhazardous and may be backfilled and left on the site. The process generally improves soil condition (e.g., increases compressive strength, decreases permeability, etc.), and there is ease of permitting for proven methods.

7.2.5   Encapsulation

Encapsulation is a remedial alternative comprised of the physical isolation and containment of the contaminated material, as is typical of a well-engineered landfill. The technique often includes a physical barrier installation at the sides of the contaminated area, since its purpose is to leave the contaminant safely in-place at a selected location.

The encapsulation technique consists of isolating the impacted soils through the use of low-permeability caps, slurry walls, grout curtains, or cut-off walls. The contaminant source is covered with low-permeability layers of synthetic textiles or a clay cap; the cap is designed to limit infiltration of precipitation and thus prevent leaching and migration of contaminants away from the site and into groundwater.

7.2.6   Incineration

Incineration is a thermal treatment/degradation process by which contaminated materials are exposed to excessive heat in a variety of incinerator types. The different types of incinerators are designed and built for different purposes. The incineration process typically involves the thermal destruction of contaminants by burning. Depending on the intensity of the heat, the contaminants of concern are volatilized and/or destroyed during the incineration process. The ash remaining from incineration of hazardous materials is usually disposed of in landfills.

Incineration of high-calorific-value wastes may be regarded as a form of recycling if the heat generated is utilized for other economic purposes. Even so, incineration is typically a very expensive treatment alternative — with the cost being very much dependent on the processes involved. Of particular interest is *catalytic conversion*, which is an incineration process that uses a catalyst to reduce the usually high incineration temperature requirements. Used in special treatment applications, catalytic units can be built for specialized contaminants, but the cost of such units tends to be rather high.

7.2.7   *In Situ* Soil Flushing

In situ soil flushing (or, *in situ* leaching) is a process by which in-place soils are flushed with water, usually mixed with a biodegradable nontoxic surfactant, in an effort to leach the compounds present in the soil into groundwater. The flushing agent is allowed to percolate into the soil and enhance the transport of contaminants to groundwater extraction wells for recovery. The groundwater is then collected downgradient or downstream of the leaching site for treatment, recycling, and/or disposal.

Water is normally used as the flushing agent in the *in situ* leaching technique. However, other solvents may be used for contaminants that are tightly held or only slightly soluble in water. Chemically enhanced *in situ* soil flushing may have extensive applications, but such applications will generally require site-specific evaluation and system design. For hydrophobic compounds (such as most hydrocarbons), flushing with surfactants is likely to be more effective than flushing with water; flushing with water generally may suffice for hydrophilic compounds. Solvents are selected on the basis of their ability to solubilize the contaminants and also on their environmental and human health effects. Thus, it is important to know the chemistry and toxicity of the surfactant. It is also important to understand the hydrogeology of the site to ensure that contaminants will be extracted once they are mobilized.

In summary, the *in situ* leaching technology is most applicable for soluble organics and metals at a low-to-medium concentration that are distributed over a wide area. Inorganic and organic contaminants are extracted from soil by flushing the soil with solvents; solvents are recovered, contaminants are extracted, and the solvents are recirculated through the soils (Figure 7.2). In general, a high water solubility, a low soil-water partition coefficient (K_{oc}), and a porous soil matrix will aid in the effective removal of chemical contaminants from soils using the soil leaching technique. The K_{oc}, which is a measure of the equilibrium between the soil organic content and water, is the leading factor in controlling the effectiveness of soil flushing; a low K_{oc} value indicates a favorable leaching tendency of the constituent from the soil.

7.2.8   *In Situ* Vitrification

In situ vitrification (ISV) is an innovative thermal treatment process that converts contaminated materials into a chemically inert, stable glass and crystalline product.

ISV uses electrical energy to melt contaminated solids at very high temperatures. The initial step in the vitrification process is to identify the boundaries of the area of contaminated soil to be treated. An array of electrodes is inserted into the ground to the desired treatment depth. The electrodes are placed vertically into the contaminated soil region and an electrical current is applied; an electrical potential of over 12 kV is applied to the electrodes which establishes the electrical current. The resultant power heats the path and surrounding soil to above fusion temperatures. The soil is melted by the resulting high temperatures. When the melt cools and solidifies, the resulting material is stable and glass-like, with the contaminants bound in the solid.

The ISV method may be used to provide a solution to mixed wastes (including organic, inorganic, and radioactive wastes) in soils at a potentially contaminated site, to about a 15-m ($\approx$ 50-ft) depth. The application of this technology can be rather expensive.

7.2.9   Landfarming

Landfarming is a land treatment process by which contaminated materials are spread over an area to enhance naturally occurring processes such as volatilization, aeration, biodegradation, and photolysis. Standard earth-moving equipment is typically used to prepare the landfarm area and to apply the material for treatment. The treatment area usually has a low-permeability liner material to prevent or minimize leaching. It usually is also surrounded by berms to control potential runoff processes and erosion from the soil during rainstorm events.

General treatment requirements for the landfarming process include periodic applications of nutrients (such as phosphorus and nitrogen), moisture control, and discing for oxygen exposure; biological treatments use bacteria and other microorganisms to degrade the contaminants.

7.2.10   Solvent Washing

Solvent washing (also known as *soil washing*) consists of excavating soils from contaminated areas and washing the contaminants from the soil using water or an aqueous solution. The contaminated effluent is then recovered, treated and recycled, or disposed of. Water is normally used as the washing agent; however, other solvents

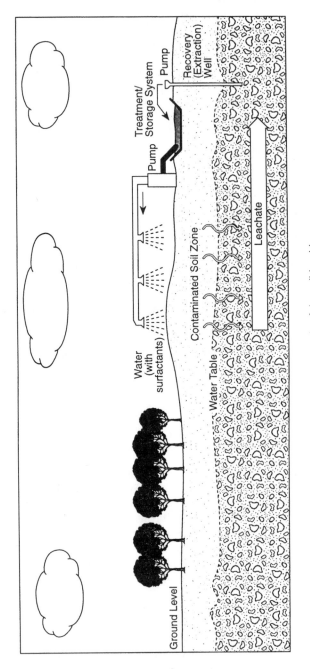

Figure 7.2 A general schematic of the soil leaching process.

may be used for contaminants that are tightly held or only slightly soluble in water. Solvents are selected on the basis of their ability to solubilize the contaminants and also on their environmental and human health effects. Thus, it is important to know the chemistry and toxicity of the solvent.

The soil-washing technique is most applicable for soluble organics and metals. Typically, inorganic and organic contaminants are extracted from soil by washing the soil with solvents; solvents are recovered, contaminants are extracted, and the solvents are recirculated through the soils.

7.2.11 Vapor Extraction Systems

Vapor extraction systems (VESs) may be applied to the removal of volatile organic chemicals (VOCs) present at contaminated sites. The VES is most applicable to the remediation of the higher volatile or lower molecular weight constituents of organic compounds only. As a general rule, heavier organic fractions (such as diesel fuel and fuel oils) are not candidates for remediation by vapor extraction. Also, vapor extraction is more effective at sites where the more volatile chemicals are still present, especially when the spill/release is reasonably recent. At most sites, the initial VOC recovery rates are relatively high and then decrease asymptotically to zero with time.

The VES technology is a particularly economical and efficient means of removing VOCs from the subsurface environment. A very important advantage for using a VES relates to the fact that there is minimal site disruption during its implementation/ operation. VESs can indeed be designed for many areas of site remediation such as contaminated soil piles and inaccessible locations (e.g., underneath buildings and similar structures).

In general, the VES may vary considerably in size and design, depending on site-specific requirements and/or conditions. Typically, a well-designed VES consists of a series of extraction/injection wells connected to a common manifold and a positive displacement air blower, among other surface equipment (such as emission controls, instrumentation, electric motor, etc.). The most common applications of VESs include soil vapor extraction, air or steam stripping, and air sparging; these systems are discussed below.

Soil Vapor Extraction

Soil vapor extraction (SVE) (also known as *in situ* soil venting, subsurface venting, vacuum extraction, or *in situ* soil stripping), is a technique that uses soil aeration to treat subsurface zones of VOC contamination in soils. In the SVE technology, VOCs are extracted from soil by using a vacuum system. This soil cleanup technique employs vacuum blowers to pull large volumes of air through contaminated soil. The airflow flushes out the vapor-phase VOCs from the soil pore spaces, disrupting the equilibrium that exists between the contaminants on the soil and in the vapor. This causes volatilization of the contaminants and subsequent removal in the air stream. Treatment rate depends on airflow through the soils and how effectively the contaminants partition into the mobile air phase (Newman et al., 1993).

Figure 7.3 depicts the pertinent design elements of a typical SVE system. The general type of SVE system consists of extraction vents/wells, air inlets or injection vents/wells (optional), air (piping) headers, vacuum pumps or air blowers, flow meters and controllers, vacuum gauges, sampling ports, air-water separator (optional), vapor treatment (optional), and an impermeable cap (optional) (Hutzler et al., 1990; USEPA,

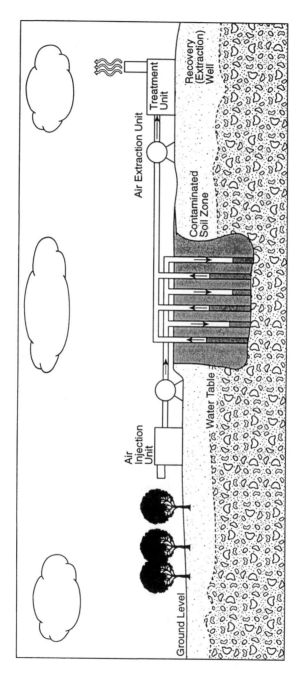

Figure 7.3 Illustrative example of a typical soil vapor extraction system.

1989). The SVE system operation is based on the extraction of VOC-laden air from contaminated soils. Inlet or injection wells, usually located at the boundaries of the contaminated area, may be used to enhance VOC-laden airflow to the extraction wells. Inlet wells are passive, with ambient air being drawn into the ground at the well locations due to pressure differentials caused by the removal of air from the extraction wells; injection wells are active and force air into the ground at the well locations. The inlet air going to the injection well may be supplied by the VOC control treatment exhaust or the vacuum pump (blower) exhaust, at the site engineer's discretion. Insulation is occasionally used on the piping and headers, especially in colder climates, to prevent condensate freezing.

The SVE process involves removing and venting VOCs from the vadose or unsaturated zone of contaminated soils by mechanically drawing or venting air through the soil matrix. Fresh air is injected or allowed to flow into the subsurface at locations in and around the contaminated soil to enhance the extraction process. This is carried out by connecting a vacuum pump or fan to one or more extraction wells, with the extraction wells typically being installed to penetrate the contaminant plume near the zone of highest VOC concentration. When suction is applied to the extraction wells, it induces a subsurface radial air flow towards perforations in the well casings. In addition, injection or ventilation wells (to facilitate the infiltration of clean air into the soil) may be placed at selected locations to help direct the flow of induced air towards the extraction wells. The VOC-laden air is withdrawn under vacuum from recovery or extraction wells which are placed in selected locations within the contaminated site. This air is then either vented directly to the atmosphere, or it is vented to an above-ground VOC treatment unit such as a carbon adsorber or a catalytic incinerator prior to being released to the atmosphere. The decision to employ VOC control system treatment is largely dependent on VOC concentrations and applicable local regulations. The selection of a particular VOC treatment option may be based in part on individual site characteristics.

SVE is effective under a wide range of site conditions. Extraction vents/wells are typically designed to fully penetrate the unsaturated soil zone or the geologic stratum to be cleaned. Spacing of extraction vents is usually based on an estimate of the radius of influence of an individual extraction vent; vent spacing typically ranges from about 4.5 to 30 m ($\approx$15 to 100 ft) (Hutzler et al., 1990). Capping the entire site with plastic sheeting, clay, concrete, or asphalt enhances horizontal movement towards the extraction vent. Impermeable caps extend the radius of influence around the extraction vent. The use of a ground surface cover will also prevent or minimize infiltration which, in turn, reduces moisture content and further chemical migration. If water should be pulled from the extraction vents, an air-water separator is required to protect the blowers or pumps and to increase the efficiency of the vapor treatment system. The overall design and operation of SVE systems depend on several factors, including (Hutzler et al., 1990):

- Contamination volume — the extent to which the contaminants are dispersed in the soil is important in deciding if vapor extraction will be cost-effective.
- Groundwater depth — SVE systems have been used in both shallow and deep unsaturated zones. However, where groundwater is more than 12 m ($\approx$40 ft) deep and contamination extends to the water table, a SVE system may be the only way to remove VOCs from the unsaturated zone.
- Soil heterogeneity — inhomogeneities influence air movements, affecting the placement of extraction and inlet vents.

- Site soil characteristics — higher contaminant removal efficiency can be anticipated in highly permeable soils. Soil moisture content or degree of saturation
  is also important because it is easier to draw air through drier soils.
- Chemical properties — in conjunction with site conditions and soil properties,
  chemical properties determine the feasibility of a SVE system. The system
  most effectively removes compounds that exhibit significant volatility at ambient subsurface temperatures. Generally, compounds with Henry's law constants >0.01, or vapor pressures >25 mmHg (or >1 in. mercury) are removable
  by vapor extraction.
- Contaminant location and area development — if, in particular, contamination
  extends across property lines, beneath buildings, or beneath an extensive utility
  trench network, then the applicability of VESs should be comprehensively
  evaluated in favor of intrusive remediation techniques. Since *in situ* vapor
  extraction generally can be conducted with minimal site disruption, this is a
  particularly favored remedial option in highly developed areas.
- Operating variables — among several factors, operating variables such as
  higher airflow rates tend to increase vapor removal because the zone of influence is increased and air is forced through more of the air-filled pores. Also, the
  water infiltration rate can be controlled by placing an impermeable cap over the
  site.
- Response variables — these consider system performance parameters that
  include air pressure gradients, VOC concentrations, and power usage; vapor
  removal rates are affected by the chemical's volatility, its sorptive capacity into
  soil, the air flow, initial distribution of chemical, soil moisture content, etc.

SVE is a cost-effective technique for VOC removal, and is finding more wide-spread applications/uses. The method is a relatively simple concept and can be used in conjunction with other soil decontamination procedures such as biological degradation. Because contaminants must volatilize and partition into air while undergoing removal, contaminated sites having significant amounts of low-volatility compounds (e.g., diesel fuel, fuel oils, or jet fuels) often are not targeted for remediation by soil venting. Sites having soil heterogeneities that result in uneven air permeation also can prevent effective remediation by conventional soil venting.

Air Stripping

Air stripping is a remediation technique that involves the physical removal of dissolved-phase contamination from a water stream. It is a separation technology that takes advantage of the fact that certain chemicals are more soluble in air than in water (Nyer, 1993). By bringing large volumes of air in contact with the contaminated water, a driving gradient from the water to the air can be created for the contaminants of concern. The air-stripping process involves pumping contaminated water containing VOCs from the ground and allowing it to trickle over packing material in an air-stripping tower. At the same time, clean air is circulated past the packing material. When the contaminated water comes into contact with the clean air, the contaminants tend to volatilize from the water into the air. The contaminated air is then released into the atmosphere, or into a granular activated carbon (GAC) system.

Air strippers are designed to maximize removal of VOCs from groundwater, leading to transfer of these contaminants into air that can be treated in a control device or discharged to the atmosphere. An air-stripping tower provides the air mechanism

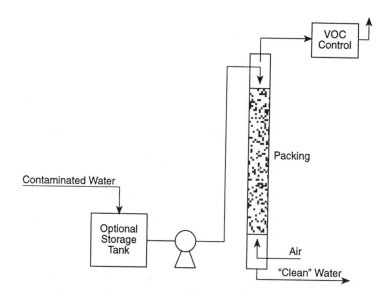

Figure 7.4 Schematic of a typical/representative air stripping process.

for, and water contact to remove the volatiles from the water by countercurrent flow through a packing material. The removal efficiency of an air stripper depends upon the volatility of the contaminants.

In a typical air stripping process, contaminated water containing VOCs is countercurrently contacted with air in a packed tower (Figure 7.4). Contaminants, usually VOCs, are transferred from the liquid phase to the gaseous phase. By contacting contaminated water with clean air, dissolved VOCs are transferred to the airstream to create equilibrium between the phases. The process takes place in a cylindrical tower packed with inert material, which allows sufficient air/water contact to remove volatiles from water. Contaminants are then removed from the airstream. An optional unit operation (e.g., catalytic oxidation or vapor-phase carbon adsorption) may be used for the control of air emissions. The VOCs are transferred to the gas phase during the intimate gas-liquid contact. The stripped water may further be treated in an optional carbon absorber polishing bed. The treated effluent water is either recycled as process water or discharged; in the case of a groundwater cleanup operation for site restoration programs, the treated water may be pumped back into the aquifer.

Air stripping of chemical contaminants from contaminated water is an effective method of removing VOCs from the contaminated water. However, this method also transfers pollutants from the water to the gas phase, and the resulting air emissions may need to be appropriately controlled. Indeed, air stripping is not a destructive technology; it merely moves contaminants from a liquid phase to an air phase, to be addressed differently.

Air Sparging

Air sparging is the highly controlled injection of air into a contaminant plume in the soil *saturated* zone. Air pumped into contaminated groundwater is used to strip volatiles from the groundwater to the soil vadose zone for capture using SVE. Air bubbles traverse horizontally and vertically through the soil column, creating a transient air-filled porosity in which volatilization can occur. In fact, air sparging effectively creates a crude air stripper in the subsurface, with the soil acting as the packing.

In a typical design of an air sparging system, an array of vents (or shallow wells) penetrating the impacted area of the vadose zone are connected via manifolding to air blowers. The blowers create a partial vacuum in the vents and pull air, including VOCs, out of the soil. The air sparging treatment involves injecting the soils with air, which flows vertically and horizontally to form an oxygen-rich zone in which VOCs are volatilized. Air bubbles that contact dissolved or adsorbed-phase contaminants in the aquifer cause the VOCs to volatilize. The volatilized organics are carried by the air bubbles into the vadose zone where they can be captured by a VES. Also, the sparged air maintains a high dissolved oxygen (DO) content, which enhances natural biodegradation. A carbon treatment system is used to treat off-gases. Careful investigation for optimal system design will result in an efficient operating system.

Used in conjunction with soil vapor extraction, air sparging is emerging as an effective treatment technology for soils and groundwater contaminated with VOCs. For instance, air sparging, in combination with SVE, has been used to remove and destroy organic solvents at difficult corrective action cleanup sites. In one case in the State of New York, trichloroethylene (TCE) and other VOCs were adsorbed onto soils below the water table at a facility located adjacent to a wetlands area. Applying SVE alone would have required pumping out groundwater to lower the water table; however, this option was unacceptable because of the location of the wetlands. Consequently, an air sparging-vapor extraction design was selected after careful site evaluation and pilot studies. With air sparging, air injected into saturated soils travels vertically and horizontally to form an oxygen-rich zone in which adsorbed and dissolved VOCs are volatilized. As vapors rise from the saturated zone to the soil vadose zone above, VOCs are captured by the SVE system. Despite the fact that sparging systems generally require very sophisticated analysis of site hydrogeology and careful engineering before implementation, overall the system can be very cost-effective.

7.2.12 Miscellaneous and Integrated Remediation Methods

There are several proven remediation technologies and processes that may be employed in the management of contaminated site problems. However, no single technology is universally applicable with equal success to all contaminant types and at all sites. Oftentimes, more than one remediation technique is needed to effectively address most contaminated site problems. In fact, treatment processes can be, and usually are, combined into process trains for more effective removal of contaminants and hazardous materials present at contaminated sites. For example, whereas biological treatment (with or without enhancement techniques) should result in the most desirable treatment scenario for petroleum-contaminated sites, the "hot spots" at such sites may best be handled by physical removal (i.e., excavation) and thermal treatment of the removed materials — rather than by biological methods. Consequently, several technologies (or combination of technologies) that can provide both efficient and cost-effective remediation should normally be reviewed as possible candidates in the remedy selection process.

Some commonly utilized techniques of general interest that are variations and/or combinations of some of the methods elaborated earlier are presented below. The contaminant treatment processes and techniques, as well as the remediation methods, processes, and technologies discussed earlier on in this chapter are by no means complete and exhaustive, and neither is the additional discussion presented below.

7.2.12.1   Bioventing

Bioventing is a variation of the VES. It comprises the delivery of oxygen to unsaturated soils by forced air movement for the purpose of enhancing biodegradation of organic contaminants. In bioventing, increased microbial activity results in the degradation of contaminants that are less easily removed by volatilization using the VES. Unlike vacuum-enhanced vapor extraction, bioventing injects air into the contaminated media at a rate designed to maximize *in situ* biodegradation and minimize or eliminate off-gassing of volatilized contaminants to the atmosphere.

Considering the fact that state and local regulations often require permitting, monitoring, and/or treatment of soil-venting off-gases that discharge to the atmosphere, a modification of the conventional vapor extraction remediation process to allow contaminants to be biologically removed *in situ* will reduce or eliminate air emissions, and therefore significantly cut down remediation costs. Bioventing also biodegrades less volatile organic contaminants and allows treatment of less permeable soils, because a reduced volume of air is required for treatment (Newman et al., 1993).

7.2.12.2   Excavation and Treatment/Disposal

Excavation is the physical process of removing soil by digging and scooping it out for treatment and/or disposal. It often is an initial step in many of the site restoration technology options available for the treatment of contaminated soils. Soil excavation, transport, and disposal processes use mechanized equipment to move contaminated soil. During excavation actions, adequate precautions and measures are necessary to minimize VOC emissions and fugitive dust generation.

Excavation and disposal at a regulated landfill is usually a preferred option only when relatively small volumes of "hot spot" soils have to be removed from the contaminated site. This is because extensive soil excavation is often too costly and disruptive to normal operations. Excavation of contaminated soil also creates increased potential for exposure to site personnel and the general public. Furthermore, the potential for long-term liabilities is always a concern for landfill disposals, since contributions of even small quantities to a disposal facility could make one a potentially responsible party (PRP) at a future date. Other *in situ* remedial alternatives should therefore be carefully evaluated before excavating large volumes of soil from a contaminated site.

7.2.12.3   Groundwater Pump-and-Treat Systems

Groundwater pump-and-treat systems involve contaminated groundwater being pumped out of the ground, treated by an appropriate treatment method to remove the contaminants of concern, and reinjected into the ground or used otherwise. Technologies for groundwater extraction and treatment are generally used to address the treatment needs of site-specific conditions and regulatory requirements. A common approach is to combine technologies to achieve effective treatment and to meet discharge criteria.

Invariably, groundwater extraction and treatment as a remedial action must address issues pertaining to the optimum design of the extraction-injection well network, as well as the selection of the proper treatment technology (such as air stripping, biodegradation, GAC adsorption, etc.) for the extracted groundwater. It is noteworthy that several types of the treatment processes utilized may be affected by a number of extraneous factors. For example, iron in groundwater can precipitate when oxidized, and may indeed have an adverse effect on air stripping and carbon adsorption treatment systems. Consequently, groundwater parameters such as total organic carbon (TOC), chemical oxygen demand (COD), biochemical oxygen demand (BOD), and iron content should be determined so as to provide input to the overall system design.

Finally, it must be recognized that treated water to be discharged may require compliance with certain pollutant discharge standards or criteria (such as compliance with the National Pollutant Discharge Elimination System [NPDES] standards, or similar criteria). Also, the use of a pump-and-treat technology would generally require the use of an air pollution control device (APCD) such as GAC columns to remove contaminants from gases released into the atmosphere. For these reasons, the exclusive use of only pump-and-treat technology often is neither a cost-effective nor an efficient approach to aquifer restoration. Pump-and-treat may be more appropriate for containing contaminant plumes, or for use in initial emergency response actions at sites having nonaqueous phase liquid releases to groundwater.

7.2.12.4 *Passive Remediation*

Passive remediation relies on natural processes (e.g., biodegradation, volatilization, photolysis, sorption, dispersion, and dilution) to remediate impacted soils and groundwater. Passive remediation may be applicable at sites where contaminant migration is limited, where potential impacts on the environment is minimum, and when health and safety considerations are insignificant.

In the application of a passive remediation alternative, continued monitoring is used to demonstrate that contamination levels are being attenuated and exposure is not occurring. Thus, following site assessment, the only activity undertaken is a progressive monitoring program to evaluate the effectiveness of the "no-action" option in the management of a contaminated site problem.

7.2.12.5 *Subsurface Control Systems*

The primary purpose of subsurface control systems is to prevent leachate migration and therefore reduce potential groundwater contamination via diversion, containment, or plume capture. Subsurface control measures include capping and top liners, seepage basins and ditches, subsurface drains, ditches and bottom liners, impermeable barriers, groundwater pumping, and interceptor trenches (USEPA, 1985).

Caps and *top liners* are generally used to reduce infiltration into a contaminated site, thereby reducing the amount of leachate that is generated. The primary objective for using *seepage basins* and *ditches* is to recharge site runoff or water withdrawn by wells or drains; it also can help improve the efficiency of plume capture by modifying groundwater flow patterns. *Subsurface drains, ditches,* and *bottom liners* are usually installed in the unsaturated zone to capture leachate before it reaches the saturated zone. *Impermeable barriers* are grout curtains, slurry walls, and sheet pilings installed in the saturated zone to divert uncontaminated groundwater around a site or to limit the migration of contaminated groundwater. Barriers can be placed in a number of locations relative to a contaminated site (e.g., upgradient, downgradient, or completely around). *Groundwater pumping* actions can have a number of configurations and design objectives. Single pumping wells or a line of well points can be used to capture a plume. Single or multiple wells can be installed to divert groundwater by lowering the water table; they can also be used to prevent unconfined aquifers from contaminating lower aquifers separated by leaky formations. The water withdrawn by pumping may be treated and subsequently reinjected through one or more wells. The reinjection wells may be used to flush contaminants toward the pumping wells or to create a hydraulic barrier to preclude further plume migration. *Interceptor trenches* are drain systems that are installed in the saturated zone; they can be used to divert groundwater by lowering the water table or to capture a plume.

An illustrative example that combines several subsurface control measures in a site restoration program is shown in Figure 7.5.

7.3   A SUMMARY OF CONTAMINATED SITE RESTORATION METHODS

Table 7.1 presents a summary recapitulation of the remedial action technologies, methods, and processes commonly used to address contaminated site problems. This partial listing includes some of the required screening and design parameters, important limitations, and also a gage of the time frame and potential costs associated with the different remedial alternatives of choice. In general, the remediation time frames and also the costs associated with remediation programs will vary significantly in accordance with site-specific conditions and the degree of cleanup attainment that is anticipated.

7.4   AN ILLUSTRATIVE EXAMPLE ANALYSIS INVOLVED IN THE SCREENING OF SITE RESTORATION OPTIONS

This section consists of the review and evaluation of remedial action alternatives considered for an inactive hazardous waste site, the S-Area landfill, located in the City of Niagara Falls, New York (Figure 7.6). The Niagara River that discharges into Lake Ontario is located approximately 120 m (≈400 ft) south of the S-Area landfill site. The site is also adjacent to the City of Niagara Falls Drinking Water Treatment Plant. The S-Area landfill lies on top of an approximately 9-m (≈30-ft) overburden (consisting of soil, clay, till, and human-made fill materials) in an area reclaimed from the Niagara River (USEPA-II, 1990). Immediately beneath the overburden is fractured bedrock. The topography of the S-Area landfill site and vicinity generally slopes gently southward towards the Niagara River. The Niagara Gorge is a major groundwater discharge zone in the Niagara Falls area. Reasonably good hydraulic connection exists between the Niagara River and an upper bedrock aquifer around the S-Area.

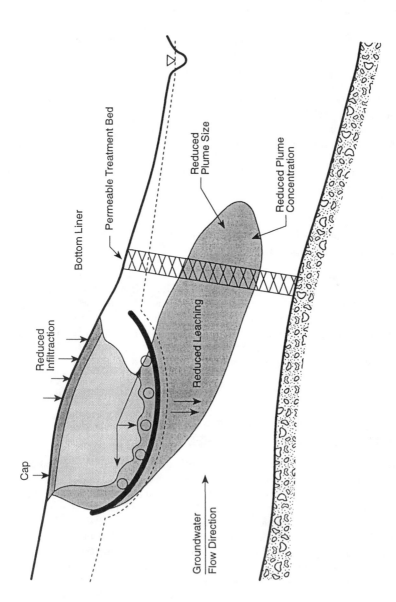

Figure 7.5 Illustrative representation of a combination subsurface control system.

Table 7.1   Summary of Site Restoration Methods and Technologies Commonly Applied to Contaminated Site Problems

Environmental Matrix of Concern	Remedial Option	Basic Technology and Process	Typical Screening and Design Parameters	Scope of Potential Applications	Limitations/ Comments	Typical Remediation Time Requirements	Estimated Remediation Costs per Volume of Impacted Matrix
Soils	Asphalt batching	Asphalt incorporation, involving the incorporation of petroleum-contaminated soils into hot asphalt mixes as a partial substitute for stone aggregate	Soil type Contaminant types Meterological factors	Economical for larger volumes of contaminated materials Dependent on climate; most asphalt plants do not operate during cold weather	May be unsuitable for clays May require offsite transportation	1–3 Months	$40–$105 per cubic meter ($30–$80 per cubic yard)
	Bioremediation	Natural or enhanced biodegradation, involving a process to degrade organic compounds into innocuous materials. Needs shallow groundwater (<15 m [or 50 ft]) or an underlying impermeable silt or clay layer	Soil type, porosity, and permeability Contaminant biodegradability/biodegradation rate Degree and extent of contamination Distribution of microorganisms Nutrient and dissolved oxygen requirements Injection/extraction well flow rates	Most cost effective for large volumes of contaminated material Most applicable when contamination extends into groundwater, and is of sufficient volume or depth below surface Minimal disruption to site operations, for in situ bioremediation	Labor intensive; requires considerable maintenance Possibility for contaminant migration Loss of efficiency in soils containing certain chemicals and having low pH	Several months to years (average time frame is 1–10 years)	$15–$195 per cubic meter ($10–$150 per cubic yard)
	Encapsulation	In situ containment and isolation process that comprises isolating contaminated soils from the surrounding environment by use of clay caps, liners, slurry walls, grout curtains, etc.	Underlying geology Presence of reactive compounds Contaminant concentration Soil type and moisture content	Used only to prevent contaminant migration Applicable to most chemicals, provided such compound does not attack containment materials Open areas preferred Example application involves capping landfills to prevent leaching by recharge water	Long-term monitoring required Does not destroy contaminants; only prevents migration Solution may not be permanent	1–6 Months	$40–$195 per cubic meter ($30–$150 per cubic yard)
	Chemical fixation	Solidification/stabilization treatment process, involving the addition of materials to decrease the mobility of the original waste constituents	Open area needed to excavate, stockpile, and treat the soil and to cure the treated materials Susceptibility to reactions Total organic carbon	Applicable to a wide variety of waste materials	Long-term monitoring required Future land use may be restricted	Up to 1 Month	$130–$390 per cubic meter ($100–$300 per cubic yard)

Table 7.1 (continued) Summary of Site Restoration Methods and Technologies Commonly Applied to Contaminated Site Problems

Environmental Matrix of Concern	Remedial Option	Basic Technology and Process	Typical Screening and Design Parameters	Scope of Potential Applications	Limitations/ Comments	Typical Remediation Time Requirements	Estimated Remediation Costs per Volume of Impacted Matrix
	Excavation/landfill disposal	Process involves the removal of contaminated soils	Underlying geology Presence of reactive compounds Contaminant concentration Soil type and moisture content	Best for removal and disposal of limited volumes of "hot spot" materials from a vast area that may otherwise be "clean" Common practice for shallow, highly contaminated soils	Potential for long-term liabilities Long-term monitoring required Trend is towards increasing disposal costs	<1 Month	$65–$390 per cubic meter ($50–$300 per cubic yard)
	Incineration	Thermal treatment	Open area needed to excavate, stockpile, and treat the soil	Economy of scale for large volumes Applicable to a broad range of organic compounds	Generally high costs Extensive permit requirements for onsite treatment Potential disposal problems for residual materials; residual ash may require further treatment	1–3 Months	$50–$390 per cubic meter ($40–$300 per cubic yard)
	In situ soil flushing (In situ soil leaching)	In situ soil leaching process that involves injecting or flushing in-place soils with water to leach chemicals in soils into groundwater; surfactants may be added to facilitate the flushing process	Soil type, porosity, and permeability Contaminant solubility Contaminant concentration and distribution Sorption properties Soil moisture content Aquifer parameters Depth to aquifer	Applicable to both organics and inorganics, but to different degrees Most applicable when contamination extends to groundwater table, requiring use in conjunction with extraction and treatment systems Site hydrogeology has strong influence	Leachate collection required Long-term monitoring required Potential problems with leaching fluid Less feasible for complex mixtures of several waste types Costs depend on site characteristics, contaminant constituents, and cleanup levels	1–2 Months (possibly longer)	$25–$160 per cubic meter ($20–$120 per cubic yard)

Method	Description	Considerations	Applicability	Limitations	Time frame	Cost
Landfarming	Land treatment process, by which affected soils are removed and spread over a treatment area in a layer so as to enhance naturally-occurring degradational processes	Soil type and porosity Soil temperature Degree of contamination Contaminant degradability/degradation rate Nutrient and dissolved oxygen requirements	Best suited for lighter organic compounds Warmer temperatures are conducive to faster degradation rates, since temperature influences the rate of degradation	Emissions control is difficult Low cleanup levels may not be practical Requires relatively large treatment areas Certain chemicals may be toxic to native microbes that could be facilitating degradation process	2–6 Months (possibly longer)	$15–$50 per cubic meter ($10–$40 per cubic yard)
Passive remediation	A "no-action" option that relies entirely on several natural processes to destroy the contaminants of concern	Underlying geology Presence of reactive compounds Contaminant concentration Soil type and moisture content	Site-specific; greater depth to groundwater, presence of aquitard, and low infiltration may minimize migration to groundwater Temperature may affect volatilization and natural degradation	Long-term monitoring required Long-term liabilities possible Low restoration levels may not be possible Effectiveness may be influenced by soil conditions Possible future landuse restrictions	10–100 Years	$0 (for remediation) — low cost (for monitoring programs)
Vapor extraction systems	*In situ* volatilization, involving the removal of volatile organic compounds from subsurface soils by mechanically drawing air through the soil matrix	Soil type, porosity, and permeability Contaminant volatility Contaminant concentration and distribution Sorption properties Well radius of influence Extraction well flow rates	Applicable to VOCs Can be applied to wide area extents Minimal disruption to site operations; conducive to both developed and undeveloped sites Most effective at higher concentrations Generally low costs	Venting and emissions difficult to control for very shallow areas Not effective below the groundwater table Performance affected by soil conditions; less effective and/or longer time frame in fine-grained soils May not work for semi-VOCs	Several weeks to a couple of years (average time frame is 1–10 years)	$15–$185 per cubic meter ($10–$140 per cubic yard)

Table 7.1 (continued) Summary of Site Restoration Methods and Technologies Commonly Applied to Contaminated Site Problems

Environmental Matrix of Concern	Remedial Option	Basic Technology and Process	Typical Screening and Design Parameters	Scope of Potential Applications	Limitations/ Comments	Typical Remediation Time Requirements	Estimated Remediation Costs per Volume of Impacted Matrix
Groundwater	Air stripping	Aeration process, effective for removal of VOCs	Contaminant volatility Groundwater temperature Influent flow rate Contaminant concentrations	Simple technology, usable in conjunction with other methods Low maintenance costs	Problems with air emissions; collection and treatment systems required for off-gas releases	Several months to years (average time frame: 3 months– 30 years)	$15–$185 per cubic meter ($10–$140 per cubic yard)
	Carbon adsorption	Activated carbon adsorption technology, based on the principle that certain organic contaminants preferentially adsorb onto organic carbon	Carbon adsorptive capacity Contaminant adsorptability Total organic carbon Influent flow rate Contaminant concentrations	Simple technology, usable in conjunction with other methods Attractive as point-of-use treatment method More appropriate for aquifer restoration Applicable to a broad range of (organic and inorganic) contaminants	Generates "spent carbon" as a waste byproduct High maintenance costs; carbon requires frequent regeneration	<1 Month	$15–$25 per cubic meter ($10–$20 per cubic yard)
	Groundwater extraction	Groundwater pump-and-treat system for the restoration of contaminated aquifers; it involves pumping groundwater for treatment at the surface	Soil type and porosity Aquifer parameters (including aquifer storage coefficient, hydraulic conductivity, aquifer saturated thickness, etc.) Depth to aquifer Number of wells Well extraction rates Contaminant sorption and solubility Contaminant distribution Presence of nonaqueous phase liquids (NAPLs)	Very common approach for the restoration of contaminated groundwater Best used for containing contaminant plumes	Treated water to be discharged requires compliance with certain pollutant discharge standards or criteria Generally requires use of air pollution control devices to remove contaminants released into air Not considered cost-effective nor efficient for aquifer restoration	Several months to years	$25–$210 per cubic meter ($20–$160 per cubic yard)
	Passive remediation	A "no-action" option that relies entirely on several natural processes to destroy the contaminants of concern	Groundwater flow regime Contaminant types	Several physical and chemical variables (e.g., temperature and alkalinity) may affect natural degradation and other attenuation processes	Long-term monitoring required Long-term liabilities possible Low restoration levels may not be possible	10–100 Years	$0 (For remediation) — low cost (for monitoring programs)

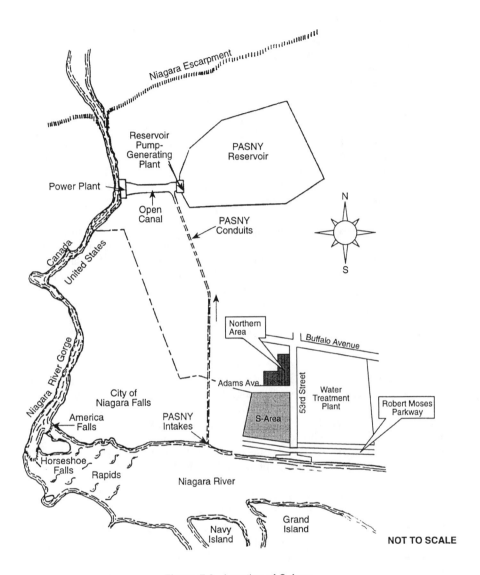

Figure 7.6 Location of S-Area.

Background Information on the Extent of Environmental Contamination Problems at the S-Area

Characterization of the nature and extent of contamination is an important input to making informed decisions on the appropriate remedial response types for the S-Area landfill site. The results of remedial investigations at the site have indicated, among other things, that (USEPA-II, 1990):

• The major contaminants of concern at the S-Area landfill site consists of hexachlorobenzene, benzene, hexachlorocyclohexanes, mirex, trichlorobenzenes,

tetrachlorobenzenes, polychlorinated biphenyls, dioxins, furans, and other similar
organic constituents.
- The groundwater flow in the overburden is to the south of the S-Area landfill
  site, toward the Niagara River.
- The groundwater flow in the bedrock is generally away from the S-Area, but
  ultimately to the lower Niagara River.
- In the overburden, NAPL chemicals have migrated in all directions.
- Aqueous-phase liquid (APL) chemicals have migrated through the overburden
  towards the Niagara River.
- Nonaqueous-phase liquid (NAPL) has migrated from the S-Area landfill site
  into the bedrock beneath the site and also beneath the Niagara River.
- There are areas of the overburden (confining layer) that lack sufficient clay/till
  layer to prevent the downward migration of NAPL.

In fact, results of several of the investigative studies for the site indicate that NAPL
and APL have migrated away from the S-Area in several directions; the extent of
these migration and spread is not fully mapped. In addition, it is apparent that
significant chemical loadings on the Niagara River could result from the erosion of
contaminated surface soils from the S-Area into the Niagara River; the long-term
erosion potential of the S-Area site can be evaluated and addressed by the S-Area
remedial systems in order to ensure that excessive loading to the Niagara River does
not occur.

Program Objective

The principal goal of any comprehensive corrective action program for this site
would be to prevent contaminant migration from the site to potential receptors, and
therefore prevent the endangerment of human health and the environment in and
around the S-Area landfill site. The specific major objective of the remedial action
program for the S-Area is to contain and/or collect and treat APL and NAPL, to be
accomplished by barrier walls and collection wells. This will ensure the protection of
human health and the environment in the vicinity of the site.

Evaluation of Feasible Remedial Options

Remediation of the S-Area landfill site may be accomplished via such processes
as waste removal (including groundwater and NAPL pumping and soil/waste excavation), waste treatment, containment and collection systems (including perimeter slurry
walls, impermeable liners, subsurface grouting, and APL/NAPL pumping), and/or
capping of the landfill site.

Groundwater has been determined to be the most significant medium of concern
in screening the remedial technologies and in evaluating the process options. The
assumption here is that much of the remediation program will be directed towards
containing, recovering, treating and disposing of APL and NAPL. Additional evaluation of remedial technologies will be required if it is determined that surface water
and sediment contamination has occurred to a point requiring remediation. For the
purpose of this illustrative evaluation, however, no surface water and sediment
remediation programs are anticipated.

Table 7.2 consists of the feasible remedial alternative development processes, and
a summary evaluation of the remedial technologies for the S-Area remedial measures
evaluation. The remedial alternative of choice will consist of a containment system

Table 7.2 Screening of Feasible Remedial Technologies and an Evaluation of Remediation Process Options for the S-Area Site

Target Media for Remediation Program	General Response Action	Remedial Technology	Preferred Process Option	Description of Remediation Program	Screening Comments
Groundwater	No Action	None	Not applicable	No action	Site remediation is required in order to achieve the remedial action objectives for this site
	Pump-and-Treat	Groundwater extraction	Extraction/injection wells	Series of wells to extract contaminated groundwater; injection wells inject uncontaminated or treated water to increase flow to extraction wells	Potentially applicable to the remediation of the overburden aquifer, but not recommended for intercepting contaminants in fractured bedrock
			Discharge	Extracted groundwater is discharged into Niagara River after treatment to "acceptable" discharge limits	Potentially applicable
	Containment	Capping (using low permeability caps and liners)	Clay	Compacted clay, covered with soil over areas of contamination	Potentially applicable as an effective alternative. Susceptible to cracking but has self-healing properties. Easily implemented. Low capital and maintenance costs
			Multimedia cap	Clay and synthetic membrane, covered with soil over areas of contamination	Potentially applicable as an effective alternative. Least susceptible to cracking. Easily implemented. Moderate capital and maintenance costs
		Vertical barriers	Slurry wall	Trench around area of contamination is filled with bentonite slurry	Potentially applicable

Table 7.2 (continued)   Screening of Feasible Remedial Technologies and an Evaluation of Remediation Process Options for the S-Area Site

Target Media for Remediation Program	General Response Action	Remedial Technology	Preferred Process Option	Description of Remediation Program	Screening Comments
			Grout curtain	Pressure injection of grout in a regular pattern of drilled holes	Potentially applicable to overburden, but most likely ineffective for fractured bedrock
			Vibrating beam	Vibrating force used to advance beams into the ground, with injection of slurry as beam is withdrawn	Potentially applicable
		Horizontal barriers	Grout injection	Pressure injection of grout at depth through closely spaced drilled holes	Potentially applicable to overburden, but most likely ineffective in fractured bedrock
			Block displacement	In conjunction with vertical barriers, injection of slurry in notched injection holes	Potentially applicable to overburden
	Gradient control		Hydraulic gradient manipulation	Use of hydraulic gradient to control flow	Potentially applicable
	Treatment	Bioremediation	Aerobic digestion	Degradation of chemicals using microorganisms in an aerobic environment	Potentially applicable
			Anaerobic digestion	Degradation of chemicals using microorganisms in an anaerobic environment	Potentially applicable

Medium	General response action	Remedial technology	Process option	Description	Screening comment
Soils and waste materials at S-Area		Physical treatment (e.g., physical separation or destruction) and chemical treatment (e.g., chemical modification or destruction)	Air stripping	Mixing large volumes of air and water in a packed column to promote transfer of VOCs to air	Potentially applicable to some contaminants
			Carbon adsorption	Adsorption of contaminants onto activated carbon by passing water through carbon column	Potentially applicable
			Reverse osmosis	Use of high pressure to force water through a membrane, leaving contaminants behind	Potentially applicable
			Incineration	Combustion	Potentially applicable
		Permeable treatment bed	Treatment bed	Treat shallow groundwater in-place by constructing permeable treatment beds that can physically and chemically remove contaminants	Potentially applicable to overburden as temporary remedial measure
	No action	None	Not applicable	No action	Site remediation is required in order to achieve the remedial action objectives for this site
	Containment	Capping (using low permeability caps and liners)	Clay	Compacted clay, covered with soil over areas of contamination	Potentially applicable as an effective alternative. Susceptible to cracking but has self-healing properties. Easily implemented. Low capital and maintenance costs
			Multimedia cap	Clay and synthetic membrane, covered with soil over areas of contamination	Potentially applicable as an effective alternative. Least susceptible to cracking. Easily implemented. Moderate capital and maintenance costs
		Vertical barriers	Slurry wall	Trench around area of contamination is filled with bentonite slurry	Potentially applicable

Table 7.2 (continued)   Screening of Feasible Remedial Technologies and an Evaluation of Remediation Process Options for the S-Area Site

Target Media for Remediation Program	General Response Action	Remedial Technology	Preferred Process Option	Description of Remediation Program	Screening Comments
			Grout curtain	Pressure injection of grout in a regular pattern of drilled holes	Potentially applicable
			Vibrating beam	Vibrating force used to advance beams into the ground, with injection of slurry as beam is withdrawn	Potentially applicable
		Horizontal barriers	Grout injection	Pressure injection of grout at depth through closely spaced drilled holes	Potentially applicable to overburden, but most likely ineffective in fractured bedrock
			Block displacement	In conjunction with vertical barriers, injection of slurry in notched injection holes	Potentially applicable to overburden
		Gradient control	Hydraulic gradient manipulation	Use of hydraulic gradient to control leachate flow	Potentially applicable
	Excavation/ treatment	Excavation	Soils excavation/ removal	Mechanical removal of materials for treatment and/or redisposal	Potentially applicable
		Immobilization/ stabilization	Sorption and/or encapsulation	Use of pozzolanic agents to help immobilize site contaminants	Potentially applicable
		Physical treatment	Incineration and pyrolysis	Thermal treatment methods	Potentially applicable, especially for the NAPL
		Chemical treatment	Neutralization	Use of chemical reagents to attenuate contaminant toxicity effects	Potentially applicable
		Biological treatment	Bioremediation	Use of microrganisms to degrade organic contaminants	Potentially applicable to some of the soil contaminants

with provisions for the collection of APL and NAPL, treatment of APL, incineration of NAPL, and monitoring programs to assess the effectiveness of the remedial system of choice. The specific elements of the feasible remedial system are (USEPA-II, 1990):

- A containment system to encompass the NAPL plume in the overburden.
- Drain tile collection system and purge/recovery wells in the overburden to collect APL and NAPL.
- Capping of the site to reduce infiltration and limit the creation of more APL.
- Overburden monitoring programs (consisting of a hydraulic system that will measure whether APL and NAPL are being contained within the overburden, and a NAPL system to evaluate wells for continued pumping of NAPL and to ensure that NAPL is contained).
- A pumping well system in the bedrock to collect APL and recover NAPL.
- Bedrock monitoring programs (that include a hydraulic program to assess if APL is being contained by the bedrock pumping system, a tracer program to evaluate the effectiveness of the bedrock remedial system at containing NAPL and APL within the existing NAPL plume in the bedrock, a NAPL program to evaluate the effectiveness of the bedrock remedial systems, and an environmental program to evaluate the effectiveness of the bedrock remedial system at protecting human health and the environment from the migration of chemicals from the S-Area landfill site).

The barrier walls are expected to present an impermeable, physical barrier to chemical migration; the material to be used for the barrier walls should therefore be tested for chemical compatibility with the S-Area chemicals. The operation of the tile collection system and collection wells in the overburden and collection wells in the bedrock will invariably be effective in collecting contaminated groundwater. NAPL collection will be effective to the extent that it is within the radius of influence of a collection well. The overburden is a porous media and as such will facilitate easier collection of NAPL than in the bedrock. NAPL will be recovered in the bedrock easier where the geology is substantially fractured. Capping of the site will invariably be effective at reducing the amount of precipitation entering the S-Area landfill site, and thus reducing the contamination leaving the site.

Although a "no-action" alternative will generally be presented during the initial screening of alternatives, it is apparent that some form of remedial action is necessary to address the conditions at the S-Area landfill site. Also, monitoring programs are included as part of the remediation program in order to verify the long-term effectiveness of the overall remedial program. Modifications to the remedial systems are implemented if monitoring program response actions are triggered.

The Selected Remedial System

The remedial alternative of choice for the S-Area landfill site should be designed and operated in such a manner as to be fully protective of human health and the environment. The S-Area remedial plan will comprise overburden and bedrock remedial systems, consisting of a site barrier wall containment system, drain tiles and collection wells within the barrier wall, a cap, groundwater pumping wells beneath the site to recover APL and NAPL, and associated monitoring programs (USEPA-II, 1990).

With an appropriate operation and maintenance (O&M) program for the selected remedial techniques, it is expected that the remedy of choice will remain effective during the course of remediation at the S-Area landfill site. The selected remedy will provide for the reduction of mobility of APL and NAPL by a physical barrier in the overburden, and by creating a hydraulic barrier in the overburden and in the bedrock through pumping wells. This will reduce the amount of contaminated groundwater flow that could otherwise continue to migrate under nonpumping conditions. Since APL is produced when water moves across a NAPL plume, the volume of APL would be reduced by reducing the NAPL amounts. APL will be removed from the S-Area landfill site and treated, and NAPL will be recovered and permanently destroyed by incineration. Consequently, the selected remedial action will reduce the toxicity, mobility, and volume of the chemicals of potential concern present at the S-Area landfill site and vicinity.

Ultimately, risks posed by the site will be reduced by containing and/or collecting and subsequently treating the APL and NAPL. Environmental monitoring programs will ensure that chemical loadings from the S-Area landfill site do not continue to escape the influence of the remedial systems. Other appropriate monitoring programs will help to evaluate the effectiveness of the remedy. Where necessary, the remedial system may be modified to ensure that the remedies are effective.

REFERENCES

ARB. 1991. Soil Decontamination. Compliance Assistance Program, Air Resources Board, Sacramento, CA.

Barnhart, M. 1992. "Bioremediation: do it yourself." *Soils,* Aug.-Sept. 1992, pp. 14–19.

Cairney, T. (Ed.). 1987. *Reclaiming Contaminated Land.* Blackie Academic and Professional, Glasgow, U.K.

Cairney, T. (Ed.). 1993. *Contaminated Land (Problems and Solutions).* Lewis Publishers, Boca Raton, FL.

Fredrickson, J.K., H. Bolton, Jr. and F.J. Brockman. 1993. "In Situ and On-Site Bioreclamation." *Environ. Sci. Technol.* Vol. 27, No. 9, 1711–1716.

Hutzler, N.J., J.S. Gierke and B.E. Murphy. 1990. "Vaporizing VOCs." *Civ. Eng. ASCE,* April, 1990, Vol. 60, No. 4, pp. 57–60.

Jolley, R.L. and R.G.M. Wang (Eds.). 1993. *"Effective and Safe Waste Management: Interfacing Sciences and Engineering With Monitoring and Risk Analysis."* Lewis Publishers, Boca Raton, FL.

Newman, B., M. Martinson, G. Smith and L. McCain. 1993. " 'Dig-and-Mix' bioventing enhances hydrocarbon degradation at service station site." *Hazmat World,* December, 1993: 34–40.

Nyer, E.K. 1993. *Practical Techniques for Groundwater and Soil Remediation.* Lewis Publishers, Boca Raton, FL.

OBG (O'Brien and Gere Engineers, Inc.). 1988. *Hazardous Waste Site Remediation: The Engineer's Perspective.* Van Nostrand-Reinhold, New York.

Purdue University. 1990. *Proc. 44th Industrial Waste Conf.* (May 9–11, 1989), Purdue University, West Lafayette, IN. Lewis Publishers, Chelsea, MI.

Purdue University. 1991. *Proc. 45th Industrial Waste Conf.* (May 8–10, 1990), Purdue University, West Lafayette, IN. Lewis Publishers, Chelsea, MI.

Purdue University. 1992. *Proc. 46th Industrial Waste Conf.* (May, 1991), Purdue University, West Lafayette, IN. Lewis Publishers, Chelsea, MI.

Purdue University. 1993. *Proc. 47th Industrial Waste Conf.* (May, 1992), Purdue University, West Lafayette, IN. Lewis Publishers, Chelsea, MI.

Sims, R.C. 1990. Soil Remediation Techniques at Uncontrolled Hazardous Waste Sites, A Critical Review. *J. Air Waste Manage. Assoc.,* Vol. 40, No. 5: 704–732 (May, 1990).

Sims, R. et al. 1986. *Contaminated Surface Soils In-Place Treatment Techniques.* Noyes Publications, Park Ridge, NJ.

USEPA. 1984. Review of in-place treatment techniques for contaminated surface soils. Volumes 1 and 2. U.S. Environmental Protection Agency, Hazardous Waste Engineering Research Laboratory. EPA-540/2-84-003a and b. Cincinnati, OH.

USEPA. 1985. Modeling Remedial Actions at Uncontrolled Hazardous Waste Sites. EPA/540/2-85/001 (April, 1985). Office of Emergency and Remedial Response, U.S. Environmental Protection Agency, Washington, D.C.

USEPA. 1988. Guidance on Remedial Actions for Contaminated Ground Water at Superfund Sites. Office of Emergency and Remedial Response, EPA/540/G-88/003. U.S. Environmental Protection Agency, Washington D.C.

USEPA. 1989. Soil Vapor Extraction VOC Control Technology Assessment. EPA-450/4-89-017. Office of Air Quality Planning and Standards, U.S. Environmental Protection Agency, Research Triangle Park, NC.

USEPA. 1990. Air Stripper Design Manual. Air/Superfund National Technical Guidance Study Series. Office of Air Quality Planning and Standards, EPA-450/4-90-003. U.S. Environmental Protection Agency, Research Triangle Park, NC.

USEPA — II (U.S. Environmental Protection Agency — Region II. 1990). Proposed RRT Remediation Plan, S-Area Landfill Site, Niagara Falls, New York. U.S. Environmental Protection Agency, Washington, D.C.

8   MANAGEMENT STRATEGIES FOR CONTAMINATED SITE PROBLEMS

Environmental contamination problems have reached an almost nightmarish level in most societies globally, with contaminated sites representing a significant portion of the overall problem. Whatever the cause of a site contamination problem, the impacted media usually must be cleaned up. Under certain circumstances, however, site cleanup may not be economically or technically feasible; in that case, risk assessment and monitoring of the situation, together with institutional control measures, may be acceptable site management alternatives in lieu of remediation.

The effective management of contaminated site problems has indeed become an important environmental priority that will remain a growing social concern for years to come. This is due in part to the numerous complexities and inherent uncertainties involved in the analysis of such problems. This chapter presents strategies and relevant protocols that can be used to aid contaminated site management decisions.

8.1 CONTAMINATED SITE RESTORATION STRATEGIES

A variety of corrective action strategies may be employed in the quest to restore contaminated sites into healthier conditions. The processes involved will generally incorporate a consideration of the complex interactions existing between the hydrogeological environment, regulatory policies, and the technical feasibility of remedial technologies. In principle, the site restoration goal and strategy selected for a contaminated site problem may vary significantly from one site to another due to several site-specific parameters. A number of extraneous factors may also affect the selection of site restoration strategies. The selection of an appropriate corrective action plan (including the choice of acceptable cleanup goals) depends on a careful assessment of both short- and long-term risks posed by the case-specific site. As a general rule, corrective action plans should provide for the removal and/or treatment of contaminants until a level necessary to protect human health and the environment is achieved.

8.1.1 Interim Corrective Actions

Whenever excessive risks exist at a contaminated site, a decision is made to implement interim corrective actions immediately in order to protect public health and the environment. Interim corrective actions are measures used to address situations which pose an imminent threat to human health or the environment, or to prevent further environmental degradation or contaminant migration pending final decisions on the necessary remedial activities. The U.S. Superfund program uses the removal authority (provided under Section 104 of CERCLA) to accomplish this same objective where expedited response and/or emergency actions are needed.

Common examples of interim corrective measures may include erecting a fence around a contaminated site in order to restrict/limit access to an impacted property, covering exposed contaminated soils with synthetic liner materials, applying dust suppressants to minimize emissions of contaminated fugitive dust, restricting use of contaminated water resources and/or providing alternative domestic/municipal water supplies, and temporal displacement or relocation of nearby residents away from hazardous waste sites.

Interim corrective actions usually will not be the ultimate containment or treatment strategy implemented as the solution for a contaminated site problem. These measures are developed primarily to minimize public exposure (to potentially acute risks/hazards) prior to developing a comprehensive corrective action program. In general, the interim measures may be relatively straightforward (such as erecting a fence or removing a small number of drums) or it may involve more elaborate control measures (such as installing a pump-and-treat system to prevent further migration of a groundwater contaminant plume).

In many situations, early in the corrective action response process, it is possible to identify measures that can and should be taken to control potential receptor exposures to contamination, or to stop further environmental degradation from occurring. Typically, where it is obvious that the final remedy will require excavation and treatment or removal of contaminated "hot spots", such actions are better initiated as interim measures rather than being deferred to a final remedy selection stage.

8.1.2 Removal Vs. Remedial Action Programs

The cleanup of inactive hazardous waste sites may involve removal and/or remedial actions. For example, U.S. environmental regulations under the Superfund program (Section 104 of CERCLA) authorizes that a removal or remedial action be taken to protect public health, welfare, or the environment when there is a release or substantial threat of release of any hazardous substance, pollutant, or contaminant that may present an imminent and substantial danger to the public health or welfare.

A *removal action program* provides a mechanism to clean up acute, short-term, immediate risks to public health and/or the environment. A removal response usually involves an immediate response to stabilize and contain a hazard situation; it normally is not a total site cleanup program. Such action is meant to reduce the threat sufficiently so that further studies can be conducted prior to any large-scale cleanup, without posing unacceptable short-term risks (Kunreuther and Gowda, 1990). The major factors considered in this case relate to the immediacy of the threat to human health and public safety. Consequently, the general criterion for completing a removal action is the determination of the level of contamination that will not pose short-term or acute risks to human health and/or the environment.

A *remedial action program* provides the framework for an organized response to inactive hazardous waste sites, generally consisting of the discovery and appraisal, prioritization, investigation, and cleanup of the contaminated site (Kunreuther and Gowda, 1990). Once preliminary assessment and site investigation programs are completed, a quasi-risk model (such as the hazard ranking system [HRS], developed by the EPA to determine whether or not a site qualifies for placement on the NPL) can be used for the prioritization; work may then proceed to fully characterize the site, and to determine the appropriate cleanup strategies.

8.1.3 General Response Actions

When remedial action objectives have been defined, general response actions (such as treatment, containment, excavation, pumping) or other actions that may be taken to satisfy those objectives can then be identified. In general, the risk posed by a contaminated site will normally be reduced by containing and/or collecting and subsequently treating contaminants.

The chemical nature of the contaminants of concern is a primary consideration in the identification of remedial alternatives that are capable of meeting the case-specific remediation goal(s). For instance, some technologies are designed primarily to control the release of heavy metals, whereas others are designed primarily to control the release of organics. When both types of contaminants are present at a site, the possibility of combining two or more technologies to meet the remediation objectives should be carefully evaluated. Potential short-term risks that may arise as a result of construction activities that are part of certain types of remedial actions should also be carefully evaluated and addressed as part of implementation package.

Oftentimes, the cleanup of contaminated sites will result in the intermedia transfer of contamination. For example, vacuum extraction and soil excavation (used to remove contaminants from soils), and air stripping in combination with granular activated carbon (used to remove contaminants from water), are typical technologies that may result in the transfer of contaminants into the air. Such releases would have to be treated before being released into the atmosphere; this requirement can indeed exert significant influence on the choice of cleanup technologies. Ultimately, however, remedial action alternatives selected for contaminated site problems should be designed and operated in a such a manner as to be fully protective of human health and the environment.

Human Health Risks Associated with Remedial Actions

As part of the overall corrective action response process, remedial alternatives are analyzed for both potential long- and short-term health effects associated with the implementation of each remediation option. Indeed, a remedial action will often increase potential short-term exposures and risks at a contaminated site above baseline conditions. For instance, the short-term public health risks posed by air emissions during excavation, handling/transportation, and disposal of contaminated soil materials could be substantial, and may well exceed the long-term benefits associated with the cleanup objectives. Nonetheless, escalated short-term risks should not become the principal determinant (or deterrent) of the corrective action decision.

Short-term health risks should generally not be used as a selection criterion for remedial alternatives, but should be used to determine appropriate management practices during implementation of the selected remedial action. For instance, a remedial option at a site may involve excavation and removal of contaminated soil. In the absence of precautionary or mitigative measures, fugitive dust generated by heavy equipment and remedial activities may create unreasonably high short-term health hazards. On the other hand, these and other temporary sources of contaminant releases associated with construction and implementation of a remedy may not be sufficiently adequate grounds for rejecting the remedial alternative. However, management practices, such as the temporary relocation of populations potentially at risk, should be considered in order to mitigate the health risks associated with such temporary source releases.

Ecological Effects of Remedial Alternatives

Remedial actions, by their nature, can alter or destroy aquatic and terrestrial habitats. The potential for the destruction or alteration of ecological habitats, and the consequences of ecosystem disturbances and other ecological effects, must be given adequate consideration during the corrective action response process. Thus, remedial alternatives, in addition to being evaluated for the degree to which they protect human health, are also evaluated for their ability to protect ecological receptors and ecosystems. Indeed, it is very important to integrate ecological investigation results and general concerns into the overall site cleanup process.

8.2 DESIGN OF CORRECTIVE ACTION PROGRAMS

Several formalized steps will generally be involved in the design of corrective action programs for contaminated site problems. Typically, the major activities carried out will comprise a remedial investigation (RI) and a feasibility study (FS).

The RI is conducted to gather sufficient data in order to characterize conditions at a contaminated site. This may involve extensive field work and site sampling used to identify site contaminants, determine their concentrations and distribution across the site, characterize migration pathways and routes of exposure to surrounding populations, and assess other site conditions that might affect cleanup options.

The FS is undertaken to identify and analyze potential site restoration alternatives. Typically, the FS uses a screening process to reduce the number of alternatives to a limited range of remedial options. The short-listed options are subjected to a detailed analysis in which various trade-offs are evaluated (such as cost effectiveness, extent of cleanup, and permanence of a cleanup action).

The following general tasks are typically carried out as part of the overall RI/FS program for contaminated site problems (Cairney, 1993; OBG, 1988; USEPA, 1988a):

- Characterize the site completely.
- Define and develop remediation objectives appropriate for the specific contamination problems.
- Identify the technologies that can achieve the site restoration goal(s).
- Develop and screen remedial alternatives, and select only those that are superior based on engineering, environmental, and economic criteria, and that concurrently meet regulatory and community expectations.
- Perform detailed analyses of remedial alternatives by scrutinizing each of the selected alternatives.
- Choose, offering justification, the preferred remedial alternative(s).

In general, once existing site information has been analyzed and a conceptual understanding of a site is obtained, potential remedial action objectives should be defined for all impacted media at the contaminated site; a preliminary range of remedial action alternatives and associated technologies should then be identified. The identification of potential technologies at this stage will help ensure that data needed to evaluate them can be collected as early as possible in the corrective action process.

Information derived from the RI will generally help determine the need for, and the extent of cleanup necessary for a contaminated site. After review of the detailed analysis and upon selection of one of a number of cleanup options identified in the FS, the project enters an engineering design phase in which plans and specification are developed for the selected remedy. Once the engineering design is completed, construction activities to deal with the actual cleanup of the site can be implemented. In some cases, such as groundwater remediation, cleanup activities may continue for several years.

8.2.1 Preliminary Remediation Goals

Preliminary remediation goals (PRGs) are usually established as cleanup objectives early in the site characterization process. The early determination of remediation goals facilitates the development of a range of feasible corrective action decisions, which in turn helps focus remedy selection on the most effective remedial alternative(s).

The development of PRGs requires site-specific data relating to the impacted media of interest, the chemicals of potential concern, and the probable future land uses. Methods for developing the appropriate PRGs were discussed previously in Chapter 6. It is noteworthy that an initial list of PRGs may have to be revised as new data become available during the site characterization activities. In fact, PRGs are refined into final remediation goals throughout the process leading up to the final remedy selection. Therefore, it is important to iteratively review and reevaluate the media and chemicals of potential concern, future land uses, and exposure assumptions originally identified at scoping.

8.2.2 Factors Affecting the Selection and Design
of Remediation Techniques

Once media cleanup standards have been established, the potential remedies can be evaluated, and then a selection made for the best demonstrated available technology (BDAT). Of utmost interest in this process is the nature of contamination. Important contaminant properties that may affect the choice and design of remediation techniques include the following:

- Retardation (which is a fundamental factor in contaminant treatment design)
- Adsorption properties of compounds on solids
- Volatility of compounds
- Henry's law constant (usually for the design of air strippers)
- Solubility
- Octanol/water partition coefficient
- Specific gravity
- Degradation rate or half-life

By using these types of information, several treatment methods can be selectively eliminated before significant sums of money are spent on a focused feasibility study.

For instance, by using the Henry's law constant data, a designer can estimate the removal efficiency that an air stripper will have on a particular compound; consequently, the designer can determine whether an air stripper should be studied further or eliminated as a possible candidate option. Carbon adsorption and biological treatment can be evaluated in a similar manner using the appropriate chemical data. Several extraneous factors may also affect the choice of remedial options. A more comprehensive review of information necessary for these types of evaluation are provided elsewhere in the literature (e.g., Nyer, 1993; OBG, 1988; Sims et al., 1986; USEPA, 1985).

In general, there is no single remedial technique that is best for all types of contamination and for all site conditions. In fact, successful remediation efforts may have to rely on proper marriages between remediation technologies.

8.2.3 Treatability Investigations in a Focused Feasibility Study

If remedial actions involving treatment are identified for a contaminated site problem, and if existing site and/or treatment data are insufficient to adequately evaluate such alternatives, then the need for treatability studies should be determined as early as possible during the site characterization activities. Treatability tests may be necessary to evaluate the effects of a particular technology on specific site contaminants. Such tests will typically involve bench-scale testing to gather information that will help assess the feasibility of selected technologies.

A treatability investigation involves performing bench or pilot treatability tests as necessary. In some situations, a pilot-scale study may be necessary to furnish perfomance data and to develop better cost estimates so that a detailed analysis can be carried out to aid the selection of a remedial action. Almost invariably, bioremediation programs usually will require treatability studies to enhance the development of the design criteria and parameters relevant to this technique.

In a typical application, the treatability study will consist of placing a small amount of a representative soil sample into a bowl or container, thereby creating a microcosm of the site, and dosing the microcosm with the selected site remedy. For example, a measured amount of a contaminated soil system can be placed into a container and innoculated with the selected remedy. After a specified time period, a sample is taken from the microcosm and analyzed for the parameters of concern. This is continued until the soil has reached the desired cleanup criteria or until the soil no longer exhibits a change in condition, indicating the endpoint of the lower limit of attainable cleanup (i.e., the asymptotic threshold level).

Treatability studies are conducted to provide sufficient data that will allow treatment alternatives to be fully developed and evaluated during a detailed analysis, and to support the remedial design of a selected alternative (USEPA, 1988a). Ultimately, it will help reduce cost and performance uncertainties for treatment alternatives to acceptable levels, resulting in an appropriate remedy being implemented.

8.3 A CONTAMINATED SITE RESTORATION DECISION FRAMEWORK

A number of indicators can create concern about possible contaminant releases into the environment. If any of the indicators of a possible release are present, then site investigations should be initiated since the indicators may not necessarily be confirmatory. In some cases, however, the evidence of release is so strong that a confirmation investigation is not necessary, in which case an investigator may move directly to the remedial investigation and mitigation studies.

The use of a structured decision framework will generally facilitate rational decision making on site restoration efforts undertaken for potentially contaminated sites. The key components of a representative site restoration decision process is comprised of the tasks discussed below.

Task 1.   *A Preliminary Site Appraisal,* consisting of the identification of possible source(s) of contaminant releases. The purpose of the preliminary site appraisal is to quickly assess the potential for a site to adversely impact the environment and/or public health. The site appraisal is initiated by the discovery of a potentially contaminated site. Conventional site reconnaissance procedures may be used in this qualitative site assessment that involve the collection and review of all available information (including an off-site reconnaissance to evaluate the source and nature of contamination, and the identification of any PRPs). Depending on the results of the preliminary survey, a site may or may not be referred for further action. In general, the preliminary site appraisal allows for site screening in order to establish a basis for more detailed site investigations.

Task 2.   *A Site Assessment,* involving the characterization of site contamination and a site categorization, where necessary. The objectives of the site assessment are to identify site contaminants and to determine site-specific characteristics that influence the migration of the contaminants. Overall, site assessment activities are undertaken to better define the characteristics of a potentially contaminated site and its neighboring areas.

Task 3.   *A Risk Appraisal,* consisting of the determination of the migration and exposure pathways, and an assessment of the environmental fate and transport of the contaminants of concern. The primary objective of the risk appraisal is to determine whether existing or potential receptors are presently, or may be in the future, at risk of adverse effects as a result of exposure to site contaminants. To determine if potential receptors are at risk, it is necessary to identify potential migration and exposure pathways as well as contaminant exposure point concentrations in relation to acceptable threshold levels.

Task 4.   *A Risk Determination,* to include an evaluation of the environmental and health impacts associated with contaminant releases, and also the development of site-specific cleanup criteria. The objective of the risk determination is to evaluate potential site risks, which can be compared against a benchmark risk; this then becomes the basis for developing cleanup goals.

Task 5.   *A Site Mitigation,* consisting of the development and implementation of a corrective action plan that may include site remediation and/or monitoring programs. Site mitigation strategies and objectives are based on the protection of both current and potential future receptors that could become exposed to site contaminants. Consequently, the site restoration is carried out so as to leave the site in such a condition as to pose no significant risks to populations potentially at risk.

Each of the above tasks may be comprised of a number of elements or subtasks. Although the tasks appear sequential in this presentation, it is noteworthy that certain aspects may be executed concurrently or in an iterative manner. In all cases of contaminated site assessments, a qualitative evaluation of site conditions identifies potential contaminant release source(s), determines the environmental media affected by each release, and broadly defines the possible extent of the release(s). For a confirmed release, the release source(s) must be stopped and hazards mitigated. Where applicable, any free product present should be removed immediately to prevent the development of health and safety hazards. Removal will also help prevent further migration of free product into soil and/or groundwater.

To meet environmental regulations and specifications for many regions, the appropriate regulatory agencies should be notified of any source release(s), and of steps taken in immediate response. In general, the initiation of free product removal should *not* have to await approval by regulatory agencies. In fact, since free product removal, if necessary, is the first step in every cleanup, there is no justification to delay its commencement.

8.3.1 Prescriptions for Cost-Effective Corrective Action Decisions

In order to design an effectual corrective action program for contaminated site problems, a number of relevant decision elements are generally evaluated as part of the corrective action assessment process for such sites (Box 8.1). The information derived from this evaluation will help define the following:

- Concentrations of contaminants of interest at and near the impacted site
- Populations that might be exposed to site contamination
- Magnitude and frequency of potential receptor exposures
- The health and environmental effects that might result from receptor exposures to site contaminants
- Site restoration strategies that might be employed at the site

Subsequently, appropriate corrective action decisions being considered to learn more about the site, and/or to mitigate any risks, can be formulated to support the corrective action program.

8.4 THE DEVELOPMENT AND SCREENING OF REMEDIAL ACTION ALTERNATIVES

The development and screening of remedial action alternatives involves identifying a range of remediation options that will ensure adequate protection of public health and the environment. Depending on the site-specific circumstances, such remedial options may result in the complete elimination or destruction of the contaminants of concern that are present at the site, the reduction of contaminant concentrations to "acceptable" risk-based levels, and/or the prevention of exposure to the contaminants of concern via engineering or institutional controls.

The broad group of remedial alternatives generally screened during corrective measure assessments for contaminated site problems include containment, removal, and treatment of contaminated materials. Site characterization data are used to identify

---

BOX 8.1
Important Decision Elements for Designing
Corrective Action Programs

- Establish a basis for contamination indicators
- Gather and review site background information for evidence of release
- Determine site history, to help identify other possible sources of contamination
- Identify potentially affected areas
- Address health and safety issues associated with the case-specific situation
- Address emergency response by mitigating release and potential hazards
- Characterize the site, to include an identification of the contaminants of potential concern and a delineation of the extent of contamination for affected matrices
- Determine contaminant behavior in the environment, and develop working hypothesis about contaminant fate and transport
- Determine contaminant migration and exposure pathways
- Identify populations potentially at risk
- Map areas where contaminants may impact human health and/or environment
- Develop realistic exposure scenarios appropriate for the specific site
- Conduct site-specific exposure assessments
- Define magnitude/severity of health and environmental impacts
- Establish remediation objectives and site-specific cleanup criteria
- Assess variables influencing selection of cleanup criteria and remedial systems
- Develop remedial action objectives
- Identify remedial action alternatives
- Develop general response actions

---

the general approach, or combination of approaches, that are likely to be most effective in addressing each impacted environmental matrix at a contaminated site. The selection of one approach over another will depend on several factors, such as ease of implementation with respect to technical feasibility and regulatory compliance, brevity of project duration, effectiveness of reducing contamination and risk to "acceptable" levels, and attainment of reasonable risk reduction within a justifiable cost-benefit framework.

In the process of developing contaminated site remediation alternatives, information on the nature and extent of contamination, applicable environmental regulations, contaminant fate and transport properties, and the toxicity of the contaminants are used to guide decisions made about the potentially feasible and appropriate remedial options. Subsequently, the remedial action alternatives and associated technologies are screened to identify those that will likely be effective for the contaminants and media of interest at the site. Information so developed is used in assembling technologies into alternatives for the site. Following the development and screening of feasible remedial alternatives, the more appropriate options can be identified and evaluated in detail for a contaminated site problem.

8.4.1   Remediation Options

Site remediation programs generally involve the use of containment or cleanup strategies in the management of contaminated site problems. Containment strategies have the goal of preventing further migration of mobile contaminants by controlling contaminant plume movements within a specified area and time frame; the primary benefit of a containment strategy is that, for some contaminants, given enough containment time, natural processes of attenuation will reduce concentrations to acceptable levels. Cleanup strategies have the goal of removing contaminants or contaminated media in a specified area until acceptable concentration levels are attained. The variety of remedial action alternatives associated with contaminated site problems may be categorized into the following broad and general groups:

- No-Action or Passive (which implicitly depends on natural processes such as volatilization, biodegradation, leaching, photolysis, and redox to destroy or attenuate the contaminants of concern).
- Containment (i.e., in-place isolation/containment).
- Removal/Disposal (i.e., contaminant removal and disposal elsewhere).
- Treatment/Restoration (i.e., contaminant treatment on- or off-site for site restoration).

The first three groups usually will be limited to problems posing low to no risks, or when the fourth option — a generally preferred alternative — is not feasible. Long-term monitoring will generally be required to supplement the use of these methods, with the extent of monitoring dependent on the category of remedial option. The principal remediation options that seem to have found widespread applications in the management of contaminated site problems have usually involved the removal of material from the case site for disposal elsewhere, or retention and isolation of material on-site using an appropriate form of cover, barrier, or encapsulation system, or physical, chemical, or biological treatment to eliminate or immobilize the contaminants, and/or lowering of the contaminant concentrations by diluting the contaminated material with clean material (Cairney, 1993).

The selection of a particular type of remedial option depends on the type of contaminants involved, cost effectiveness, practicability, site conditions and accessibility, and applicable regulations that must be met in the site restoration process. For instance, specific characteristics of diesel fuel (such as its low volatility, low health risk, and moderate viscosity) make certain remediation options better suited for diesel-contaminated soils, whereas the same remedial techniques may not be as effective in addressing other petroleum hydrocarbons. This is because the moderate viscosity of diesel makes remedial options such as containment or fixation reasonable candidate techniques to adopt for diesel-contaminated soils. On the other hand, the low diesel volatility makes vapor extraction systems (which may work reasonably

---

BOX 8.2
Requirements for Developing Remediation Alternatives

- Establish remedial action objectives
- Identify potential treatment technologies and containment or disposal requirements for the contaminants of concern (that will satisfy remedial action objectives)
- Determine process options and general response actions (that will satisfy remedial action objectives)
- Identify volumes or areas of impacted media (to which general response actions might be required)
- Identify regulatory limits and compare with removal efficiencies of remedial techniques
- Prescreen remedial technologies and process options based on their effectiveness, implementability, and cost
- Assemble technologies and their associated containment or disposal requirements into alternatives for the contaminated media
- Establish preliminary remediation goals and/or cleanup levels
- Determine the area of attainment (i.e., area over which cleanup levels will be achieved for the contaminated site, encompassing the area outside the site boundary and up to the boundary of the contaminant plume)
- Estimate the restoration time frame (i.e., the period of time required to achieve selected cleanup levels at all locations within the area of attainment)
- Formulate remedial alternatives (i.e., compile and group technologies into appropriate remedial alternatives)

---

well for the remediation of gasoline-contaminated soils, in most cases) ineffective for diesel fuels.

8.4.2 Development of Remedial Alternatives

The development of remedial alternatives for contaminated site problems involves compiling a limited number of options for source control and/or remedial action. This often includes the evaluation of a "no-action" alternative as a possibly suitable candidate option.

Remedial alternatives can be developed to address a specific contaminated medium (e.g., groundwater), specific areas of a contaminated site (e.g., a waste lagoon or "hot spots"), or an entire contaminated site problem. Typically, the development of a remedial action alternative will consist of the set of activities indicated in Box 8.2, with the remedial action objectives made up of medium-specific or site-specific goals for protecting human health and the environment (USEPA, 1988a, 1988b). They are developed following a site characterization program in order to specify the area of attainment, the restoration time frame, and cleanup levels.

General response actions that describe those actions required to satisfy the remedial action objectives are an important aspect of the remedial alternative development process. The general response actions may include treatment, containment, excavation, extraction, disposal, institutional controls, or a combination of these. Similar to remedial action objectives, these are indeed medium- and site-specific. Process options are determined using effectiveness, implementability, and cost criteria. At this stage, however, the evaluation will usually focus on effectiveness factors, with less effort directed at implementability and cost evaluation.

8.4.3 Screening of Remedial Alternatives

In situations where numerous potential remedial options are initially developed, it usually is necessary to screen out some of the available options in order to reduce the number of alternatives that will be analyzed in detail. The screening process involves evaluating alternatives with respect to their effectiveness, implementability, and cost. This is usually done on a general basis and with limited effort (relative to the detailed analysis) because the information necessary to fully evaluate the alternatives may not be complete at this point in the process. Also, because the screening process addresses approaches to remediation rather than specific remedial technologies, the evaluation is more qualitative, rather than quantitative. However, the screening analysis uses the quantitative site characterization data to recommend an approach to remediation.

It is important during the screening of remedial action measures to include as many alternatives as possible. This will ensure that the most cost-effective technique is not excluded from consideration. However, it is impractical or uneconomical to conduct extensive full-scale prototype investigations on every alternative during planning and preliminary design stages. Thus, the first step is to determine the potentially feasible alternatives that can be evaluated further, based on technical and economic factors.

8.4.4 In-Place Remediation as a Method of Choice

An in-place remedial technology or process will generally be given preference as a method of choice in the development of remedial options for a potentially contaminated site. In-place treatment techniques may include extraction (e.g., soil washing), immobilization (e.g., sorption, ion exchange, precipitation), chemical degradation (e.g., degradation involving oxidation, reduction, and polymerization reactions), biodegradation (using natural microorganisms or genetically engineered organisms), photolysis (that may include enhanced photodegradation achieved by addition of proton donors in the form of polar solvents), attenuation (e.g., mixing of contaminated surface soil with clean soil), and reduction of volatilization effects (Sims et al., 1986). An in-place treatment can be used to contain the source of contamination, or to remove contamination through treatment processes.

A primary consideration in the identification of appropriate site-specific in-place technologies for potentially contaminated sites relates to the chemical nature of the contaminants of concern. This is because, as an example, some technologies are designed primarily to control the release of heavy metals, whereas others are designed primarily to control the release of VOCs or other organics. In fact, the design and implementation of an in-place treatment process requires information on characteristics of the contaminant/soil systems as a whole, especially regarding the following important variables (Sims et al., 1986; USEPA, 1984):

- *Depth to Contamination.* If contamination is limited to the upper 15–20 cm (≈6–8 in.) of the soil and is well above the water table, in-place treatment techniques may be much more easily applied than if the contamination extends well below the ground surface and into a seasonally high water table.
- *Contaminant Concentrations and Quantities.* The efficiency and effectiveness of an in-place process depends on both contaminant concentration levels and quantity of each contaminant present in a given area.
- *Treatability.* Based on the contaminant, soil, and system characteristics, an analysis can be made of the pathways and rates of contaminant migration, and the potential for damage to human health and the environment as a function of time under conditions of "no-action".

Contaminant and soil characteristics are used in prescreening in-place alternatives for potential applicability in meeting remedial action objectives. For instance, contaminant, soil, and system characteristics can be used in the evaluation of a *"no-action" alternative* vs. an *extraction alternative* vs. a *degradation alternative.* In the *"no-action" alternative*, the inherent assimilative capacity of the soil for the contaminants is assessed in order to determine if the intrinsic properties of the soil are adequate to block migration of chemical constituents along pathways of concern; this will likely be the most cost-effective method, if applicable. Even if naturally occurring degradation, immobilization, and attenuation processes are deemed inadequate to meet remedial action objectives, an assessment of these processes generally provides a useful baseline for designing in-place remedial measures. In the *extraction alternative,* extraction techniques actually remove the undesired contaminants from the soil by dissolution in a fluid which is subsequently recovered and treated either on-site or off-site. In the *degradation alternative*, degradation techniques, generally applicable to organic compounds, convert the contaminant species into innocuous or less toxic compound(s). When several types of contaminants are of concern, the possibility of combining two or more in-place technologies to achieve the overall remediation goals becomes imminent.

8.5   THE DETAILED EVALUATION OF REMEDIAL ACTION ALTERNATIVES

After an initial screening of remedial alternatives, a detailed evaluation process is used to identify the remedial technology most likely to be successful from among the remedial approaches previously compiled during the screening analysis. The number of technologies selected for a detailed evaluation are determined based on knowledge of the alternatives that have proven to be successful under conditions similar to those at the case site; untested technologies and those known to perform inadequately under similar site conditions are not evaluated.

Purpose and Scope

The detailed analysis of remedial alternatives is conducted with the principal objective of providing decision makers with sufficient information to compare alternatives in a technically justifiable and socioeconomically acceptable manner. It follows the development and screening of feasible alternatives and precedes the actual selection of a remedy. The evaluation of both short-term and long-term risks are an important part of the detailed analyses. The detailed evaluation of the applicable alternative technologies being considered for a contaminated site problem will typically incorporate the following information:

- Successful application of the technology under similar site conditions, supported by identification of project locations, dates, and managing entity.
- Total project cost, supported by an estimate itemizing technology testing, capital equipment, operating and maintenance labor, equipment, environmental testing and monitoring, and closure costs.
- Risk reduction, supported by numeric estimates of risk posed to site workers or other receptors during remediation and the risk posed by any contaminants remaining after remediation.
- Project duration, supported by an estimated schedule showing major milestones, including any permitting activities that may be required.
- Manageability of data gaps, supported by the identification of any environmental testing or treatability studies necessary to determine the effectiveness of a remedial technology under site conditions.

In the detailed evaluation process, one or more of the screened remedial measures undergo detailed analyses that may involve field/prototype investigations to identify the most cost-effective alternative. An important component of the detailed evaluation of remedial alternatives involves the assessment of design parameters for remedial technologies. The design of many remedial technologies requires data that may not generally be collected during routine site characterization or remedial investigation. On the other hand, it is important to consider data needs for such a design during scoping so as to minimize the amount of time needed to select and implement the remedy. The important data needs for the evaluation and design of various remedial technologies are enumerated in the literature (e.g., USEPA, 1988a, 1988b).

Evaluation Criteria

In general, the detailed analysis of remedial alternatives is completed with consideration given to several evaluation criteria stipulated under federal, state, and local regulatory requirements. Under the U.S. federal environmental programs, the detailed evaluation of remedial alternatives is guided by the application of the following specific set of evaluation criteria: protection of human health and the environment; short-term effectiveness; long-term effectiveness; compliance with regulatory standards; reduction of toxicity, mobility, or volume; technical and administrative implementability; cost; state acceptance; and community acceptance (USEPA, 1988b, 1989a). The process helps determine the respective strengths and weaknesses of alternative remedial measures, and to identify the key trade-offs that must be balanced for a contaminated site problem.

The results of the detailed evaluations will comprise a recommended technology or combination of technologies to remediate each impacted medium posing "unacceptable" risks. If such a determination cannot be made with the available information, data gaps are identified and a program capable of providing the missing information is implemented.

8.5.1   Risk Evaluation of Remedial Alternatives

Risk assessment plays a very important role in the development of remedial action objectives for contaminated sites, in the identification of feasible remedies that meet the remediation objectives, and in the selection of a protective and balanced remedial alternative. The risk evaluation of remedial alternatives involves the same general steps as a baseline risk assessment (see Chapter 5). However, the baseline risk assessment typically is more refined and requires a greater degree of effort than the risk comparison of remedial alternatives. Much of the data collected during the baseline risk assessment can also be used to calculate the long-term residual risk associated with remedial alternatives.

In analyzing exposures associated with remedial actions, it is noteworthy that treatment processes and technologies used as part of a remediation strategy may facilitate the transfer of contaminants into unimpacted environmental matrices. On the other hand, well-engineered remedial alternatives are not expected in themselves to cause a net increase in exposures and impacts as the direct result of contaminant releases into the environment. In any case, the risks associated with a remedial action should be evaluated so as to include the risks created by implementing a remedial alternative as well as the postremediation risks associated with residual contamination that remains at the site. The difference between the site risks in the absence of remedial action (i.e., the baseline risk) and the risks associated with a remedial alternative will generally help define the net benefits associated with a given remedial option. Overall, it is crucial to ensure that the projected risks posed by a remedial option do not offset any benefits associated with reducing site contamination in order to achieve an established remediation or risk reduction goal.

8.6   SELECTION AND IMPLEMENTATION OF REMEDIAL OPTIONS

Typically, once a list of remedial alternatives is developed, these alternatives are analyzed in detail so that the most appropriate for the site-specific problem can be selected. The analyses usually involve an initial screening and then a detailed evaluation. The initial screening of alternatives is designed to eliminate alternatives which are clearly inappropriate to the given situation or are clearly inferior to other alternatives. Alternatives which remain after the initial screening are subjected to more detailed evaluation; additional data gathering may be required to complete this level of analysis. In addition, laboratory or pilot-scale studies may be required at this stage, especially with respect to treatment technologies. Based on the results of the detailed analysis, the appropriate remedial alternative(s) can be selected.

The selection of an appropriate corrective action plan for a contaminated site problem depends on a careful assessment of both short- and long-term risks posed by the site. Risk assessment, among other tools, can be used to aid this process of selecting among remedial options for such sites. Risk assessment techniques can indeed be used to quantify the human health risks and environmental hazards created by implementing specific remedial options at contaminated sites. These procedures can help determine whether a particular remedial alternative will pose unacceptable risks following implementation, and to determine the specific remedial alternatives that will create the least risk with respect to the cleanup goals or remedial action

---

BOX 8.3
Remedy Selection Decision Criteria

- Overall protection of human health and the environment
- Compliance with applicable laws and regulations
- Short- and long-term effectiveness, and permanence of corrective actions
- Potential short- and long-term liability issues
- Amount of contaminated media (such as soils, groundwater, etc.)
- Ease of implementation (i.e., the technical and administrative feasibility of a remedy)
- Cost effectiveness and cost efficiency of plans
- Reduction of toxicity, mobility, or volume of contaminated materials
- Regulatory and community acceptance of the program

---

objectives for the site. By using such an approach, the remedy selection process ensures that remedies satisfy the following pertinent conditions:

- Be protective of human health and the environment
- Attain media cleanup standards specified by regulatory requirements
- Control the source(s) of releases so as to reduce or eliminate, to the extent practicable, further releases that may pose a threat to human health or the environment
- Comply with all relevant corrective action program specifications and standards

The selection criteria, as well as other extraneous factors affecting the selection and implementaion of a number of remediation options for contaminated sites are discussed below; further details can be found elsewhere in the literature (e.g., Cairney, 1993; Calabrese and Kostecki, 1991; Nyer, 1993; OBG, 1988; Sims et al., 1986; USEPA, 1984, 1985, 1988b).

8.6.1   Remedy Selection Criteria

Remedial alternatives considered for potentially contaminated sites will generally be evaluated based on several criteria, such as required by the National (Oil and Hazardous Substances Pollution) Contingency Plan (NCP). Box 8.3 contains the particularly important remedy selection decision factors that should be considered during the remedial action evaluation (USEPA, 1988a, 1988b). Ultimately, the selected remedy will be the alternative found to provide the best balance of trade-offs among alternatives in terms of these evaluation criteria. This will generally satisfy several important requirements, such as the NCP requirements of providing the lowest-cost alternative that is technologically feasible and reliable, and which effectively mitigates and minimizes environmental damages and provides adequate protection of public health, welfare, or the environment (40 CFR 300.68(j)). Any remedies not meeting these criteria are eliminated from further consideration as preferred alternatives.

In general, the remedial alternative selected following the detailed evaluation should attain or exceed pertinent regulatory standards that apply to the site, and should also realize sustained effectiveness. An alternative that does not meet the applicable standards may, however, be selected in situations where (OBG, 1988):

- The selected alternative is not a final remedy, but is part of a more comprehensive remedy package
- No alternative that attains or exceeds the standards is feasible
- All alternatives that attain or exceed the requirements will result in other significant adverse environmental effects

Indeed, a number of other extraneous but important site-specific features may also affect the selection of the ultimate corrective measure. These include several site characteristics pertaining to surface features, subsurface conditions, populations potentially at risk, climate, adjacent land uses, cultural and social situations, and regulatory climate.

8.6.2 Choosing between Remediation Options

The selection of a specific remedial alternative for a contaminated site problem depends, to a great extent, on the required cleanup criteria established for the site. Once the cleanup criteria has been determined, a variety of techniques can be evaluated for containing and treating impacted media (such as soils, groundwater, etc.) associated with a contaminated site problem. The remediation approach may comprise the use of physical containment techniques and/or physical, chemical, and biological treatment processes. In a representative situation, the remedial action will include the use of techniques to contain the contamination plume, and to recover and treat the impacted matrices.

It is noteworthy that, oftentimes, the best solution for remediation of potentially contaminated sites is not necessarily one specific technology, but rather a combination of several technologies or remedial options capable of addressing the site-specific concerns of the case-site. For instance, soil vapor extraction (SVE) can be successfully coupled with bioremediation and air sparging technologies in comprehensive site cleanups. Furthermore, treatment processes can be, and usually are, combined into process trains for more effective removal of contaminants.

Successful and long-term site restoration requires adequate mitigation strategies to remove the contaminant source(s). Source control technologies which involve treatment of contaminated materials, or that otherwise do not rely on containment structures or systems to prevent future releases, should be strongly preferred to those that offer temporary, or less reliable, controls. As far as practicable, corrective action remedies must ensure, with a high level of confidence, that environmental damage and health impacts from the sources of contamination will not occur in the future.

Groundwater Remediation Techniques
Methods for containing/mitigating contaminated groundwater include the use of physical containment systems (such as impermeable barriers, hydraulic barriers, and subsurface collection systems) and leachate controls. The impermeable barriers may consist of slurry walls, grout curtains, and sheet piling; these can be used to contain, capture, or redirect groundwater flow for the remediation/mitigation of contaminated

groundwater. The hydraulic barriers (e.g., recovery wells, interceptor trenches, etc.) are used to modify hydraulic gradients around contained waters; this can be used to manipulate, through pumping/injection strategies, the movement and size of a contaminant plume — given the proper subsurface conditions). Leachate controls may include capping (to prevent or minimize rainwater from infiltrating through contaminated soil into groundwater) and/or the use of subsurface drains (consisting of buried conduits that collect and convey leachate). The containment system acts to interrupt contaminant transport mechanisms in order to prevent or minimize the continuing spread of contamination.

Groundwater pump-and-treat systems, consisting of extraction/injection well networks, have been a very commonly used remediation technique. In the remedial design for groundwater contamination problems, the recovery well systems are designed to intercept the contaminant plume so that no further degradation of the impacted aquifer occurs. Modeling is a very useful tool in the design of such systems (USEPA, 1985). Groundwater treatment technologies applicable to the recovered groundwaters include physical treatment processes (such as phase-separated hydrocarbon recovery, air stripping, activated carbon adsorption, and filtration — which are all processes generally applied without the aid of chemical or biological agents), chemical processes (such as coagulation-precipitation, oxidation-reduction [redox], neutralization, etc.), and biological methods (such as suspended-growth and fixed-film reactors, as well as *in situ* biodegradation).

The Challenges of Contaminated Aquifer Restoration Programs

The restoration of contaminated aquifer systems is one of the most challenging problems in corrective action planning decisions. Oftentimes, far too much attention is given to pump-and-treat remedial technologies. However, this technique tends to leave a great part of contaminant residues in the capillary fringe or vadose zone, unaffected by groundwater pumping. In fact, a number of studies have been carried out to assess the feasibility and effectiveness of such a remedial strategy. For instance, Yin (1988) used a two-dimensional random-walk solute transport model to simulate aquifer restoration processes for groundwater that is contaminated by dissolved petroleum constituents. The model simulates proposed aquifer restorations by pumping out the contaminated groundwater. The aquifer restoration process consists of a single pumping well located at the center of the contaminated area. The effectiveness of pumping is evaluated in terms of duration of pumping and concentration reduction. The simulation results are then used to evaluate the feasibility and effectiveness of aquifer restoration that can be attained by pumping out the contaminated water. The modeling results indicate that the time of pumping required to reduce concentrations of benzene and xylene to acceptable levels is likely to extend over several years — to the point of affecting the efficacy of the approach.

Removal of the contaminated soil at the case site, followed by treatment and/or disposal, will generally be a better strategy to adopt in order to address source elimination. In fact, in most aquifer contamination problems, containment of the aquifer contaminants is the most immediate concern. This can be achieved through the use of a physical barrier (such as a grout curtain, slurry cut-off wall, or sheet piles), or by the creation of a hydraulic barrier resulting from a network of extraction (pumping) and injection (recharge) wells. The next requirement will be the removal of the mobile contaminants from the aquifer system. This may include free product recovery (e.g., petroleum products floating on a water table), air stripping and vacuum extraction of VOCs, bioremediation, and pump-and-treat processes for soluble constituents. Insoluble constituents may remain adsorbed onto soil particles in the aquifer or vadose zone, to be addressed differently. The details of commonly used technologies and processes employed for preventing and/or cleaning up groundwater contamination were previously discussed in Chapter 7.

8.6.3 Monitoring Programs

In addition to adopting appropriate remedial actions, monitoring programs will normally be implemented to verify the long-term effectiveness of an overall corrective action program. Monitoring is indeed considered a very important component of corrective action programs, and can serve as a useful tool for evaluating the performance of remedial actions.

Several different monitoring programs may be implemented to evaluate the effectiveness of a remedy. For instance, an environmental monitoring program is normally used to ensure that chemical loadings from contaminated sites do not continually escape the influence of applicable remedial systems. Under a biomonitoring program, measurements of toxicity (through bioassays) and bioaccumulation can be used to assess the nature and extent of potential biological impacts in off-site areas. Where necessary, the remedial system is modified to ensure that the remedies are effective.

Several monitoring parameters are important to the design of an effectual monitoring program. Of special interest is the selection of monitoring constituents. In general, the selection of monitoring constituents should consider the possibility for chemical constituents to be transformed over time and space. For instance, knowledge about the degradation of contaminants can be an extremely important factor in identifying monitoring constituents. Thus, specific monitoring constituents and indicator parameters may have to be modified as an investigation progresses in time, in order to account for transformation products. This is because physical, chemical, and biological degradation may transform certain constituents as the release ages or advances. In fact, despite the notion that most chemicals usually degrade into less toxic, more stable species, this is not universally true; for example, one of the degradation products of trichloroethylene is vinyl chloride, both of which are carcinogenic chemicals. Consequently, the potential for physical, chemical, or biological transformations of constituents should be given adequate consideration in identifying monitoring constituents.

Finally, all other factors being equal, siting and placement of monitoring networks becomes a critical factor in monitoring programs. The success of the network will very much depend on whether or not representative samples or measurements can be obtained. Thus, it is vital that the finite number of monitoring locations are placed such that they produce an accurate description of the prevailing site conditions.

8.6.4 Performance Evaluation

The overall remedial action objective for a contaminated site is to protect human health, the environment, and public and private properties in the vicinity of the site. Risks posed through contaminant migration pathways are generally eliminated, reduced, or controlled through treatment, engineering measures, or institutional controls. The amount of reduction in toxicity, mobility, or volume offered by treatment processes give a measure of the anticipated performance of the treatment technologies that a remedy may employ. This becomes an important factor in the decision regarding the remedial action alternative that should be selected and implemented for the contaminated site problem.

The adequacy of protection afforded by a remedial option to human health and the environment is measured by whether or not the remedy meets the remediation goals. If, after remedy implementation, it is determined that a remedial alternative does not (or will not) meet stipulated remediation goals for all contaminants in the media of concern, then the residual risk remaining after implementation should be examined to determine whether other measures are necessary to assure protectiveness.

The performance goal of remedial alternatives may be evaluated based on a cleanup criteria and a time period for the restoration of the contaminated site. The favored alternatives are compared based on the trade-offs between the time to attain an "acceptable" cleanup level and the costs associated with the remedial actions. However, the complexities in the fate and transport mechanisms at contaminated sites often make it difficult to predict, with any degree of accuracy, the performance of site remedial actions. Consequently, there should be reasonable flexibility in the remediation process so that, where necessary and possible, appropriate changes can be made to the remedial program to improve the performance of the selected remedial action alternative.

8.7 MANAGEMENT OF PETROLEUM-CONTAMINATED SITES AS AN EXAMPLE

The release of chemical substances from leaking underground storage tanks (USTs) is a common occurrence. The subsequent impacts of such releases on the environment and public health is a particularly important environmental issue. Of an even greater interest/concern is the management of sites contaminated as a result of petroleum product releases from USTs.

Leakage of petroleum hydrocarbon products and other chemicals from underground storage tanks and pipelines is a frequent occurrence in commercial, industrial, and even domestic activities. Such leakages can result in the contamination of several environmental media, especially soils and groundwater. When petroleum products enter the soil, gravitational forces act to draw the fluid in a downward direction. Other forces act to retain it; some product is either adsorbed to soil particles or trapped in soil pores. Petroleum hydrocarbon products can therefore become a major source of long-term contamination of soils and groundwater. The amount of product retained in the soil is of importance, because this could determine both the degree of contamination and the likelihood of subsequent contaminant transport into other environmental compartments.

Preventing soil and groundwater contamination by petroleum products is important because many communities depend on groundwater as their primary source of drinking water. Furthermore, under Subtitle I of the *Hazardous and Solid Waste Amendments of 1984* (HSWA), the EPA is required to develop and implement a comprehensive regulatory program for USTs. Consequently, the EPA and many states have promulgated regulations pertaining specifically to the management of USTs containing petroleum products and other hazardous materials. Regulations covering petroleum products are codified in Title 40, Part 280 of the *Code of Federal Regulations.*

8.7.1   Composition of Petroleum Products

Specific petroleum products commonly found in USTs include gasoline, diesel, heating oil, aviation fuel, waste oils, and other related petroleum hydrocarbons. Typically, motor fuel alone is a mixture of over 200 petroleum-derived chemicals plus a few synthetic products that are added to improve the fuel performance (LUFT, 1989). Several of these chemical constituents can potentially affect human and eco-logical receptors if released into the environment.

Petroleum fuel contaminants of major health concern include benzene, toluene, ethylbenzene, and xylene — commonly referred by the acronym, BTEX. These BTEX compounds have the potential to move through soils to contaminate groundwaters. In fact, groundwater contamination from petroleum products, and particularly from leaking USTs, is a growing concern especially because of the potential carcinogenic and neurotoxic effects exhibited by the BTEX compounds. For instance, benzene is a known human carcinogen, whereas the others (i.e., toluene, ethylbenzene, and xylene) are noncarcinogens but are known to possess neurotoxicity effects. This means that the BTEX compounds generally constitute preselected target/indicator chemicals of con-cern for petroleum product releases. As BTEX constitute the most toxic and environ-mentally mobile constituents, their selection as indicator chemicals assures that any corrective action cleanup criteria developed based on these will also adequately address the less toxic or the less mobile constituents.

In addition to an evaluation of BTEX, analysis for total petroleum hydrocarbons (TPH) is commonly carried out for most petroleum-contaminated sites. This latter analysis detects aliphatic and aromatic constituents contained in the fuel. Detection is reported as the sum total of all hydrocarbons in the sample — rather than as individual chemical constituents. Because the lighter fractions (such as BTEX) are more mobile, these constituents can migrate or dissipate away from the main body of contamination. Less mobile hydrocarbons (such as those detected in TPH analysis) may give a more accurate indication of the actual contamination. As a consequence, soils are preferably analyzed for both the BTEX and TPH as indicators of contamination.

Further to BTEX and TPH analyses, and because of its extreme toxicity, the possible presence of organo-lead would generally be investigated where significant leaks of leaded motor fuel have occurred, or where an investigator feels that there may be potential danger of exposure to organo-lead.

It is noteworthy that diesel fuels consist primarily of aliphatic hydrocarbons (though they may also contain some limited quantities of aromatic constituents, including benzene, depending on the source and the refining process). Consequently, TPH analysis is usually the only one required for leakage and spills from diesel storages.

8.7.2   The Subsurface Behavior of Petroleum
Constituent Releases

Typically, in the event of contamination from a leaking UST used for petroleum products that have densities less than that of water (e.g., motor fuel), the leaking product first enters the unsaturated soil below the tank. This eventually forms a floating layer on any underlying water table if the leaking quantity is large. Also, product vapor will enter the unsaturated soils around the tank and above the water table. Water-soluble fractions of the petroleum product, such as benzene and xylene, will eventually form a plume within the groundwater. The soluble plume will spread within the groundwater by diffusion and move with the groundwater as it flows downgradient.

In the unsaturated zone, petroleum constituents tend to move downward under the influence of gravity, and laterally due to capillary forces and the heterogeneous characteristics of soil. The movement also depends on the viscosity of the product and the rate of product release. If the amount of petroleum constituents is large enough, it will pass through the capillary zone and reach the water table. For motor fuels which are immiscible with water and also less dense than water, a layer of petroleum constituents will lie on and slightly depress the water table. Only a small amount of hydrocarbons will dissolve in the water.

Flow in the saturated zone generally transports the contamination in the direction of decreasing hydraulic potential. Petroleum constituents that reach the groundwater table will dissolve in water and be transported by groundwater. Transport of dissolved constituents in the groundwater is governed by advection and dispersion of groundwater, and by attenuation mechanisms (such as biodegradation and adsorption) of the soil media.

In general, the migration of contaminants such as petroleum constituents through the subsurface environment is governed by four major factors: the quantity or volume of release, the physical properties of the contaminants, the physical properties of the subsurface material (e.g., the adsorptive capacity of the earth materials), and the subsurface flow, such as the rates and directions of groundwater movement (Hunt et al., 1988; Yin, 1988). In addition, all processes which attenuate contaminant concentrations and/or limit the area of the contaminated zones will affect the fate of the source release.

The quantity of the contaminant determines whether it will reach the water table. Physical properties of the leaked substance that are important to migration include solubility, specific gravity, viscosity, and surface tension; biodegradability should also be considered important to long-term contaminant migration. Physical properties of subsurface materials important to the migration of petroleum constituents include porosity, permeability, and homogeneity. Other important physical parameters would include soil organic carbon content available for partitioning and also available oxygen to aid aerobic biodegradation. All these parameters affect the subsurface behavior of the contaminants of concern.

The Fate and Transport Processes for Source Releases

An understanding of the fate and transport processes are important in determining the likelihood of medium-specific releases and exposures that result from leaking USTs. A variety of environmental fate models may be used to predict the transport and fate of contaminants. These models may range from a simple mass balance equation to multidimensional numerical solution of coupled differential equations. In any case, the basic equation governing fate and transport is based on the principle of conservation of mass for the contaminant.

Several factors affect the fate and transport of the mobile liquid, vapor, and dissolved hydrocarbon phases. Releases of petroleum constituents from leaking USTs can contaminate soils and groundwater by the migration of free product through the unsaturated zone and by dissolution of certain constituents into groundwater. It is noteworthy that although petroleum fuel odor may be reported during soil sampling in the unsaturated zone in a site investigation/appraisal, at some site locations no hydrocarbons may be detected. This situation suggests that the mechanisms of ground-water contamination away from the gasoline spill are vaporization of gasoline com-ponents into the soil gas, migration of denser-than-air soil gas downward to the water table, followed by radial spreading and then component partitioning into the ground-water (Hunt et al., 1988).

In a typical situation when petroleum products have leaked from an UST, a nonaqueous phase liquid is released that can move through soil pores. Some of this liquid is left behind as disconnected fluid, called ganglia because of strong capillary forces (Hunt et al., 1988). These ganglia make up the residual saturation in the soil and are the long-term source of contaminants released to the air and groundwater. Even where no regional groundwater flow may be evident, a far field transport of selective gasoline components may indicate groundwater contamination by subsurface vapor migration away from the spill. In the unsaturated zone, denser gas is produced by volatilization of liquid gasoline, and this gas sinks toward the water table and then spreads out over the capillary fringe. Air saturated with gasoline and in contact with capillary water and groundwater generally will lose the more water-soluble compo-nents as it spreads out radially. Thus, the more volatile, less water-soluble hydrocar-bons are expected to move greater distances than the more water-soluble compounds such as benzene. Indeed, available investigation data strongly suggest that groundwater contaminants detected in wells away from a petroleum fuel spill arrive mostly via the vapor phase and not by groundwater flow (Hunt et al., 1988). Thus, it is important, in several situations, to model subsurface vapor transport in order to adequately assess the potential for contamination from leaking USTs. Relevant information from this type of evaluation is also important in the choice of appropriate remediation technolo-gies and/or cleanup strategies.

8.7.3   The Corrective Action Decision Process for UST Release Sites

It is almost a certainty that the result of most cleanup actions undertaken at petroleum-contaminated sites will leave some residual contamination. In fact, cleanup of all contaminated soil and dissolved product in groundwater is not always necessary to protect public health and the environment (LUFT, 1989). Consequently, it is important to develop cleanup goals attainable for appropriate remedial actions, based on estimates of health and environmental risks associated with the contaminants resulting from UST releases. However, generic cleanup levels for contaminated soil and dissolved product are undesirable, since conditions vary from one region to another. Instead, site-specific cleanup levels are usually recommended for corrective action decisions (LUFT, 1989). Indeed, it is believed that the use of site-specific cleanup criteria could result in significant cost savings, because such an approach can allow the use of corrective actions which are most effective in risk reduction. As part of a UST release site investigation, therefore, one should determine appropriate cleanup levels and the attainable remediation goals prior to implementing any correc-tive action plans.

Site Categorization

To facilitate a site-specific and phased approach to cost-effective corrective action decisions at petroleum-contaminated sites, a site categorization scheme may be adopted in the investigation of UST releases. The different categories will reflect the seriousness or hazardous nature of the potential release situation.

A site designation process may be used to categorize sites after observing the presence of one or more of a number of suspect conditions (such as tank closure, reported nuisance conditions, monitoring problems, and observed leakages). Existence of any of the suspect conditions provides justification to initiate a preliminary investigation that will confirm or disprove the suspected situation. Based on the preliminary site investigation and site history, the site can be assigned to one of a number of categories. For the purpose of this discussion, sites are classified into the following three categories:

- *Category 1: LOW-RISK SITES (LrS)*   ≈ *No Suspected Soil Contamination*
- *Category 2: MEDIUM-RISK SITES (MrS)*   ≈ *Suspected or Known Soil Contamination*
- *Category 3: HIGH-RISK SITES (HrS)*   ≈ *Known Soil and Groundwater Contamination*

Moving from the LrS category to the HrS sends one from a less serious to a more serious and hazardous scenario. For instance, the HrS category presents a case where both soil and groundwater contamination is confirmed. If the field personnel suspect that more serious contamination has occurred than was anticipated, then a site may be reclassified from a lower risk category to an intermediary risk category, or to a higher risk category, as appropriate. Conversely, the discovery of a less serious situation may result in reclassifying a site from HrS to MrS, or to LrS category. Relevant and standard fuel leak detection and screening methods can be used to aid field personnel in the classification/categorization tasks.

Contamination Assessment

Generally, the LrS will require a field TPH test only. This method is recommended only when there is likelihood that a problem exists and a quick, qualitative confirmation is desired. Thus, at LrS, the field personnel may use a field TPH test to confirm the absence of soil contamination. In fact, only sites showing no evidence of possible soil contamination can use this qualitative form of analysis. The field TPH test can indeed prove helpful as a gross guidance tool.

If a field inspection or background check indicates that a site is suspected or known to have contaminated soil, or if a site has failed the field TPH test under category 1, then *quantitative* laboratory analysis of soil samples is needed. To ensure quality results, standard procedures should be followed in the design of the sample collection and analysis procedures. For instance, soil samples should be quantitatively analyzed for BTEX using an appropriate method (e.g., EPA Method 8020). If any of these constituents is detected above the minimum practical quantitation limit, then the site investigation should proceed through a more detailed analysis that includes a general risk appraisal.

If a general risk appraisal shows that groundwater is at risk, then further evaluation is required that will help determine a HrS category. The necessary procedures involve more detailed investigations and decisions above what is performed under a

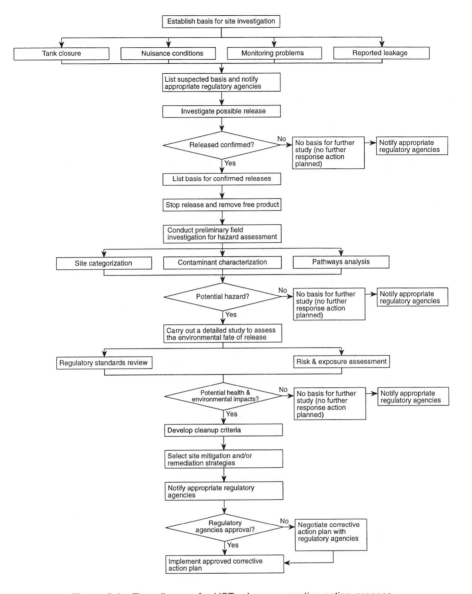

Figure 8.1 Flow diagram for UST release corrective action process.

MrS. Among other things, the groundwater gradient is determined using piezometers, which show the direction of groundwater and contaminant plume movements.

A Corrective Action Decision Framework

Figure 8.1 is a diagrammatic representation of a process flow that can be utilized in the investigation of potentially contaminated sites that are the result of petroleum hydrocarbon releases from USTs.

Decision Parameters

Table 8.1 presents a comprehensive suite of decision elements typically employed in the assessment and restoration of sites contaminated by leaking USTs.

The Site Restoration Program

If levels of contamination present at the spill site and impacted media are determined to exceed "acceptable" risk levels, then corrective actions, which may include site remediation, will generally be required. It must be recognized, however, that most cleanup actions cannot achieve a "zero" contamination level for petroleum-contaminated sites — or for any contaminated site problem, for that matter. The development of site-specific cleanup objectives necessary for the case-specific site will therefore aid in the selection of appropriate site restoration strategies.

During the initial planning stages of the restoration program, it is recommended that soils at the contaminated site be screened to determine if it is possible for bioremediation to meet designated cleanup specifications for the site. As a general guide, cleanup levels of TPH ≤ 100 ppm and BTEX ≤ 10 ppb have been determined to be typical attainable cleanup limits for the bioremediation of gasoline- and diesel-contaminated sites (Barnhart, 1992). Bioremediation programs usually will require bench-scale tests and/or treatability studies to enhance the development of the design criteria and parameters. After the treatability study has demonstrated that the soil system will indeed support it, a full-scale program can then be implemented.

Cleanup of petroleum fuel spill sites may generally be carried out to a "realistic" level only. This is because once the major chemical constituents (such as benzene) have reached approximately a 100-ppb level in groundwater, for example, continued use of most groundwater remediation technologies may not yield any substantial improvement over that achieved through natural processes. In fact, trace residual hydrocarbons left in the subsurface at such "threshold" levels will degrade through natural processes in an equally timely manner. Risk assessment procedures can be used to establish cleanup objectives in this regard. The site-specific risk assessment will guide the development of appropriate cleanup objectives with reference to site conditions, land uses, and exposure scenarios pertaining specifically to the case-site and its vicinity. Such an assessment will provide directions to develop effectual corrective action and risk management plans.

8.7.4 Illustration of the Site Restoration Assessment Process

A hypothetical problem is discussed here to illustrate the site restoration requirements for a petroleum-contaminated site problem. This illustrative problem consists of the development of a corrective action plan in response to hydrocarbon releases from USTs at an operating facility.

Background Information

A limited environmental site assessment conducted for the Petro-X Fueling and Automotive Services facility located in East London indicated a high degree of soil and groundwater contamination within the site boundaries. This occurrence is the result of UST releases, and also spills of organic solvents associated with the automotive service station. For the purpose of this illustration, it is assumed that analytical

Table 8.1 Typical Decision Elements Involved in the Investigation of UST Releases

Task (and purpose)	Sub-tasks/ Action Items	Major Decision Elements	Typical Investigation Rationale and/or Requirements
Preliminary site appraisal (to investigate indicated or suspected UST releases)	Establish basis for release indicators	Inventory variance/ discrepancy	Discrepancy of 0.5% of throughput over ≈ 30-day period
		Tank closure	Cleaning of tank reveals corrosion (i.e., potential leak)
			Removal of tank reveals soil contamination or tank corrosion
		Nuisance conditions	Vapors in adjacent structures or utility conduits
			Nonaqueous phase liquid (NAPL) in adjacent structures or utility conduits
		Reported leakage	Documented release from tank or lines
		Piping problems	Loss of suction at dispensers
		Monitoring problems	Detection of NAPL in groundwater monitoring wells
			Detection of NAPL with interstitial monitoring devices
			Detection of vapors in tank pit monitoring wells
	Gather and review site background information for evidence of release	Geographic and geologic information	Topographic maps (for preliminary determination of groundwater flow direction, potential receiving water bodies, and residential communities)
			Geologic maps (for preliminary determination of subsurface conditions controlling the transport of contaminants)
			Soil survey (to determine characteristics of near surface soils with respect to contaminant transport)
			Local hydrogeologic records (for preliminary determination of groundwater depth and direction of groundwater flow)
			Well permits (for preliminary determination regarding the potential for impacting potable water wells)
		Engineering drawings	Delineate subsurface structures that may function to provide preferential contaminant transport pathways
		Interview of site personnel regarding storage tanks information, transportation mode, monitoring methods, etc.	Determine the number, size, and location of subsurface storage tanks (gasoline, heating and waste oil)
			Determine the method of product transfer from the tank to the dispenser
			Piping construction, location, and configuration
			Fill port location and configuration
			Determine if automated and/or manual monitoring devices are in use, i.e., interstitial monitoring (tanks and piping), tank pit monitoring wells, perimeter monitoring wells, leak detectors, etc.

Table 8.1 (continued)   Typical Decision Elements Involved in the *I*nvestigation
                         of UST Releases

Task (and purpose)	Sub-tasks/ Action Items	Major Decision Elements	Typical Investigation Rationale and/or Requirements
		Site history	Determine property location with respect to environmentally sensitive areas (e.g., wildlife sanctuary) Determine previous land use to assess the potential for subsurface contamination from past site activities Assess the potential for contaminant transport from off-site sources
	Perform contamination assessment	Release dates	If documented release(s) has occurred, defining the date is of importance when delineating the potential extent of contaminant migration within the subsurface
		Duration of release	If a leak rate is established during tank testing activities, then the duration of the release can be used to approximate the volume of product lost
		Quantities of liquid hydrocarbon lost	The volume of lost product dictates the severity of subsurface impact (i.e., large volumes indicate extensive soil degradation and most likely groundwater contamination; minor loss event may only impact a localized area of soil)
		Hydrocarbon type	Gasoline: vapor problems, high dissolved constituent (i.e., BTEX) concentrations in the groundwater, and rapid contaminant transport Diesel and/or fuel oils: minimal vapor problems, low dissolved constituent (i.e., BTEX) and high polynuclear aromatic hydrocarbons (PAHs) concentrations in the groundwater, and moderate to slow contaminant transport
		Origin of hydrocarbon	Source location to determine potential impact area
		Subsurface access points	Monitoring wells, storm drain catch basins, french drain systems, utility manholes, tank access ports, dry wells, basements, etc.
	Identify potentially affected areas	Underground utilities	The presence of NAPL or hydrocarbon vapors in structures such as sanitary sewers, storm sewers, water lines, gas lines, dry wells, septic systems, power lines
		Wells	Complaints of taste, odor, and sheen in the potable water of adjacent residences. The presence of NAPL and/or dissolved-phase contaminants (i.e., BTEX) in potable wells

		Surface water	The presence of NAPL on adjacent surface water bodies (i.e., streams, ponds, marsh land, etc.)
		Residential properties	Presence of NAPL or hydrocarbon vapors in basements of adjacent residential structures
	Review, verify, and reconcile inventory records/data	Examination of overall physical facility (including calibration of dispenser meters, pressure testing of piping systems, tightness test of storage systems, etc.)	Review record keeping procedures for potential discrepancy areas Look for obvious release characteristics (i.e., soil staining at fill ports, product in storm drains, etc.) Verify accuracy of dispenser meters Implement precision testing procedures on piping system to determine integrity of pipe and fittings (e.g., leak rate >0.05 gallon/hour indicates a release) Implement precision tests on individual subsurface storage tanks (e.g., leak rate >0.05 gallon/hour indicates a release) Reevaluate inventory records with tank/line test and dispenser calibration information
	Investigate other possible sources of hydrocarbon releases	Adjacent land uses	Review historic aerial photographs and records
		Other nearby tanks	Identify potential source areas based upon assumed direction of groundwater movement and product (contaminant) identification
	Address health and safety issues	Health and safety requirements	Evaluate the potential adverse health and safety effects associated with the release to the environment
		Regulatory requirements	Once a release is confirmed, identify all appropriate regulatory agencies/personnel (e.g., state regulatory agency and local fire marshal)
	Address emergency response	Release source control	If subsurface tank(s) are responsible for product discharge, evacuate the tank contents and remove the tank from service
		Free product removal	Implement emergency response product recovery efforts, gaining access to the subsurface through pit excavations and/or wells, to minimize the extent of environmental impact
		Imminent hazard mitigation	Address the potential for the development of an explosive environment Mitigate the presence of vapors in adjacent residential structures Provide alternate potable water supplies for impacted wells
Site assessment (to confirm suspected UST releases)	Implement subsurface investigation to characterize site	Identify contaminated soils and determine extent of soil contamination	Establish area of contaminated soil utilizing a soil vapor survey and/or soil borings Determine the magnitude of soil contamination through the collection of depth-specific samples for total

Table 8.1 (continued)   Typical Decision Elements Involved in the *In*vestigation of UST Releases

Task (and purpose)	Sub-tasks/ Action Items	Major Decision Elements	Typical Investigation Rationale and/or Requirements
			petroleum hydrocarbon (TPH), BTEX, and gasoline additive analyses
		Identify free product on water table; identify dissolved contaminants in groundwater; and determine extent of groundwater contamination	Install groundwater monitoring wells that are screened across the water table to detect the presence of NAPL
			Conduct periodic liquid level measurements to determine the depth of groundwater and the thickness of NAPL accumulations
			Conduct product recovery (numeric and/or field) tests to delineate the "true" product thickness
			Develop a groundwater monitoring well array designed to define the background water quality and the downgradient extent of dissolved contaminant transport
			Collect groundwater samples to facilitate quantitative laboratory analysis for BTEX and gasoline additive constituents
		Determine presence and distribution of vapor accumulations	Implement a soil vapor investigation and/or conduct periodic vapor monitoring of adjacent utility conduits
		Determine sources and directions of contaminant migration	Utilize the monitoring well array to establish the direction of contaminant transport and define potential source areas
			Conduct periodic liquid level measurements to determine the hydraulic gradient and groundwater velocity
		Determine hydraulic properties controlling contaminant movement	Conduct hydraulic conductivity tests (slug and/or pump tests) to establish aquifer hydraulic properties
		Identify hydrogeologic conditions	Determine the following: depth to groundwater, type of aquifer (i.e., water table, confined or perched), thickness of the water bearing strata, permeability and porosity of the formation
		Determine regional recharge	Define the recharge area, typically upgradient topographically, for the aquifer of concern
		Determine discharge areas	Look for local streams, seeps, springs and surface water bodies that function as groundwater discharge points. Withdrawal of groundwater from large-volume production wells can have significant control over the direction of localized groundwater flow

	Determine outlet points where contamination can impact public health and safety	Conduct a record search for potable water wells within the immediate area of the site
Characterize petroleum hydrocarbons	Product type	Utilize visual observations, specific gravity, and gas chromatograph (GC) fingerprint analyses to characterize the type of product (such as gasoline, middle distillates [diesel, kerosene, jet fuels, and lighter fuel oils], heavier fuel oils and lubricating oils, asphalts and tars)
	Physical properties	Viscosity, which will control migration and adsorption of contaminants Volatility, which will control the production of hazardous vapors and the ratio of dissolved to NAPL-phase contamination within the groundwater
Verify presence of hydrocarbons	Chemical properties	Organic carbon partitioning coefficient, half-life, solubility, etc.
	Delineation of residual liquid hydrocarbons	Collect depth-specific soil samples above the water table for TPH and BTEX analysis to define the quantity of adsorbed contaminants. The percentage of organic carbon within the soil matrix must also be evaluated
	Delineation of liquid-phase hydrocarbons	Measure "apparent" product thickness from monitoring wells and conduct NAPL bail-down recovery tests to delineate the "true" product thickness and, therefore, provide more accurate estimates regarding the volume of product within the geologic formation
	Delineation of dissolved-phase hydrocarbons in vadose zone	Conduct a soil gas survey utilizing a portable gas chromatograph for constituent identification Verify results and calibrate survey with quantitative laboratory analysis of soil samples
	Delineation of dissolved-phase hydrocarbons in groundwater zone	Conduct BTEX and gasoline additive analysis in all nonproduct-bearing wells
	Delineation of vapor phase	Conduct soil gas investigations
	Delineation of adsorbed-phase	Soil sampling to determine soil partitioning coefficient
Confirm presence of free-phase hydrocarbons	Remove free-phase hydrocarbons	Implement manual (bailing) recovery of NAPL from all product-bearing wells until a centralized, automated, recovery program is initiated.
Confirm presence of dissolved-phase contamination	Sample for indicator parameters	Sample potentially impacted potable wells and surface waters for indicator parameters (such as BTEX, methyl-t-butylether [MTBE], ethylene dibromide [EDB], lead, and other applicable additives)

Table 8.1 (continued)   Typical Decision Elements Involved in the Investigation of UST Releases

Task (and purpose)	Sub-tasks/ Action Items	Major Decision Elements	Typical Investigation Rationale and/or Requirements
	Confirm presence of vapor-phase contamination	Delineate preferential vapor migration pathways	Review utility conduit drawings and engineering plans to delineate potential vapor migration pathways
		Ventilate vapors to eliminate explosive environment	Evacuate hydrocarbon vapors from all structures, including utility conduits and residences, to reduce vapor concentrations below the lowest explosive limit (LEL)
		Institute remedial action program to reduce vapor concentrations to within acceptable risk-based levels	Establish either positive pressure within or negative pressure outside of adjacent structures possessing vapor contamination to mitigate the migration of vapors into the structure
	Perform contamination pathways analysis	Develop working hypothesis about migration of liquid, vapor, and dissolved phases beneath and near release sites	Evaluate all information regarding indigenous stratigraphy and subsurface structures to assess preferential contaminant migration pathways, if any
		Predict timing and concentration levels of mobile phases reaching pathway outlets	Utilize aquifer hydraulic parameters and contaminant characteristics to implement a contaminant transport analysis (computer modeling)
		Determine populations potentially at risk	Having defined both spatial and temporal controls on contaminant transport, evaluate the potential impact area and associated receptors
		Identify potential exposure routes	Map areas where hydrocarbon phases may impact human health and/or environment
			Map distribution of hydrocarbon phases and all potential pathway outlets
Risk appraisal (for contamination assessment of UST releases)	Determine factors affecting contaminant fate and transport	Adsorptive capacity of earth materials	Organic carbon content, soil/water partitioning coefficient, solubility of constituents
		Perching horizons and interconnected void spaces	Porosity and permeability
		Relative conductivity of earth materials to water and other hydrocarbon fluids	Hydraulic conductivity, grain size analysis, matrix sorting

		Rates and directions of groundwater movement	Hydraulic gradient, permeability, and porosity
		Processes that attenuate concentrations and limit area of contaminated zones	Dispersion, dilution, adsorption, biodegradation, and biotransformation
	Determine acceptable cleanup criteria to qualify risks	Establish liquid hydrocarbon criteria that confirms removal	Cessation of liquid hydrocarbon discharges into underground structures or openings. Attainment of practical limits for hydrocarbon recovery at wells and drains. Presence of only traces of free hydrocarbons in monitoring wells
		Establish dissolved hydrocarbon criteria (acceptable levels of BTEX), confirming removal	Background levels of offsite contaminant levels. Constituents reach asymptotic contaminant levels
		Establish vapor hydrocarbon criteria	Explosion limit for total vapors in air. Offensive odors. Background levels for key constituents
		Establish residual hydrocarbon criteria	Leaching potential of key constituents to groundwater or soil. Proximity of potential discharge points connecting leaching process locations
	Perform exposure assessment	Indicator/target chemical selection	For gasoline, common indicators are BTEX
		Select significant exposure pathways	Selection based on information on probable exposure scenarios
		Address fate and transport modeling	Assess modeling needs and select appropriate model
	Define magnitude/ severity of health and environmental impacts	Potential for exposure, via contaminant mobility and pathways analyses	Determine concentrations, toxicity, sensitive receptors, future land uses
Site remediation (consisting of an evaluation of the applicability of remedial alternatives)	Implement emergency response	Soil excavation	Remove or minimize contaminant source(s)
		Hydrocarbon vapor control	Eliminate explosive environment and detrimental health conditions
		NAPL recovery	Manual recovery utilizing hand bailing or vacuum truck without water table depression
		Dissolved hydrocarbon-groundwater control	Groundwater withdrawal creating a localized cone of depression to reverse contaminant migration
	Establish remediation objectives and cleanup criteria	Cleanup objectives and criteria	Negotiate cleanup standards with regulatory agency based on results of risk analysis

Table 8.1 (continued) Typical Decision Elements Involved in the *In*vestigation of UST Releases

Task (and purpose)	Sub-tasks/ Action Items	Major Decision Elements	Typical Investigation Rationale and/or Requirements
	Establish limits of applicability of corrective action program for vadose zone soil remediation	Soil excavation	Limited areal extent of contaminated soils
			Limited depth of contaminant migration
			Moderate to high concentrations of adsorbed contaminants
			Soils away from foundations and below ground structures
		Soil venting	Volatile contaminants
			Permeable soils
			Moderate to high concentrations of adsorbed contaminants
			Absence of NAPL
		Enhanced biodegradation	Indigenous microbial population
			Nonreactive soil matrix
			Neutral pH groundwater
			Permeable soils
		Surfactant flushing	Permeable soils
			Low to moderate viscosity product
		Landfilling	Low to high concentration of adsorbed contaminants
			Low to high viscosity products
			Classified soils
		On-site aeration treatment	Volatile contaminants
			Limited depth of contaminant migration
			Low to high concentration of adsorbed contaminants
			Moderate to high permeability soils
			Air emission monitoring
		Asphalt batching/ incorporation	Low to high concentration of adsorbed contaminants
			Moderate to high viscosity product
			Moderate to low volatile content
	Establish limits of applicability of corrective action program to vapor-phase remediation	Soil venting	Volatile contaminants
			Permeable soils
			Moderate to high concentrations of adsorbed contaminants
			Absence of NAPL
	Establish limits of applicability of corrective action program to NAPL recovery	Interceptor trenches and drains	NAPL on perched water system
			Low permeability formations
			Shallow water table aquifers
			Limited volume of fugitive product
		Skimming systems	Shallow to moderate depth water table aquifers
			Perched water systems
			Minimal NAPL accumulations
			Low to moderate permeability formations
			Plume of large areal extent
			Low to moderate viscosity product
		Single-pump (total fluids) systems	Shallow to deep aquifers
			Low to moderate permeability formations
			Minimal NAPL accumulations

	Two-pump systems	Low to moderate viscosity product
		Requires phase separation (i.e., oil/water separator)
		Shallow to deep aquifers
		Substantial NAPL accumulations
		Moderate to high permeability formations
Establish limits of applicability of corrective action program to dissolved hydrocarbon recovery/ treatment	Air stripping	Low to moderate viscosity products
		Volatile contaminants (i.e., BTEX)
		Low to high solubility constituents
		Large throughput volume
		Low to high VOC concentration
		Low ambient iron concentration
		Air emissions
	Activated carbon adsorption	Low to moderate solubility compounds
		Volatile contaminants
		Low to moderate throughput volume
		Low to moderate VOC concentration
		Low ambient iron and microbial concentration
		Carbon regeneration or disposal
	Chemical oxidation	Oxidizable contaminants
		Low to high throughput volume
		Low ambient iron concentration
	Combined air-stripping and carbon adsorption	Moderate to highly volatile contaminants
		Stringent air emission and/or water discharge requirements
	Spray irrigation	Large surface area
		Liberal air emission and groundwater discharge requirements
		Indigenous microbial population
	In situ enhanced biodegradation	Indigenous microbial population
		Neutral pH groundwater
		Nonreactive aquifer matrix
		Permeable formation
	Natural (passive) biodegradation	Indigenous microbial population
		Low to moderate contaminant concentrations
		Plume of limited areal extent
		Low to moderate contaminant transport velocity

results from the site investigations indicate BTEX as the only chemical constituents of significant concern for this site.

Objective

The overall objective of the corrective action assessment for the Petro-X facility is to determine the type of remedial systems necessary to abate potential risks posed by the site.

Assessment of the Corrective Action Program Needs

Site conditions at the facility should be fully characterized, to help establish the vertical and lateral extents of hydrocarbon contaminations from this site. Based on the information available from the previous site assessments for the Petro-X facility, it is apparent that petroleum product releases from the gasoline station have significantly

impacted both soils and groundwater at the site. Additional sampling and characterization will provide the information necessary to properly characterize the site, to
perform a risk assessment that will help establish cleanup criteria for the soil and
groundwater matrices, and to develop cost-effective corrective action plans for the
facility. Supplementary site information needed to complete a comprehensive site
characterization for the Petro-X facility will, at a minimum, include the following:

- Sediment samples from drainage systems at and near the site.
- Representative soil samples across the site, in particular within areas with
  visually obvious contamination, to help identify potential "hot spots" at the
  facility.
- Groundwater monitoring wells, placed such that upgradient and downgradient
  water quality relative to the potential "hot spots" can be fully investigated.
- Limited background soil samples representative of the general vicinity of the
  site. This will become an important basis for comparison when developing
  cleanup levels for the site.

The additional investigations should also include an assessment of the potential
for hydrocarbon vapors to migrate along or within human-made conduits at and near
the Petro-X facility location. This is because, among other things, such vapors may
pose health hazards and threats of explosion or fire if concentrations reach explosive
levels and an ignition source is present.

Identification of Feasible Remedial Options
    Typically, the most frequently used treatment methods for groundwater contami-
nated with BTEX are air stripping and granular activated carbon (GAC) adsorption.
For greater removal efficiency, air stripping is used in tandem with GAC (i.e., a
GAC/air stripping system). As necessary, remediation of the petroleum hydrocarbon
plumes using a pump/treat/reinject method may be recommended for groundwater at
the Petro-X facility. Reinjection of the treated water is considered necessary in order
to maintain the water supply capacity of the aquifer at this site. The preferred approach
will be to pump from the edges of the mapped plume and reinject (after treatment) at
the center of the plume. A remedial plan developed around four pumping wells and
two injection wells is considered to be a reasonable starting point for an average-sized
program for a contaminated gasoline fueling station.
    An important site restoration technique for hydrocarbon-contaminated soils would
be bioremediation; clearly identified "hot spots" may, however, be removed for
disposal or for treatment by other processes such as incineration or asphalt batching.
Yet other favored remediation techniques would consist of the use of *in situ* soil
venting systems.

The Corrective Action Plan
    The corrective action plan for the Petro-X facility is divided into soil and
groundwater remediation tasks. It has been decided to use vacuum extraction systems for the impacted soils, and a pump-and-treat technology for the affected
groundwater. The soil remediation tasks will consist of the design of a vapor
extraction system (VES), the installation and operation of the VES (after obtaining
applicable permits to operate the system), and laboratory analyses of verification

borings (for monitoring purposes and performance evaluation). The groundwater remediation tasks will consist of the design of a groundwater treatment system (GTS), the installation and operation of the GTS, water sampling and analysis during operation, and laboratory analyses of verification samples (for monitoring purposes and performance evaluation).

Since the Petro-X facility is still operational, the VES should be designed, installed, and operated in such a manner that will minimize potential disruption to normal activities at the facility and in the vicinity. The VES in this case will utilize an array of vapor extraction wells and air injection wells located in areas of impacted soils as determined from the prior site characterization, and designed so as to achieve maximum efficiency for contaminant removal; the use of the complementing air injection wells will generally improve the performance of the VES. The VES should be operated until the acceptable cleanup goals are achieved for the key contaminants of concern in the contaminated soils. Subsequently, verification samples (to ascertain the performance of the VES) should be collected and analyzed prior to closure.

The GTS for the Petro-X facility should also be designed, installed, and operated in such a manner that will minimize potential disruption to normal activities at the facility and in the vicinity. The GTS, a pump-and-treat system, will consist of both extraction and injection wells appropriately located so as to achieve maximum efficiency for contaminant removal from the groundwater. Pumped groundwater will be treated prior to reinjection or disposal. The GTS should be operated until the acceptable cleanup goals are achieved for the key contaminants of concern in groundwater. Subsequently, verification samples (to ascertain the performance of the GTS) should be collected and analyzed prior to closure; furthermore, long-term monitoring wells should be installed downgradient of the contaminated areas.

It is noteworthy that in a number of situations a detailed assessment may indicate that no extensive and expensive remediation or cleanup program is necessary for a contaminated site. Thus, prior to implementing any remediation plan for the Petro-X facility, it is recommended that appropriate risk-based cleanup criteria be developed and compared with the levels of contamination that presently exist at the site (i.e., the baseline conditions). Based on such criteria, it may be determined from the types of exposure scenarios relevant to the site that only limited cleanup is warranted; this could save Petro-X substantial amounts of money as well as other potential problems and liabilities that the implementation of a remedial action program could carry. In fact, even if it is determined that some degree of cleanup is required, the cleanup criteria developed will aid in optimizing the efforts involved so as to arrive at more cost-effective solutions.

8.8 DEVELOPMENT OF A SITE CLOSURE PLAN FOR A CONTAMINATED SITE PROBLEM

The purpose of this section is to present pertinent information relevant to the preparation of a decommissioning or closure plan for an abandoned industrial facility. This hypothetical facility has been used for a multitude of operations, including machine components cleaning, electroplating, sandblasting, painting, and vehicle maintenance. The site is located within an industrial estate in the State of California. Based on current zoning plans, it is anticipated that this land parcel will be used for residential developments in the near future.

Introduction and Background

The former industrial facility, located at A2Z in an industrially zoned area, operated for about two decades before being permanently closed. Site facilities include a main plant building, office buildings, storage tanks, and post-closure areas (that consist of surface impoundments for wastewater treatment operations, sludge ponds, etc.). Past operations at the plant required the storage of raw materials in above-ground tanks, the distribution of raw materials in pipelines, and the storage of chemicals, fuels, and waste materials in USTs. Historical uses of the site included component cleaning (in which acids, caustics, and chlorinated hydrocarbon-based solvents were used) and electroplating (for which major associated chemicals included cadmium, nickel, and chromium). Other significant activities included sandblasting of unpainted metal parts, painting, and vehicle maintenance.

Due to the sandblasting activities, incidental spillage during materials handling, and possible leakage of underground storage and transmission/distribution systems, soils and groundwater underlying the A2Z plant site are expected to be significantly impacted; this is the result of releases of chemical materials that were used in the industrial processes and related activities carried out at this facility. Preliminary remedial activities have been implemented to remove buried drums and storage tanks, and to remove soil materials from some of the most heavily contaminated areas.

The soil materials at the A2Z site consist mostly of sand and silty sand, underlain by silts and clays that overlie some cherty shales and siltstones. An extensive program has been udertaken to define the nature and extent of the soil and groundwater contamination within the site boundary. The past sampling activities indicate that a number of chemicals of potential concern associated with the site have impacted soils and groundwater, with the possibility to affect surface water in the vicinity of the site if timely corrective measures are not implemented.

Objective and Scope

The objective of this illustrative example is to develop a preliminary decommissioning/closure plan for the A2Z site. This will normally include the preparation of an initial documentation (consisting of a general description of the facilities to be decommissioned, a site plan drawn to scale and also showing surrounding land uses and natural features, the approximate time frame envisaged for the decommissioning or site cleanup, and the current official plan designation and zoning of the site) as well as a preliminary inventory (which provides an initial understanding of the potential range and quantity of contaminants possibly present at the site), a sampling and analysis program, a risk assessment, and a corrective action program.

For the purpose of this illustrative example, only the sampling requirements, a risk assessment, and a proposed corrective action plan are discussed (Figure 8.2); a general descriptive presentation of the facility layout and inventory data are not included here.

Sampling Requirements for the Decommissioning Plan

The general types of site data required as part of the decommissioning plan for the A2Z site include:

- The contaminant identities
- Chemical concentrations in the key sources and media of concern

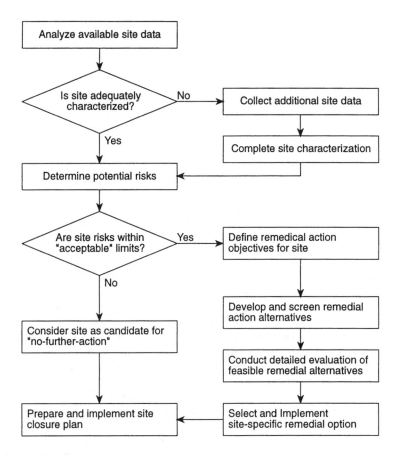

Figure 8.2 Technical elements of the process for developing a decommissioning plan for the A2Z facility.

- Characteristics of sources, in particular, information on chemical release potentials
- Characteristics of the environmental setting (including groundwater flow direction) that may affect the fate, transport, and persistence of the chemicals of potential concern present at this site

Typically, a preliminary sampling program for soils, groundwater, surface water, and sediments is undertaken to identify on-site contamination. If it is apparent that contaminants may have or could migrate off-site, the initial sampling program should be extended to off-site locations for verification.

A preliminary sampling program is required for the A2Z site in order to completely characterize the site conditions, as well as to determine the nature and extent of contamination present at the facility. This is achieved through sampling and analysis of all contaminant sources and all potentially impacted media. Since soil often is the major source of chemical releases to other media, the number, location, and type of soil samples collected could have significant impacts on the overall investigations.

Consequently, a systematic grid pattern that covers the areal extent of this site is preferably used for the soil sampling program. Additionally, where higher degrees of contamination are visually apparent or suspected, a greater number of samples should be collected to better define any existing "hot spots" at the facility.

The sampling and analysis program for the A2Z site was designed to produce representative samples to support the closure plan. Chemicals found in soils and groundwater at the A2Z site included both organic and inorganic constituents (Table 8.2).

Evaluation of Potential Risks

By virtue of the physical setting and the type of surface materials present at the site, the primary release mechanisms for the contaminants of concern at the A2Z facility have been determined to be via infiltration and migration into groundwater, and also by wind erosion and fugitive dust transport. Surface water runoff and overland transport could become an important contributor to the spread of contamination if timely corrective actions are not taken. For the purpose of this discussion, the significant exposure pathways to potential receptors that are evaluated here relate to the air, groundwater, and soils only.

The following represents an evaluation of the potential impacts of the chemicals of potential concern (COCs) present at the A2Z site. Residential exposure scenarios involving possible exposure of potential receptors to contaminated soils and to contaminated groundwater were evaluated using the methods of approach discussed in Chapters 5 and 6; the results are presented in Tables 8.3 and 8.4, respectively. The 95% upper confidence level (UCL) concentrations of the COCs in the environmental samples were used as the exposure point concentrations in the evaluation of potential risks. Also, case-specific exposure parameters obtained from the literature (viz., DTSC, 1994; OSA, 1993; USEPA, 1989a, 1989b, 1991, 1992) were used in the modeling effort. Toxicity values used pertain to those promulgated into California regulations, since this facility is located in the State of California, and in some cases those found within federal EPA guidance, as appropriate. It is assumed, for the sake of simplicity, that all the noncarcinogenic effects of the COCs are attributable to the same toxicological endpoint.

Table 8.3 consists of an evaluation of the potential risks associated with a hypothetical residential development at A2Z, assuming the contaminated soils present at the site remain in place. It is assumed that potential receptors may be exposed via inhalation of airborne contamination (consisting of particulates and/or volatile emissions), through the ingestion of contaminated soils, and by dermal contact with the contaminated soils at the site. Default exposure parameters indicated in the literature (viz., DTSC, 1994; OSA, 1993; USEPA, 1989a, 1989b, 1991, 1992) were used for the calculations shown in this spreadsheet. Based on this scenario, it is determined that risks exceed the generally accepted benchmark of 10^{-6}. The noncarcinogenic hazard index also exceeds the reference level of unity. In particular, acceptable risk-based soil criteria are exceeded for arsenic, beryllium, hexavalent chromium, and TCE at the 10^{-6} risk level.

Table 8.4 consists of an evaluation of the potential risks associated with a hypothetical population exposure to impacted groundwater originating from the A2Z site, assuming the contaminated water is not treated before going into a public water supply system. It is assumed that potential receptors may be exposed through the inhalation of volatiles during domestic usage of contaminated water, from the ingestion of contaminated water, and by dermal contact to contaminated waters. Default exposure parameters indicated in the literature (viz., DTSC, 1994; OSA, 1993; USEPA,

Table 8.2 Summary of the Chemicals of Potential Concern at the A2Z Site

Chemicals of Potential Concern in Soils	Important Synonyms or Trade Names, or Chemical Formula	Chemical Abstracts Service number (CAS No.)	95% UCL Soil Concentration (mg/kg)
Inorganic chemicals			
Antimony	Sb	7440–36-0	4.0
Arsenic	As	7440-38-2	17.2
Beryllium	Be	7440-41-7	1.4
Cadmium	Cd	7440-43-9	4.5
Chromium	Cr	16065-83-1	92.2
Chromium VI	Cr(VI)	7440-47-3	4.6
Cobalt	Co	7440-48-4	7.8
Manganese	Mn	7439-96-5	512
Mercury	Hg	7439-97-6	0.25
Molybdenum	Mo	7439-98-7	16.6
Nickel	Ni	7440-02-0	90.9
Selenium	Se	7782-49-2	1.4
Vanadium	V	7440-62-2	62.9
Zinc	Zn	7440-66-6	413
Organic compounds			
Chloroform	Trichloromethane	67-66-3	0.008
Trichloroethene	TCE	79-01-6	32
1,1,2-Trichloroethane	1,1,2-TCA	79-00-5	0.020
Tetrachloroethene	PCE	127-18-4	0.033
Ethylbenzene	EB; phenylethane	100-41-4	0.009
Xylenes (mixed)	Dimethylbenzene	1330-20-7	0.044

Chemicals of Potential Concern in Groundwater	Important Synonyms or Trade Names, or Chemical Formula	Chemical Abstracts Service number (CAS No.)	95% UCL Groundwater Concentration (μg/L)
Inorganic chemicals			
Antimony	Sb	7440-36-0	3.4
Arsenic	As	7440-38-2	7.6
Cadmium	Cd	7440-43-9	4.6
Chromium	Cr	16065-83-1	8,690
Chromium VI	Cr(VI)	7440-47-3	435
Manganese	Mn	7439-96-5	299
Molybdenum	Mo	7439-98-7	164
Nickel	Ni	7440-02-0	82.2
Vanadium	V	7440-62-2	54
Zinc	Zn	7440-66-6	64.6
Organic compounds			
Vinyl chloride	VC	75-01-4	0.729
1,1-Dichloroethene	1,1-DCE	75-35-4	3.31
trans-1,2-Dichloroethene	1,2-trans-DCE	156-60-5	11.8
cis-1,2-Dichloroethene	cis-1,2-DCE	540-59-0	1,220
Trichloroethene	TCE	79-01-6	548

Table 8.3 Risk Screening for a Hypothetical Residential Population Exposure to Soils at the A2Z Site

Chemical of Potential Concern	95% UCL Soil Concentration (mg/kg)	Chemical-Specific Dermal Absorption (ABSs)	Oral RfD (mg/kg-day)	Oral SF (1/mg/kg-day)	Inhalation RfD (mg/kg-day)	Inhalation SF (1/mg/kg-day)	Risk for Air	Hazard for Air	Risk for Soil	Hazard for Soil	Total Risk (Air + Soil)	Total Hazard (Air + Soil)	Risk-Based Soil Criteria (mg/kg)	Soil Criteria Exceeded?
Inorganic chemicals														
Antimony	4.0	0.01	4.00E−04				0.00E+00	3.20E−04	0.00E+00	1.41E−01			28	No
Arsenic	17.2	0.03	3.00E−04	1.75E+00		1.20E+01	1.54E−06	1.83E−03	6.41E−05	9.54E−01			0.26	Yes
Beryllium	1.4	0.01	5.00E−03	4.30E+00		8.40E+00	8.76E−08	8.95E−06	1.06E−05	3.94E−03			0.13	Yes
Cadmium	4.5	0.001	5.00E−04			1.50E+01	5.03E−07	2.88E−04	0.00E+00	1.16E−01			9	No
Chromium (total)	92.2	0.01	1.00E+00				0.00E+00	2.95E−06	0.00E+00	1.30E−03			70,862	No
Chromium (VI)	4.6	0.00	5.00E−03	4.20E−01		5.10E+02	1.75E−05	2.94E−05	3.03E−06	1.18E−02			0.22	Yes
Cobalt	7.8	0.01	2.90E−01				0.00E+00	8.59E−04	0.00E+00	3.79E−01			21	No
Manganese	512	0.01	1.40E−01		2.90E−04		0.00E+00	1.49E−01	0.00E+00	5.15E−02			2,557	No
Mercury	0.25	0.01	3.00E−04		1.10E−04		0.00E+00	9.29E−05	0.00E+00	1.17E−02			21	No
Molybdenum	16.6	0.01	5.00E−03		8.60E−05		0.00E+00	1.06E−04	0.00E+00	4.67E−02			354	No
Nickel	90.9	0.01	2.00E−02		5.00E−03	9.10E−01	6.16E−07	1.45E−04	0.00E+00	6.40E−02			148	No
Selenium	1.4	0.01	5.00E−03				0.00E+00	8.95E−06	0.00E+00	3.94E−03			354	No
Vanadium	62.9	0.01	7.00E−03				0.00E+00	2.87E−04	0.00E+00	1.27E−01			496	No
Zinc	413	0.01	3.00E−01				0.00E+00	4.40E−05	0.00E+00	1.94E−02			21,259	No
Organic compounds														
Chloroform	0.008	0.10	1.00E−02	3.10E−02	1.00E−02	1.90E−02	1.32E−09	2.98E−05	8.53E−10	2.05E−05			9	No
Trichloroethene	32	0.10	6.00E−03	1.50E−02	6.00E−03	1.00E−02	2.01E−06	1.44E−01	1.65E−06	1.37E−01			19	Yes
1,1,2-Trichloroethane	0.020	0.10	4.00E−03	5.70E−02	4.00E−03	5.60E−02	3.54E−09	6.77E−05	3.92E−09	1.28E−04			5	No
Tetrachloroethene	0.033	0.10	1.00E−02	5.10E−02	1.00E−02	5.10E−02	5.18E−09	4.36E−05	5.79E−09	8.45E−05			6	No
Ethylbenzene	0.009	0.10	1.00E−01		2.90E−01		0.00E+00	5.47E−07	0.00E+00	2.30E−06			3,901	No
Xylenes	0.044	0.10	2.00E+00		2.00E−01		0.00E+00	3.16E−06	0.00E+00	5.63E−07			78,028	No
							2.22E−05	0.30	7.94E−05	2.1	1.02E−04	2.4		

Notes: The computational formulas and models used in this evaluation are discussed in Chapters 5 and 6, and further elaborated in Appendices F, I, and J.

Risk/hazard for air accounts for the volatilization effects (for volatile organic compounds) and airborne emissions of contaminated particulates (for nonvolatile chemicals) present at the site (see Section 5.8.1 and DTSC, 1994).

Case-specific exposure parameters used in the calculations were obtained from the following sources — DTSC (1994), OSA (1993), and USEPA (1989a, 1989b, 1991, 1992).

Table 8.4 Risk Screening for a Hypothetical Residential Population Exposure to Groundwater at the A2Z Site

Chemical of Potential Concern	95% UCL Water Concentration (μg/L)	Chemical-Specific K_p (cm/hr)	Toxicity Criteria				Risk for Water	Hazard for Water	Risk-Based Water Criteria (μg/L)	MCL Values as Water Criteria (μg/L)	Water Criteria Exceeded?
			Oral RfD (mg/kg-day)	Oral SF (1/mg/kg-day)	Inhalation RfD (mg/kg-day)	Inhalation SF (1/mg/kg-day)					
Inorganic chemicals											
Antimony	3.4	1.60E − 04	4.00E − 04				0.00E + 00	5.43E − 01	6.26	6	No
Arsenic	7.6	1.60E − 04	3.00E − 04	1.75E + 00		1.20E + 01	1.98E − 04	1.62E + 00	0.04	50	Within MCL
Cadmium	4.6	1.60E − 04	5.00E − 04			1.50E + 01	0.00E + 00	5.88E − 01	8	10	No
Chromium (total)	8,690	1.60E − 04	1.00E + 00				0.00E + 00	5.55E − 01	15,647	50	Within risk-based criteria
Chromium VI	435	1.60E − 04	5.00E − 03	4.20E − 01		5.10E + 02	2.72E − 03	5.56E + 00	0.16	na	Yes
Manganese	299	1.60E − 04	5.00E − 03		1.10E − 04		0.00E + 00	3.82E + 00	78	50	Yes
Molybdenum	164	1.60E − 04	5.00E − 03		5.00E − 03		0.00E + 00	2.10E + 00	78	na	Yes
Nickel	82.2	1.60E − 04	2.00E − 02			9.10E − 01	0.00E + 00	2.63E − 01	313	100	No
Vanadium	54	1.60E − 04	7.00E − 03				0.00E + 00	4.93E − 01	110	na	No
Zinc	64.6	1.60E − 04	3.00E − 01				0.00E + 00	1.38E − 02	4,694	5,000	No
Organic compounds											
Vinyl chloride	0.729	7.30E − 03		2.70E − 01		2.70E − 01	5.91E − 06	0.00E + 00	0.12	0.5	Yes
1,1-Dichloroethene	3.31	1.60E − 02	9.00E − 03	6.00E − 01	9.00E − 03	1.20E + 00	8.98E − 05	4.74E − 02	0.04	6.0	Within MCL
trans-1,2-Dichloroethene	11.8	1.00E − 02	2.00E − 02		2.00E − 02		0.00E + 00	7.58E − 02	156	10.0	Within risk-based criteria
cis-1,2-Dichloroethene	1,220	1.00E − 02	1.00E − 02		1.00E − 02		0.00E + 00	1.57E + 01	78	6.0	Yes
Trichloroethene	548	1.60E − 02	6.00E − 03	1.50E − 02	1.00E − 02	1.50E − 02	2.49E − 04	9.43E + 00	2.20	5.0	Yes
							3.27E − 03	40.8			

Notes: The computational formulas and models used in this evaluation are discussed in Chapters 5 and 6, and further elaborated in Appendices F, I, and J.

Risk/hazard for water account for both volatile chemical emissions and nonvolatile chemical contributors present at the site (see Section 5.8.1 and DTSC, 1994).

Case-specific exposure parameters used in the calculations were obtained from the following sources — DTSC (1994), OSA (1993), and USEPA (1989a, 1989b, 1991, 1992).

K_p = chemical-specific dermal permeability coefficient for water; MCL = maximum contaminant level (for California); na = not available.

1989a, 1989b, 1991) were used for the calculations presented in this spreadsheet. Based on this scenario, it is apparent that risks exceed the generally accepted benchmark risk level of 10^{-6}. The reference hazard index of 1 is also exceeded for receptor exposures to raw/untreated groundwater from the A2Z site. The major contributors to the carcinogenic risk is from the general population exposure to arsenic, hexavalent chromium, 1,1-DCE, and TCE in groundwater. The most significant contributors to noncarcinogenic effects are hexavalent chromium, manganese, cis-1,2-DCE, and TCE in groundwater.

A Site Restoration Strategy

The risk evaluation presented above indicates that the COCs present at the A2Z site may pose significant risks to human receptors potentially exposed via the soil and groundwater media. Consequently, if the A2Z site is to be used for future residential developments, or if raw/untreated groundwater from this site is to be used as a potable water supply source, then a comprehensive site restoration program is required to abate the imminent risks that the site poses. This should comprise an integrated soil and groundwater remediation program. Thus, general response actions should be developed for each of the impacted environmental medium associated with the site (i.e., for both soils and groundwater). For other than a "no-action" situation, the site-specific risk-based criteria can be used to guide and support the development, screening, and selection of potentially feasible remedial alternatives.

The Corrective Action Plan

General response actions to consider and evaluate for the A2Z site include containment (e.g., capping and slurry walls), removal of "hot spots" (e.g., excavation and disposal), treatment (e.g., by soil venting/vapor extraction, thermal treatment, biological treatment/bioremediation, soil washing, immobilization) for the soils, and pump-and-treat systems (including air/steam stripping, GAC adsorption, etc.) for groundwaters. Where necessary, treatability investigations should be conducted following the development and screening of the remedial alternatives for the site.

The remedial alternatives selected as the most technically feasible will subsequently become the focus for more detailed analyses. Several important requirements for designing an effectual corrective action plan, as discussed earlier in this chapter, will be used to support the processes involved. The detailed evaluation ranks the potential response actions with particular attention given to several regulatory compliance requirements, community relations activities, effectiveness, implementability, and associated costs. Depending on what chemicals are determined to drive the remediation, one particular remedial technology/process may be preferred over another; a combination of various process options should be adopted if that is determined to be more cost effective and technically justifiable/defensible in its application.

REFERENCES

Barnhart, M. 1992. "Bioremediation: do it yourself." *Soils,* August-Sept. 1992, pp. 14–19.

Cairney, T. (Ed.). 1993. *Contaminated Land (Problems and Solutions).* Lewis Publishers, Boca Raton, FL.

Calabrese, E.J. and P.T. Kostecki (Ed.). 1991. *Hydrocarbon Contaminated Soils,* Vol. 1. Lewis Publishers, Chelsea, MI.

DTSC. 1994. Preliminary Endangerment Assessment Guidance Manual (A guidance manual for evaluating hazardous substance release sites). California Environmental Protection Agency, Department of Toxic Substances Control, Sacramento.

Hunt, J.R., J.T. Geller, N. Sitar and K.S. Udell. 1988. Subsurface Transport Processes for Gasoline Components. In Specialty Conf. Proc., Joint CSCE-ASCE Natl. Conf. Environ. Eng., Vancouver, B.C., Canada, July 13–15, 1988, pp. 536–543.

Kunreuther, H. and M.V. Rajeev Gowda (Eds.). 1990. *Integrating Insurance and Risk Management for Hazardous Wastes.* Kluwer Academic, Boston.

LUFT, 1989. Leaking Underground Fuel Tank Field Manual: Guidelines for Site Assessment, Cleanup, and Underground Storage Tank Closure. State of California, Leaking Underground Fuel Tank Task Force, Sacramento.

Nyer, E.K. 1993. *Practical Techniques for Groundwater and Soil Remediation.* Lewis Publishers, Boca Raton, FL.

OBG (O'Brien and Gere Engineers, Inc.). 1988. *Hazardous Waste Site Remediation: The Engineer's Perspective.* Van Nostrand Reinhold, New York.

OSA, 1993. Supplemental Guidance for Human Health Multimedia Risk Assessments of Hazardous Waste Sites and Permitted Facilities. California Environmental Protection Agency, The Office of Scientific Affairs (OSA), Sacramento.

Sims, R. et al. 1986. *Contaminated Surface Soils In-Place Treatment Techniques.* Noyes Publications, Park Ridge, NJ.

USEPA. 1984. Review of in-place treatment techniques for contaminated surface soils. Vol. 1 and 2. U.S. Environmental Protection Agency, Hazardous Waste Engineering Research Laboratory, EPA-540/2-84-003a and b. Cincinnati, OH.

USEPA. 1985. Modeling Remedial Actions at Uncontrolled Hazardous Waste Sites. EPA/540/2-85/001. Office of Emergency and Remedial Response, U.S. Environmental Protection Agency, Washington, D.C.

USEPA. 1988a. Guidance for Conducting Remedial Investigations and Feasibility Studies Under CERCLA. EPA/540/G-89/004. OSWER Directive 9355.3-01, Office of Emergency and Remedial Response, U.S. Environmental Protection Agency, Washington, D.C.

USEPA. 1988b. Guidance on Remedial Actions for Contaminated Ground Water at Superfund Sites. EPA/540/G-88/003. Office of Emergency and Remedial Response, U.S. Environmental Protection Agency, Washington, D.C.

USEPA. 1989a. Risk Assessment Guidance for Superfund. Vol. I. Human Health Evaluation Manual (Part A). EPA/540/1-89/002. Office of Emergency and Remedial Response, U.S. Environmental Protection Agency, Washington, D.C.

USEPA. 1989b. Exposure Factors Handbook, EPA/600/8-89/043. Office of Health and Environmental Assessment, U.S. Environmental Protection Agency, Washington, D.C.

USEPA. 1991. Risk Assessment Guidance for Superfund, Vol. I. Human Health Evaluation Manual. Supplemental Guidance. "Standard Default Exposure Factors" (Interim Final). Office of Emergency and Remedial Response. OSWER Directive: 9285.6-03. U.S. Environmental Protection Agency, Washington, D.C.

USEPA. 1992. Dermal Exposure Assessment: Principles and Applications. EPA/600/8-91/011B, U.S. Environmental Protection Agency, Washington, D.C.

Yin S.C.L. 1988. Modeling Groundwater Transport of Dissolved Gasoline. In Specialty Conf. Proc., Joint CSCE-ASCE Natl. Conf. Environ. Eng. Vancouver, B.C. Canada, July 13–15, pp. 544–551.

ADDITIONAL BIBLIOGRAPHY

Ahmad, Y.J., S.E. El Serafy and E. Lutz. 1989. *Environmental Accounting for Sustainable Development.* World Bank, Washington, D.C.

Alberta Environment and Alberta Labor. 1989. Subsurface Remediation Guidelines for Underground Storage Tanks. Alberta MUST (Management of Underground Storage Tanks) Project, A Joint Project of the Departments of Environment and Labour, Edmonton, Alberta, Canada.

Ananichev, K. 1976. *Environment: International Aspects.* Progress Publishers, Moscow.

Andelman, J.B. and D.W. Underhill. 1988. *Health Effects From Hazardous Waste Sites.* 2nd ed., Lewis Publishers, Chelsea, MI.

Asante-Duah, D.K., D.S. Bowles and L.R. Anderson. 1991. Framework for the Risk Analysis of Hazardous Waste Facilities. In, Proc. Sixth Int. Conf. Appl. Statistics and Probability in Civil Eng. CERRA/ICASP 6, Mexico.

Baasel, W.D. 1985. *Economic Methods for Multipollutant Analysis and Evaluation.* Marcel Dekker, New York.

Barnard, R. and G. Olivetti. 1990. "Rapid Assessment of Industrial Waste Production Based on Available Employment Statistics." *Waste Manage. Res.* 8, 2, 139–44.

Bartell, S.M., R.H. Gardner and R.V. O'Neill. 1992. *Ecological Risk Estimation.* Lewis Publishers, Chelsea, MI.

Batstone, R., J.E. Smith, Jr. and D. Wilson (Eds.). 1989. *The Safe Disposal of Hazardous Wastes — The Special Needs and Problems of Developing Countries,* Vol. I, II, & III. A Joint Study Sponsored by the World Bank, the World Health Organization (WHO), and the United Nations Environment Programme (UNEP). World Bank Technical Paper 0253-7494, No. 93. World Bank, Washington, D.C.

Berthouex, P.M. and L.C. Brown. 1994. *Statistics for Environmental Engineers.* Lewis Publishers/CRC Press, Boca Raton, FL.

Bhatt, H.G., R.M. Sykes and T.L. Sweeney (Eds.). 1986. *Management of Toxic and Hazardous Wastes.* Lewis Publishers, Chelsea, MI.

Binder, S., D. Sokal and D. Maughan. 1986. Estimating the Amount of Soil Ingested by Young Children through Tracer Elements. *Arch. Environ. Health,* 41, 341–345.

Blumenthal, D.S. (Ed.). 1985. *Introduction to Environmental Health.* Springer, New York.

Boulding, J.R. 1994. *Description and Sampling of Contaminated Soils (A Field Manual),* 2nd ed. Lewis Publishers/CRC Press, Boca Raton, FL.

Bregman, J.I. and K.M. Mackenthun. 1992. *Environmental Impact Statements.* Lewis Publishers, Chelsea, MI.

275

Bretherick, L. 1979. *Handbook of Reactive Chemical Hazards*. 2nd ed. Butterworth Publishers, Wolburn, MA.

Brown, C. 1978. Statistical Aspects of Extrapolation of Dichotomous Dose Response Data. *J. Natl. Cancer Inst.,* 60, 101–8.

Brown, K.W., G.B. Evans, Jr. and B.D. Frentrup (Eds.). 1983. *Hazardous Waste and Treatment*. Butterworth Publishers, Boston, MA.

Brusick, D.J. (Ed.). 1994. *Methods for Genetic Risk Assessment*. Lewis Publishers/CRC Press, Boca Raton, FL.

Buchel, K.H. 1983. *Chemistry of Pesticides*. John Wiley & Sons, New York.

Byrnes, M.E. 1994. *Field Sampling Methods for Remedial Investigations*. Lewis Publishers/CRC Press, Boca Raton, FL.

Cairns, J., Jr. and T.V. Crawford (Eds.). 1991. *Integrated Environmental Management*. Lewis Publishers, Chelsea, MI.

Calabrese, E.J. 1984. *Principles of Animal Extrapolation*. John Wiley & Sons, New York.

Calabrese, E.J. and P.T. Kostecki. 1988. *Soils Contaminated by Petroleum: Environment and Public Health Effects*. John Wiley & Sons, New York.

Calabrese, E.J. and P.T. Kostecki. 1989. *Petroleum Contaminated Soils,* Vol. 2. Lewis Publishers, Chelsea, MI.

Calabrese, E.J. and P.T. Kostecki (Eds.). 1991. *Hydrocarbon Contaminated Soils,* Vol. 1. Lewis Publishers, Chelsea, MI.

Calabrese, E.J. and P.T. Kostecki. 1992. *Risk Assessment and Environmental Fate Methodologies*. Lewis Publishers/CRC Press, Boca Raton, FL.

Calabrese, E.J., R. Barnes, E.J. Stanek, III, H. Pastides, C.E. Gilbert, P. Veneman, X. Wang, A. Lasztity and P.T. Kostecki. 1989. How Much Soil Do Young Children Ingest: An Epidemiologic Study. *Regul. Toxicol. Pharmacol.* 10, 123–137.

Canter, L.W., R.C. Knox and D.M. Fairchild. 1988. *Ground Water Quality Protection*. 3rd ed. Lewis Publishers, Chelsea, MI.

CAPCOA. 1989. Air Toxics Assessment Manual. California Air Pollution Control Officers Association, Draft Manual, August, 1987 (amended, 1989), Sacramento, CA.

Carson, W.H. (Ed.). 1990. *The Global Ecology Handbook — What You Can Do About the Environmental Crisis*. The Global Tommorrow Coalition. Beacon Press, Boston.

Casarett, L.J. and J. Doull. 1975. *Toxicology: The Basic Science of Poisons*. Macmillan Publishing, New York.

CCME. 1991. Interim Canadian Environmental Quality Criteria for Contaminated Sites. Report CCME EPC-CS34, The National Contaminated Sites Remediation Program, Canadian Council of Ministers of the Environment, Winnipeg, Manitoba.

Chatterji, M. (Ed.). 1987. *Hazardous Materials Disposal: Siting and Management*. Gower Publishing, Avebury, U.K.

Chiu, H.S. and K.L. Tsang. 1990. "Reduction of Treatment Cost by Using Communal Treatment Facilities." *Waste Manage. Res.* 8, 2, 165–7.

Chrostowski, P.C., L.J. Pearsall and C. Shaw. 1985. Risk assessment as a management tool for inactive hazardous materials disposal sites. *Environ. Manage.,* 9, 5, 433–442.

Clausing, O., A.B. Brunekreef and J.H. van Wijnen. 1987. A Method for Estimating Soil Ingestion by Children. *Int. Arch. Occup. Environ. Health,* 59, 73–82.

Clayson, D.B., D. Krewski and I. Munro (Eds.). 1985. *Toxicological Risk Assessment*, Vols. 1 and 2. CRC Press, Boca Raton, FL.

Clement Associates, Inc. 1988. Multi-Pathway Health Risk Assessment Impact Guidance Document. South Coast Air Quality Management District, Sacramento, CA.

Clewell, H.J. and M.E. Andersen. 1985. Risk Assessment Extrapolations and Physiological Modeling. *Toxicol. Ind. Health,* 1, 111–132.

CMA. 1984. *Risk Management of Existing Chemicals.* Chemical Manufacturers Association, Washington, D.C.

CMA. 1985. *Risk Analysis in the Chemical Industry.* Government Institutes, Inc., Rockville, MD.

Cohen, Y. 1986. Organic Pollutant Transport. *Environ. Sci. Technol.,* Vol. 20, No. 6, pp. 538–45.

Cohrssen, J.J. and V.T. Covello. 1989. Risk Analysis: A Guide to Principles and Methods for Analyzing Health and Environmental Risks. National Technical Information Service, U.S. Department of Commerce, Springfield, VA.

Cole, G.M. 1994. *Assessment and Remediation of Petroleum-Contaminated Sites.* Lewis Publishers/CRC Press, Boca Raton, FL.

Conway, R.A. (Ed.). 1982. *Environmental Risk Analysis of Chemicals.* Van Nostrand Reinhold Co., New York.

Conway, M.F. and S.H. Boutwell. 1987. The Use of Risk Assessment to Define a Corrective Action Plan for Leaking Underground Storage Tanks. In *Proc. NWWA/API Conf. Petroleum Hydrocarbons Org. Chem. Ground Water — Prevention, Detection, Restoration,* National Groundwater Association, Dublin, OH, pp. 19–40.

Cothern, C.R. (Ed.). 1993. *Comparative Environmental Risk Assessment.* Lewis Publishers/CRC Press, Boca Raton, FL.

Cothern, C.R. and N.P. Ross (Eds.). 1994. *Environmental Statistics, Assessment, and Forecasting.* Lewis Publishers/CRC Press, Boca Raton, FL.

Covello, V.T., J. Menkes and J. Mumpower (Eds.). 1986. Risk Evaluation and Management. *Contemporary Issues in Risk Analysis,* Vol. 1. Plenum Press, New York.

Covello, V.T. and J. Mumpower. 1985. Risk Analysis and Risk Management: An Historical Perspective. *Risk Anal.* 5, 103–120.

Covello, V.T. et al. 1987. Uncertainty in Risk Assessment, Risk Management, and Decision Making. *Advances in Risk Analysis,* Vol. 4. Plenum Press, New York.

Cowherd, C.M., G.E. Muleski, P.J. Englehart and D.A. Gillette. 1984. Rapid Assessment of Exposure to Particulate Emissions From Surface Contamination Sites. Midwest Research Institute, Kansas City, MO.

Crandall, R.W. and B.L. Lave (Eds.). 1981. *The Scientific Basis of Risk Assessment.* Brookings Institute, Washington, D.C.

Crouch, E. A. C., R. Wilson and L. Zeise. 1983. The Risks of Drinking Water. *Water Resour. Res.,* 19(6), 1359–1375.

Crump, K.S. 1981. An Improved Procedure for Low-Dose Carcinogenic Risk Assessment from Animal Data. *J. Environ. Toxicol.* 5, 339–46.

Crump, K.S. and R.B. Howe. 1984. The Multistage Model with Time-dependent Dose Pattern: Applications of Carcinogenic Risk Assessment. *Risk Anal.* 4, 163–176.

Daniels, S.L. 1978. Environmental Evaluation and Regulatory Assessment of Industrial Chemicals. 51st Annu. Conf. Water Pollut. Control Fed., Anaheim, CA.

Davis, C.E. 1993. *The Politics of Hazardous Waste.* Prentice-Hall, Englewood Cliffs, NJ.

Diesler, P.F. (Ed.). 1984. *Reducing the Carcinogenic Risks in Industry.* Marcel Dekker, New York.

Doherty, N., P. Kleindorfer and H. Kunreuther. 1990. "An Insurance Perspective on an Integrated Waste Management Strategy." In, *Integrating Insurance and Risk Management for Hazardous Wastes,* H. Kunreuther and M.V. Rajeev Gowda (Eds.) Kluwer Academic, Boston, pp. 271–302.

Dunster, H.J. and W. Vinck. 1979. The Assessment of Risk, its Value and Limitations".
    Eur. Nuclear Conf., Hamburg.
Erickson, M.D. 1993. *Remediation of PCB Spills*. Lewis Publishers/CRC Press, Boca
    Raton, FL.
Eschenroeder, A., R.J. Jaeger, J.J. Ospital and C. Doyle. 1986. Health Risk Assessment of
    Human Exposure to Soil Amended with Sewage Sludge Contaminated with Polychlo-
    rinated dibenzodioxins and dibenzofurans. *Vet. Hum. Toxicol.* 28, 356–442.
Ess, T.H. 1981a. Risk Acceptability. In, Proc. Natl. Conf. Risk Decision Anal. Hazardous
    Waste Disposal, Aug. 24–27, Baltimore, MD, p. 164–74.
Ess, T.H. 1981b. Risk Estimation. In, Proc. Natl. Conf. Risk Decision Anal. Hazardous
    Waste Disposal, Aug. 24–27, Baltimore, MD, p. 155–63.
Evans, L.J. 1989. Chemistry of Metal Retention by Soils. *Environ. Sci. Technol.,* Vol. 23,
    No. 9, 1047–56.
Fitchko, J. 1989. *Criteria for Contaminated Soil/Sediment Cleanup*. Pudvan Publishing,
    Northbrook, IL.
Forester, W.S. and J.H. Skinner (Eds.). 1987. *International Perspectives on Hazardous
    Waste Management — A Report from the International Solid Wastes and Public
    Cleansing Association (ISWA) Working Group on Hazardous Wastes*. Academic
    Press, London.
FPC. 1986. Benzene in Florida Groundwater: An Assessment of the Significance to Human
    Health. Florida Petroleum Council, Tallahassee, FL.
Freese, E. 1973. Thresholds in toxic, teratogenic, mutagenic, and carcinogenic effects.
    Environ. Health Perspect., 6, 171–178.
Glickman, T. S. and M. Gough (Eds.). 1990. *Readings in Risk*. Resources for the Future,
    Washington, D.C.
Gordon, S. I. 1985. *Computer Models in Environmental Planning*. Van Nostrand Reinhold,
    New York.
Gorelick, S.M., R.A. Freeze, D. Donohue and J.F. Keely. 1993. *Groundwater Contamination
    (Optimal Capture and Containment)*. Lewis Publishers/CRC Press, Boca Raton, FL.
Gots, R.E. 1993. *Toxic Risks*. Lewis Publishers/CRC Press, Boca Raton, FL.
Grasso, D. 1993. *Hazardous Waste Site Remediation (Source Control)*. Lewis Publishers/
    CRC Press, Boca Raton, FL.
Gregory, R. and H. Kunreuther. 1990. "Successful Siting Incentives." *Civ. Eng.,* 60, 4,
    73-5.
Guswa, J.H., W.J. Lyman, A.S. Donigian, Jr., T.Y.R. Lo and E.W. Shanahan. 1984.
    Groundwater Contamination and Emergency Response Guide. Noyes Publications,
    Park Ridge, NJ.
Haimes, Y.Y., L. Duan and V. Tulsiani. 1990. Multiobjective Decision-Tree Analysis. *Risk
    Anal.* Vol. 10, No. 1, 111–129.
Hallenbeck, W.H. and K.M. Cunningham-Burns. 1985. *Pesticides and Human Health*.
    Springer-Verlag, New York.
Haun, J.W. 1991. *Guide to the Management of Hazardous Waste*. Fulcrum Publishing,
    Golden, CO.
Hawken, P. 1993. *The Ecology of Commerce: A Declaration of Sustainability*. Harper
    Business, New York.
Hawley, G.G. 1981. *The Condensed Chemical Dictionary*, 10th ed. Van Nostrand Reinhold,
    New York.
Hawley, J.K. 1985. Assessment of health risks from exposure to contaminated soil. *Risk
    Anal.* 5, 4, 289–302.
Hayes, A.W. (Ed.). 1982. *Principles and Methods of Toxicology*. Raven Press, New York.

Henderson, M. 1987. *Living with Risk: The Choices, The Decisions.* John Wiley & Sons, New York.

Hinchee, H.J., H.J. Reisinger, D. Burris, B. Marks and J. Stepek. 1986. Underground Fuel Contamination, Investigation, and Remediation: A Risk Assessment Approach to How Clean Is Clean. In *Proc. NWWA/API Conf. Petroleum Hydrocarbons Org. Chem. Ground Water — Prevention, Detection and Restoration,* National Ground Water Association, Dublin, OH.

Hoddinott, K.B. (Ed.). 1992. *Superfund Risk Assessment in Soil Contamination Studies.* ASTM Publ. STP 1158, American Society for Testing and Materials, Philadelphia, PA.

Hoel, D.G., D.W. Gaylor, R.L. Kirschstein, U. Saffiotti and M.A. Schneiderman. 1975. Estimation of risks of irreversible, delayed toxicity. *J. Toxicol. Environ. Health,* 1, 133–151.

Honeycutt, R.C. and D.J. Schabacker (Eds.). 1994. *Mechanisms of Pesticide Movement into Groundwater.* Lewis Publishers/CRC Press, Boca Raton, FL.

HSE (Health and Safety Executive). 1989. Risk Criteria for Land-Use Planning in the Vicinity of Major Industrial Hazards. Her Majesty's Stationery Office, London.

HSE (Health and Safety Executive). 1989b. Quantified Risk Assessment — Its Input to Decision Making. Her Majesty's Stationery Office, London.

IJC. 1986. Literature Review of the Effects of Persistent Toxic Substances on Great Lakes Biota. Report of the Health of Aquatic Communities Task Force. International Joint Commission.

IRPTC. 1978. Attributes for the Chemical Data Register of the International Register of Potentially Toxic Chemicals. Register Attribute Series, No. 1. International Register of Potentially Toxic Chemicals, UNEP, Geneva, Switzerland.

IRPTC. 1985. International Register of Potentially Toxic Chemicals, Part A. International Register of Potentially Toxic Chemicals, UNEP, Geneva, Switzerland.

IRPTC. 1985. Industrial Hazardous Waste Management. Industry and Environment Office and the International Register of Potentially Toxic Chemicals. United Nations Environment Programme, Geneva, Switzerland.

IRPTC. 1985. Treatment and Disposal Methods for Waste Chemicals: IRPTC File. Data Profile Series No.5. International Register of Potentially Toxic Chemicals. United Nations Environment Programme, UNEP, Geneva, Switzerland.

J.C. Consultancy Ltd, London. 1986. *Risk Assessment for Hazardous Installations.* Pergamon Press, Oxford.

Jennings, A.A. and P. Suresh. 1986. Risk Penalty Functions for Hazardous Waste Management. *ASCE J. Environ. Eng.,* Vol. 112, No. 1, Feb., p. 105–22.

Johnson, B.B. and Covello, V.T. 1987. *Social and Cultural Construction of Risk: Essays on Risk Selection and Perception.* Kluwer Academic, Norwell, MA.

Jorgensen, E.P. (Ed.). 1989. *The Poisoned Well — New Strategies for Groundwater Protection.* Sierra Club Legal Defense Fund. Island Press, Washington, D.C.

Kastenberg, W.E., T.E. McKone and D. Okrent. 1976. On Risk Assessment in the Absence of Complete Data. UCLA Rep. No. UCLA-ENG-677, University of California, Los Angeles, CA.

Kastenberg, W.E. and H.C. Yeh. 1993. Assessing public exposure to pesticides — contaminated ground water. *J. Ground Water,* Vol. 31, No. 5, 746–752.

Kates, R.W. 1978. Risk Assessment of Environmental Hazard. SCOPE Rep. 8, John Wiley & Sons, Chichester.

Keeney, R.L. 1990. "Mortality Risks Induced by Economic Expenditures." *Risk Anal.* 10, 1, 147–59.

Kempa, E.S. (Ed.). 1991. *Environmental Impact of Hazardous Wastes.* PZITS Publishing, Warsaw, Poland.

Kimmel, C.A. and D.W. Gaylor. 1988. Issues in Qualitative and Quantitative Risk Analysis for Developmental Toxicology. *Risk Anal.* 8, 15–20.

Kleindorfer, P.R. and H.C. Kunreuther (Ed.). 1987. *Insuring and Managing Hazardous Risks: From Seveso to Bhopal and Beyond.* Springer-Verlag, Berlin.

Kostecki, P.T. and E.J. Calabrese (Eds.). 1989. *Petroleum Contaminated Soils,* Vol. 1, 2, and 3. Lewis Publishers, Chelsea, MI.

Kostecki, P.T. and E.J. Calabrese (Eds.). 1991. *Hydrocarbon Contaminated Soils and Groundwater,* Vol. 1. Lewis Publishers, Chelsea, MI.

Kostecki, P.T. and E.J. Calabrese (Eds.). 1992. *Contaminated Soils (Diesel Fuel Contamination).* Lewis Publishers/CRC Press, Boca Raton, FL.

Krewski, D., C. Brown and D. Murdoch. 1984. Determining Safe Levels of Exposure: Safety Factors or Mathematical Models. *Fundam. Appl. Toxicol.* 4, S383–S394.

LaGoy, P. K. 1987. Estimated Soil Ingestion Rates for Use in Risk Assessment. *Risk Anal.* 7 (3), 355–359.

Larsen, R.J. and M.L. Marx. 1985. *An Introduction to Probability and its Applications.* Prentice-Hall, Englewood Cliffs, NJ.

Lave, L.B. (Ed.). 1982. *Quantitative Risk Assessment in Regulation.* Brookings Institute, Washington, D.C.

Lave, L.B. and A.C. Upton (Eds.). 1987. *Toxic Chemicals, Health, and the Environment.* Johns Hopkins University Press, Baltimore, MD.

Lee, J.A. 1985. *The Environment, Public Health, and Human Ecology — Considerations for Economic Development.* A World Bank Publ. Johns Hopkins University Press, Baltimore, MD.

Lees, F.B. 1980. *Loss Prevention in the Process Industries.* Vol. 1. Butterworths, Boston.

Lind, N.C., J.S. Nathwani and E. Siddall. 1991. *Managing Risks in the Public Interest.* Institute for Risk Research, University of Waterloo, Waterloo, Ontario.

Lindsay, W.L. 1979. *Chemical Equilibria in Soils.* Wiley-Interscience, New York.

Lippit, J., J. Walsh, M. Scott and A. DiPuccio. 1986. Cost of Remedial Actions at Uncontrolled Hazardous Waste Sites: Worker Health and Safety Considerations. Project Summary Report No. EPA/600/S2-86/037. U.S. Environmental Protection Agency, Washington, D.C.

Liptak, S.C., J.W. Atwater and D.S. Mavinic (Eds.). 1988. Proc. 1988 Joint CSCE-ASCE Natl. Conf. Environ. Eng. July 13–15, 1988, Vancouver, B.C., Canada.

Loehr, R.C. (Ed.). 1976. *Land as a Waste Management Alternative.* Ann Arbor Science, Ann Arbor, MI.

Long, W.L. 1990. "Economic Aspects of Transport and Disposal of Hazardous Wastes." *Mar. Policy Int. J.* 14, 3 (May, 1990) 198–204.

Long, F. A. and G. E. Schweitzer (Eds.). 1982. *Risk Assessment at Hazardous Waste Sites.* American Chemical Society, Washington, D.C.

Lowrance, W.W. 1976. *Of Acceptable Risk: Science and the Determination of Safety.* William Kaufman, Los Altos, CA.

Lu, F.C. 1985. *Basic Toxicology.* Hemisphere, Washington, D.C.

Lu, F.C. 1985. Safety Assessments of Chemicals with Threshold Effects. *Regul. Toxicol. Pharmacol.* 5, 121–32.

Lu, F.C. 1988. Acceptable Daily Intake: Inception, Evolution, and Application. *Regul. Toxicol. Pharmacol.* 8, 45–60.

MacCarthy, L.S. and D. Mackay. 1993. Enhancing ecotoxicological modeling and assessment. *Environ. Sci. Technol.* Vol. 27, No. 9, 1719–1728.

MacDonald, J.A. and B.E. Rittmann. 1993. Performance standards for in situ bioremediation. *Environ. Sci. Technol.* Vol. 27, No. 10, 1974–79.

Mahmood, R.J. and R.C. Sims. 1986. Mobility of organics in land treatment systems. *J. Environ. Eng.* 112, 2, 236–245.

Malina, J.F., Jr. (Ed.). 1989. *Environmental Engineering: Proceedings of the 1989 Specialty Conference,* Austin, TX (July 10–12, 1989). American Society of Chemical Engineers, New York.

Manahan, S. 1992. *Toxicological Chemistry.* Lewis Publishers/CRC Press, Boca Raton, FL.

Manahan, S.E. 1993. *Fundamentals of Environmental Chemistry.* Lewis Publishers/CRC Press, Boca Raton, FL.

Mansour, M. (Ed.). 1993. *Fate and Prediction of Environmental Chemicals in Soils, Plants, and Aquatic Systems.* Lewis Publishers/CRC Press, Boca Raton, FL.

Maurits la Riviere, J.W. 1989: "Threats to the World's Water." Managing Planet Earth. *Scientific American.* Sept. 1989, Special Issue.

Martin, E.J. and J.H. Johnson, Jr. (Ed.). 1987. *Hazardous Waste Management Engineering.* Van Nostrand Reinhold, New York.

Mathews, J.T. 1991. *Preserving the Global Environment: The Challenge of Shared Leadership.* W.W. Norton, New York.

McColl, R.S. (Ed.). 1987. *Environmental Health Risks: Assessment and Management.* University of Waterloo Press, Waterloo, Ontario, Canada.

McKone, T.E. and J.I. Daniels. 1991. Estimating human exposure through multiple pathways from air, water, and soil. *Regul. Toxicol. Pharmacol.* 13, 36–61.

McKone, T.E., W.E. Kastenberg and D. Okrent. 1983. The use of landscape chemical cycles for indexing the health risks of toxic elements and radionuclides. *Risk Anal.* Vol. 3, No. 3, 189–205.

Merck. 1989. *The Merck Index: An Encyclopedia of Chemicals, Drugs and Biologicals.* 11th ed. Merck & Co., Rahway, NJ.

Meyer, C.R. 1983. Liver Dysfunction in Residents Exposed to Leachate from a Toxic Waste Dump. *Environ. Health Perspect.* 48, 9–13.

Miller, D.W. (Ed.). 1980. *Waste Disposal Effects on Ground Water.* Premier Press, Berkeley, CA.

Mitsch, W.J. and S.E. Jorgesen (Eds.). 1989. *Ecological Engineering, An Introduction into Ecotechnology.* John Wiley & Sons, New York.

Monahan, D.J. 1990. "Estimation of Hazardous Wastes from Employment Statistics — Victoria, Australia." *Waste Manage. Res.* 8, 2, 145–9.

Moore, A.O. 1987. *Making Polluters Pay — A Citizen's Guide to Legal Action and Organizing.* Environmental Action Foundation, Washington, D.C.

Mulkey, L.A. 1984. Multimedia fate and transport models: an overview. *J. Toxicol. Clin. Toxicol.* 21, 1–2, 65–95.

Munro, I.C. and Krewski, D.R. 1981. Risk assessment and regulatory decision making. *Food Cosmet. Toxicol.* 19, 549–560.

Nathwani, J., N.C. Lind and E. Siddall. 1990. "Risk-Benefit Balancing in Risk Management: Measures of Benefits and Detriments." Presented at the Annu. Meet. Soc. Risk Analysis, 29th Oct.–1 Nov. 1989, San Francisco, CA.

Neely, W.B. 1980. *Chemicals in the Environment (Distribution, Transport, Fate, Analysis).* Marcel Dekker, New York.

Neely, W.B. 1994. *Introduction to Chemical Exposure and Risk Assessment.* Lewis Publishers/CRC Press, Boca Raton, FL.

NIOSH. 1982. Registry of Toxic Effects of Chemical Substances. R.L. Tatken and R.J. Lewis (Eds). U.S. Department of Health and Human Services (DHHS) National Institute for Occupational Safety and Health (NIOSH) Publ. No. 83-107, Cincinnati, OH.

Norris, R.D. et al., 1994. *Handbook of Bioremediation.* Lewis Publishers/CRC Press, Boca Raton, FL.

NRC. 1972. Specifications and Criteria for Biochemical Compounds. 3rd ed. National Academy of Sciences, Washington, D.C.

NRC. 1977. Drinking Water and Health, Vol. 1. National Academy Press, Washington, D.C.

NRC. 1977. Environmental Monitoring, Analytical Studies for the U.S. Environmental Protection Agency, Vol. IV. National Academy Press, Washington, D.C.

NRC. 1977. Drinking Water and Health. Safe Drinking Water Committee, Advisory Center on Toxicology, National Academy of Sciences, Washington, D.C.

NRC. 1980. Drinking Water and Health, Vol. 2. National Academy Press, Washington, D.C.

NRC. 1980. Drinking Water and Health, Vol. 3. National Academy Press, Washington, D.C.

NRC. 1981. Prudent Practices for Handling Hazardous Chemicals in Laboratories. National Academy Press, Washington, D.C.

NRC. 1982. Drinking Water and Health, Vol. 4. National Academy Press, Washington, D.C.

NRC. 1982. Risk and Decision-Making: Perspective and Research. Committee on Risk and Decision-Making. National Academy Press, Washington, D.C.

NRC. 1983. Drinking Water and Health, Vol. 5. National Academy Press, Washington, D.C.

NRC. 1988. Hazardous Waste Site Management: Water Quality Issues. National Academy Press, Washington, D.C.

NRC. 1989. Ground Water Models: Scientific and Regulatory Applications. National Academy Press, Washington, D.C.

NRC (National Research Council). 1993. In Situ Bioremediation: When Does it Work? National Academy Press, Washington, D.C.

NRC (National Research Council). 1994. Alternatives for Ground Water Cleanup. Committee on Ground Water Cleanup Alternatives. National Academy Press, Washington, D.C.

OECD. 1986. Existing Chemicals — Systematic Investigation, Priority Setting and Chemical Review. Organization for Economic Cooperation and Development, Paris.

OECD. 1986. Report of the OECD Workshop on Practical Approaches to the Assessment of Environmental Exposure, Organization for Economic Cooperation and Development, 14–18 April, 1986, Vienna.

O'Hare, M., L. Bacow and D. Sanderson. 1983. *Facility Siting and Public Opposition.* Van Nostrand Reinhold, New York.

Onishi, Y., A.R. Olsen, M.A. Parkhurst and G. Whelan. 1985. Computer-Based Environmental Exposure and Risk Assessment Methodology for Hazardous Materials. *J. Hazardous Mater.* 10, 389–417.

OSHA. 1980. Identification, Classification, and Regulation of Potential Occupational Carcinogens. *Federal Register*, 45, 5002–5296.

OTA. 1981. Assessment of Technologies for Determining Cancer Risks from the Environment. Office of Technology Assessment, Washington, D.C.

Ott, W.R. 1995. *Environmental Statistics and Data Analysis.* Lewis Publishers/CRC Press, Boca Raton, FL.

Overcash, M.R. and D. Pal. 1979. *Design of Land Treatment Systems for Industrial Wastes — Theory and Practice.* Ann Arbor Science, Ann Arbor, MI.

Paasivirta, J. 1991. *Chemical Ecotoxicology.* Lewis Publishers, Chelsea, MI.

Park, C.N. and R.D. Snee. 1983. Quantitative Risk Assessment: State of the Art for Carcinogenesis. *Am. Stat.* 37(4), 427–441.

Peck, Dennis L. (Ed.). 1989. *Psychosocial Effects of Hazardous Toxic Waste Disposal on Communities.* Charles C Thomas, Springfield, IL.

Pedersen, J. 1989. Public Perception of Risk Associated with the Siting of Hazardous Waste Treatment Facilities. European Foundation for the Improvement of Living and Working Conditions, Dublin, Eire.

Petak, W. J. and A. A. Atkisson. 1982. *Natural Hazard Risk Assessment and Public Policy: Anticipating the Unexpected.* Springer-Verlag, New York.

Pickering, Q.H. and C. Henderson. 1966. "The Acute Toxicity of Some Heavy Metals to Different Species of Warm Water Fish". *Air Water Pollut. Int. J.* 19–453.

Piddington K.W. 1989. "Sovereignty and the Environment." *Environment,* Vol. 31, No. 7, Sept, 1989, pp. 18–20, 35–39.

Postel S. 1988. *Controlling Toxic Chemicals.* State of the World, 1988. Worldwatch Institute, New York.

Prins, G. and R. Stamp. 1991. *Top Guns & Toxic Whales: The Environment & Global Security.* Earthscan Publications, London.

Rail, C.C. 1989. *Groundwater Contamination: Sources, Control and Preventive Measures.* Technomic Publishing , Lancaster, PA.

Ramamoorthy, S. and E. Baddaloo. 1991. Evaluation of Environmental Data for Regulatory and Impact Assessment. *Studies in Environmental Science 41,* Elsevier, Amsterdam.

Ray, D.L. 1990. *Trashing the Planet.* Regnery Gateway, Washington, D.C.

Ricci, P.F. (Ed.). 1985. *Principles of Health Risk Assessment.* Prentice-Hall, Englewood Cliffs, N.J.

Ricci, P.F. and M.D. Rowe (Eds.). 1985. *Health and Environmental Risk Assessment,* Pergamon Press. New York.

Rodricks, J.V. 1984. Risk assessment at hazardous waste disposal sites. *Hazardous Waste Hazardous Mater.,* 1, 3, 333–362.

Rodricks, J.V. and R.G. Tardiff (Eds.). 1984. *Assessment and Management of Chemical Risks.* ACS Symp. Ser. 239, American Chemical Society, Washington, D.C.

Rodricks, J. and M.R. Taylor. 1983. Application of risk assessment to food safety decision making. *Regul. Toxicol. Pharmacol.* 3, 275–307.

Rowe, W.D. 1983. *Evaluation Methods for Environmental Standards.* CRC Press, Boca Raton, FL.

Rowland, A.J. and P. Cooper. 1983. *Environment and Health.* Edward Arnold, London.

Royal Society of London. 1983. *Risk Assessment: A Study Group Report.* The Royal Society, London.

Ruckelshaus, W.D. 1985. Risk, Science, and Democracy. *Issues in Science & Technology,* Spring, 1985, pp. 19–38.

Russell, M. and M. Gruber. 1987. "Risk Assessment in Environmental Policy-Making." *Science,* Vol. 236, 286–90.

Sara, M.N. 1993. *Standard Handbook of Site Assessment for Solid and Hazardous Waste Facilities.* Lewis Publishers/CRC Press, Boca Raton, FL.

Sax, N.I. 1979. *Dangerous Properties of Industrial Materials,* 5th ed. Van Nostrand Reinhold, New York.

Saxena, J. and F. Fisher (Eds.). 1981. *Hazard Assessment of Chemicals.* Academic Press, New York.

Schleicher, K. (Ed.). 1992. *Pollution Knows No Frontiers: A Reader.* Paragon House Publishers, New York.

Schroeder, R.L. 1985. Habitat Suitability Index Models: Northern Bobwhite. U.S. Department of Interior, Fish and Wildlife Service, Washington, D.C.

Schwartz, S.I. and W.B. Pratt. 1990. *Hazardous Waste from Small Quantity Generators — Strategies and Solutions for Business and Government.* Island Press, Washington, D.C.

Schwing, R.C. and W.A. Albers, Jr. (Eds.). 1980. *Societal Risk Assessment: How Safe is Safe Enough.* Plenum Press, New York.

Searle, C.E., (Ed.) 1976. *Chemical Carcinogens.* ACS Monogr. 173. American Chemical Society, Washington, D.C.

Sebek, V. (Ed.). 1990. "Maritime Transport, Control and Disposal of Hazardous Waste." Spec. Iss. *Mar. Policy Int. J.* Vol. 14, No. 3, May, 1990.

Sedman, R.M. 1989. The Development of Applied Action Levels for Soil Contact: A Scenario for the Exposure of Humans to Soil in a Residential Setting. *Environ. Health Perspect.* Vol. 79, pp. 291–313.

Shrader-Frechette, K.S. 1985. *Risk Analysis and Scientific Method.* D. Reidel Publishing, Boston, MA.

Sims, R.C. and J.L. Sims. 1986. Cleanup of contaminated soils. In: *Utilization, Treatment, and Disposal of Waste on Land*, pp. 257-278. Soil Science Society of America, Madison, WI, pp. 257–278.

Sitnig, M. 1985. *Handbook of Toxic and Hazardous Chemicals and Carcinogens.* Noyes Data Corp., Park Ridge, NJ.

Smith, A.H. 1987. Infant Exposure Assessment for Breast Milk Dioxins and Furans Derived from Waste Incineration Emissions. *Risk Anal.* 7, 347–53.

Spencer, E.Y. 1982. Guide to the Chemicals Used in Crop Protection, 7th ed. Research Institute, Agriculture Canada. Publ. No. 1093. Information Canada, Ottawa.

Splitstone, D.E. 1991. "How clean is clean .... statistically?" *Pollution Engineering,* pp. 90–96.

Sposito, G., C.S. LeVesque, J.P. LeClaire and N. Sensi. 1984. Methodologies to predict the mobility and availability of hazardous metals in sludge-amended soils. California Water Resource Center, Contrib. No. 189. University of California, Los Angeles.

Starr, C., R. Rudman and C. Whipple. 1976. Philosophical Basis for Risk Analysis. Annu. Rev. Energy, 1, 629–662.

States, J.B., P.T. Hang, T.B. Shoemaker, L.W. Reed and E.B. Reed. 1978. A System Approach to Ecological Baseline Studies. FQS/DBS-78/21. USFWS. Washington, D.C.

Suess, M.J. and J.W. Huismans (Eds.). 1983. Management of Hazardous Wastes: Policy Guidelines and Code of Practice. WHO Regional Publ. Eur. Ser. No. 14, World Health Organization, Geneva.

Tardiff, R.G. and J.V. Rodricks (Eds.). 1987. *Toxic Substances and Human Risk.* Plenum Press, New York, 445 pp.

Tasca, J.J., M.F. Saunders and R.S. Prann. 1989. Terrestrial Food-Chain Model for Risk Assessment. In Superfund '89: Proc. 10th Natl. Conf.

Testa, S.M. 1994. *Geological Aspects of Hazardous Waste Management.* Lewis Publishers/ CRC Press, Boca Raton, FL.

Testa, S.M. and D.L. Winegardner. 1990. *Restoration of Petroleum-Contaminated Aquifers.* Lewis Publishers/CRC Press, Boca Raton, FL.

The Conservation Foundation. 1985. *Risk Assessment and Risk Control.* The Conservation Foundation, Washington, D.C.

Theiss, J.C. 1983. The ranking of chemicals for carcinogenic potency. *Regul. Toxicol. Pharmacol.*, 3, 320–328.

Theodore, L., J.P. Reynolds and F.B. Taylor. 1989. *Accident and Emergency Management.* Wiley-Interscience, New York.

The World Bank. 1985. *Manual of Industrial Hazard Assessment Techniques.* Office of Environment and Scientific Affairs, Washington, D.C., October, 1985.

Thibodeaux, L.J. 1979. *Chemodynamics: Environmental Movement of Chemicals in Air, Water and Soil.* John Wiley & Sons, New York.

Tolba, M.K. (Ed.). 1988. Evolving Environmental Perceptions: From Stockholm to Nairobi. United Nations Environment Programme, UNEP, Nairobi.

Tolba, M.K. 1990. "The Global Agenda and the Hazardous Waste Challenge." *Mar. Policy Int. J.* 14, 3 (May), 205–9.

Travis, C.T. and A.D. Arms. 1988. Bioconcentration of organics in beef, milk, and vegetation. *Environ. Sci. Technol.* 22, 271.

Travis, C. C. and H.A. Hattemer-Frey. 1988. Determining an Acceptable Level of Risk. *Environ. Sci. Technol.* 22 (8).

USBR. 1986. Guidelines to Decision Analysis. ACER Techn. Memo. No.7, U.S. Bureau of Reclamation, Denver, CO.

USEPA. 1982. Test methods for evaluating solid waste: Physical/chemical methods, 1st ed., SW-846. U.S. Environmental Protection Agency, Washington, D.C.

USEPA. 1983. Hazardous waste land treatment. Rev. ed. SW-8974. U.S. Environmental Protection Agency, Cincinnati, OH.

USEPA. 1984. Approaches to risk assessment for multiple chemical exposures. Environmental Criteria and Assessment Office, U.S. Environmental Protection Agency, EPA-600/9-84-008. Cincinnati, OH.

USEPA. 1984. Proposed Guidelines for Carcinogen, Mutagenicity, and Developmental Toxicant Risk Assessment. *Federal Register*, 49, 46294–46331.

USEPA. 1985. Chemical, Physical, and Biological Properties of Compounds Present at Hazardous Waste Sites. Report Prepared by Clement Associates for the USEPA, U.S. Environmental Protection Agency, Washington, D.C.

USEPA. 1985. Development of Statistical Distribution or Ranges of Standard Factors Used in Exposure Assessments. Office of Health and Environmental Assessment, U.S. Environmental Protection Agency, Washington, D.C.

USEPA. 1985. Practical Guide to Ground-Water Sampling. Robert S. Kerr Environmental Research Lab., Office of Research and Development, EPA/600/2-85/104. U.S. Environmental Protection Agency, Ada, OK.

USEPA. 1985. Rapid Assessment of Exposure to Particulate Emissions From Surface Contamination Sites, EPA/600/8-85/002, NTIS PB85-192219. Office of Health and Environmental Assessment, U.S. Environmental Protection Agency, Washington, D.C.

USEPA. 1986. Ecological Risk Assessment. Hazard Evaluation Division Standard Evaluation Procedure. U.S. Environmental Protection Agency, Washington, D.C.

USEPA. 1986. Guidelines for Carcinogen Risk Assessment. *Federal Register* 51(185): 33992-34003, CFR 2984, September 24, 1986.

USEPA. 1986. Methods for Assessing Exposure to Chemical Substances. Vol. 8. Methods for Assessing Environmental Pathways of Food Contamination. Exposure Evaluation Division, Office of Toxic Substances. EPA 560/5-85-008, U.S. Environmental Protection Agency, Washington, D.C.

USEPA. 1986. Registry of Toxic Effects of Chemical Substances. U.S. Environmental Protection Agency, Research Triangle Park, NC.

USEPA. 1986. Superfund Public Health Evaluation Manual. EPA/540/1-86/060. Office of Emergency and Remedial Response, U.S. Environmental Protection Agency, Washington, D.C.

USEPA. 1986. Superfund Risk Assessment Information Directory. EPA/540/1-86/061. Office of Emergency and Remedial Response, U.S. Environmental Protection Agency, Washington, D.C.

USEPA. 1987. Data Quality Objectives for Remedial Response Activities: Example Scenario. EPA/540/G-87/004. U.S. Environmental Protection Agency, Washington, D.C.

USEPA. 1987. Handbook for Conducting Endangerment Assessments. U.S. Environmental Protection Agency, Research Triangle Park, NC.

USEPA. 1987. Quality Assurance Program Plan. EPA/600/X-87/241. Quality assurance Management Staff, U.S. Environmental Protection Agency, Las Vegas, NV.

USEPA. 1987. Technical Guidance for Hazard Analysis. U.S. Environmental Protection Agency, Washington, D.C.

USEPA. 1988. A Workbook of Screening Techniques for Assessing Impacts of Toxic Air Pollutants. EPA-450/4-88-009. Office of Air Quality Planning and Standards, U.S. Environmental Protection Agency, Research Triangle Park, NC.

USEPA. 1988. Estimating Toxicity of Industrial Chemicals to Aquatic Organisms Using Structure Activity Relationships. EPA/560/6-88/001. Office of Toxic Substances, U.S. Environmental Protection Agency, Washington, D.C.

USEPA. 1988. Review of Ecological Risk Assessment Methods. Office of Policy, Planning and Evaluation, U.S. Environmental Protection Agency, Washington, D.C.

USEPA. 1989. Ecological Assessments of Hazardous Waste Sites: A Field and Laboratory Reference Document. Office of Research and Development — Corvallis Environmental Research Laboratory, EPA/600/3-89/013. U.S. Environmental Protection Agency, Corvallis, OR.

USEPA. 1989. Estimating Air Emissions from Petroleum UST Cleanups. Office of Underground Storage Tanks, U.S. Environmental Protection Agency, Washington, D.C.

USEPA. 1989. Exposure Assessment Methods Handbook. Office of Health and Environmental Assessment, U.S. Environmental Protection Agency, Cincinnati, Ohio.

USEPA. 1989. Methods for Evaluating the Attainment of Cleanup Standards. Vol. I. Soils and Solid Media. EPA/230/2-89/042. Office of Policy, Planning and Evaluation, U.S. Environmental Protection Agency, Washington, D.C.

USEPA. 1990. Emission Factors for Superfund Remediation Technologies. Draft. Office of Air Quality Planning and Standards, U.S. Environmental Protection Agency, Research Triangle Park, NC.

USEPA. 1990. Hazard Ranking System. *Federal Register,* Vol. 55, No. 241, EPA, 40 CFR Part 300, pp. 51532–51666. December 14, 1990.

USEPA. 1991. Guidance for Performing Site Inspections Under CERCLA — Interim Version. Draft Publication, OSWER Directive 9345.1-06, Office of Emergency and Remedial Response, U.S. Environmental Protection Agency, Washington, D.C.

USEPA. 1991. Risk Assessment Guidance for Superfund. Vol. I. Human Health Evaluation Manual (Part B, Development of Risk-based Preliminary Remediation Goals). PB92-963333, OSWER Directive: 9285.7-01B, Interim. Office of Emergency and Remedial Response, U.S. Environmental Protection Agency, Washington, D.C.

USEPA. 1991. Risk Assessment Guidance for Superfund. Vol. I. Human Health Evaluation Manual (Part C, Risk Evaluation of Remedial Alternatives). PB90-155581, OSWER Directive: 9285.7-01C, Interim. Office of Emergency and Remedial Response, U.S. Environmental Protection Agency, Washington, D.C.

USEPA. 1991. The Role of Baseline Risk Assessment in Superfund Remedy Selection Decisions. Office of Solid Waste and Emergency Response. OSWER Directive: 9355.0-30. U.S. Environmental Protection Agency, Washington, D.C.

USEPA. 1992. Framework for Ecological Risk Assessment. EPA/630/R-92/001, U.S. Environmental Protection Agency, Washington, D.C.

Van Ryzin, J. 1980. Quantitative risk assessment. *J. Occup. Med.*, 22, 321–326.

Verschueren, K. 1983. *Handbook of Environmental Data on Organic Chemicals*, 2nd ed. Van Nostrand Reinhold, New York.

Volpp, C. 1988. "Is It Safe or Isn't It ?": An Overview of Risk Assessment In *Water Resource News,* Department of Environmental Protection, Division of Water Resources, Trenton, NJ.

Weast, R.C. (Ed.) 1984. *Handbook of Chemistry and Physics*, 65th ed. CRC Press, Boca Raton, FL.

WHO. 1983. Management of Hazardous Waste. WHO Regional Eur. Ser. No. 14, World Health Organization, Geneva.

Wilson, R. and E. A. C. Crouch. 1987. Risk Assessment and Comparisons: An Introduction. *Science,* 236–267.

Wilson, L.G., L.G. Everett and S.J. Cullen (Eds.). 1995. *Handbook of Vadose Zone Characterization and Monitoring.* Lewis Publishers/CRC Press, Boca Raton, FL.

Whipple, C. 1987. De Minimis Risk. *Contemporary Issues in Risk Analysis,* Vol. 2. Plenum Press, New York.

Whipple, W., Jr. 1994. *New Perspectives in Water Supply.* Lewis Publishers, Boca Raton, FL.

Worster, D. 1993. *The Wealth of Nature: Environmental History and the Ecological Imagination.* Oxford University Press, New York.

Yakowitz, H. 1989. Monitoring and Control of Transfrontier Movements of Hazardous Wastes: An International Overview. Technical Paper W/0587M, OECD, Paris.

Yakowitz, H. 1990: "Monitoring and Control of Transfrontier Movements of Hazardous Wastes: An International Overview." In, *The Management of Hazardous Substances in the Environment.* K.L. Zirm and J. Mayer, Eds. Elsevier, London, pp. 139–162.

Yang, J.T. and W.E. Bye. 1979. A guidance for protection of ground water resources from the effects of accidental spills of hydrocarbons and other hazardous substances. EPA-570/9-79-017, U.S. Environmental Protection Agency, Washington, D.C.

Zirm, K.L. and J. Mayer. (Eds.). 1990. *The Management of Hazardous Substances in the Environment.* Elsevier, London.

Zogg, H.A. 1987. *"Zurich" Hazard Analysis.* Zurich Insurance Group, Risk Engineering, Zurich, Switzerland.

APPENDIX A
LIST OF SELECTED
ABBREVIATIONS AND ACRONYMS

ACL	alternate concentration limit
ADD	average daily dose
ADI	acceptable daily intake (or allowable daily intake)
AL	action level
ANOVA	analysis-of-variance
APCD	air pollution control device
APL	aqueous phase liquid
ARAR	applicable or relevant and appropriate requirement
ASC	acceptable soil concentration
AWQC	ambient water quality criteria
BCF	bioconcentration (or bioaccumulation) factor
B(D)AT	best (demonstrated) available technology
BTEX	benzene, toluene, ethylbenzene, and xylene
CAA	Clean Air Act
CAG	Carcinogen Assessment Group
CAP	corrective action plan
CAS	Chemical Abstracts Service
CDI	chronic daily intake
CERCLA	Comprehensive Environmental Response, Compensation, and Liability Act (also, Superfund)
CFR	Code of Federal Regulations
COC	chemical of potential concern
COEC	chemical of potential ecological concern
CPF	carcinogenic potency factor (see also, SF)
CSM	conceptual site model
CWA	Clean Water Act
DDT	dichlorodiphenyl trichloroethane
DL	detection limit (of analyte)
DQO	data quality objective
EA	endangerment assessment
EPA	Environmental Protection Agency
ERA	environmental (or ecological) risk assessment
ESA	Endangered Species Act
EQ	ecological quotient

FIFRA	Federal Insecticide, Fungicide, and Rodenticide Act
FS	feasibility study
FSP	field sampling plan
FWPCA	Federal Water Pollution Control Act
GAC	granular activated carbon
GC/PID	gas chromatograph/photoionization detector
HEAST	Health Effects Assessment Summary Tables (USEPA)
HI	hazard index
HMTA	Hazardous Materials Transport Act
HQ	hazard quotient
HRS	hazard ranking system
HSP	health and safety plan
HSWA	Hazardous and Solid Waste Amendments (of 1984, to RCRA)
IARC	International Agency for Research on Cancer
IDL	instrument detection limit
IDW	investigation-derived waste
IRIS	Integrated Risk Information System
IRPTC	International Register of Potentially Toxic Chemicals
ISV	*in situ* vitrification
LADD	lifetime average daily dose
LC_{50}	mean lethal concentration
LD_{50}	mean lethal dose
LOAEL	lowest observed adverse effect level
LOEL	lowest observed effect level
MCL	maximum contaminant level
MCLG	maximum contaminant level goal
MDD	maximum daily dose
MDL	method detection limit
MEL	maximum exposure level
MSW	municipal solid waste
(M)USLE	(modified) universal soil loss equation
NAAQS	National Ambient Air Quality Standard
NAPL	nonaqueous phase liquid
NCP	National Contingency Plan
ND	non-detect (for analytical results)
NESHAP	National Emission Standards for Hazardous Air Pollutants
NFA	no further action
NFRAD	no-further-response-action decision
NOAEL	no observable adverse effect level
NOEL	no observable effect level
NPDES	National Pollutant Discharge Elimination System
NPL	National Priorities List
NTIS	National Technical Information Service
OHEA	Office of Health and Environmental Assessment (USEPA)
OSHA	Occupational Safety and Health Administration
OSWER	Office of Solid Waste and Emergency Response (USEPA)
OTA	Office of Technology Assessment (Congress of the United States)
OTS	Office of Toxic Substances (USEPA)
OVA/GC	organic vapor analyzer/gas chromatograph
PA	preliminary assessment

PAH	polyaromatic hydrocarbon
PAR	population-at-risk
PCB	polychlorinated biphenyl
PCE	perchloroethylene
PCP	pentachlorophenol
PEL	permissible exposure limit
ppb	parts per billion
ppm	parts per million
PQL	practical quantitation limit (see also, SQL)
PRG	preliminary remediation goal
PRP	potentially responsible party
QAPP	quality assurance project plan
QA/QC	quality assurance/quality control
RAP	remedial action plan
RBAG	risk-based action goal
RBC	risk-based concentration
RCRA	Resource Conservation and Recovery Act
RFA	RCRA facility assessment
RfC	reference concentration
RfD	reference dose
RfD_s	subchronic reference dose
RFI	RCRA facility investigation
RI	remedial investigation
RI/FS	remedial investigation/feasibility study
RMCL	recommended maximum contaminant level (renamed, MCLG)
RME	reasonable maximum exposure
RSD	risk-specific dose
RSCL	recommended site cleanup limit
RTECS	Registry of Toxic Effects of Chemical Substances
SAP	sampling and analysis plan
SARA	Superfund Amendments and Reauthorization Act
SDI	subchronic daily intake
SDWA	Safe Drinking Water Act
SF	slope factor (see also, CPF)
SI	site inspection
SQL	sample quantitation limit (see also, PQL)
SVE	soil vapor extraction
SWMU	solid waste management unit
TBC	to-be-considered (as a nonpromulgated advisory or guidance material)
TCE	trichloroethylene (or trichloroethene)
TLV	threshold limit value
TPH	total petroleum hydrocarbon
TSCA	Toxic Substances Control Act
TSDF	treatment, storage, and disposal facility
UCL	upper confidence level
UCR	unit cancer risk
UF	uncertainty factor (also, safety factor)
UR	unit risk
USEPA	United States Environmental Protection Agency

UST	underground storage tank
VES	vapor extraction system
VOC	volatile organic compound/chemical
VSD	virtually safe dose
WHO	World Health Organization
WQA	Water Quality Act
WQC	Water Quality Criteria (with respect to CWA)

Absorption — The transport of a substance through the outer boundary of a medium. Oftentimes used to refer to the uptake of a chemical by a cell or an organism, including the flow into the bloodstream following exposure through the skin, lungs, and/or gastrointestinal tract.

Absorbed dose — The amount of a chemical substance actually entering an exposed organism via the lungs (for inhalation exposures), the gastrointestinal tract (for ingestion exposures), and/or the skin (for dermal exposures). It represents the amount penetrating the exchange boundaries of the organism after contact. It is calculated from the intake and the absorption efficiency, expressed in mg/kg-day.

Absorption factor — The percent or fraction of a chemical in contact with an organism that becomes absorbed into the receptor.

Acceptable daily intake (ADI) — An estimate of the maximum amount of a chemical (in mg/kg body weight/day) to which a potential receptor can be exposed on a daily basis over an extended period of time — usually a lifetime — without suffering a deleterious effect, or without anticipating an adverse effect.

Acceptable risk — A risk level generally deemed by society to be acceptable.

Action level (AL) — The level of a chemical in selected media of concern above which there are potential adverse health and/or environmental effects. It represents the contaminant concentration above which some corrective action (e.g., monitoring or remediation) is required by regulation.

Activated carbon — A highly adsorbent form of carbon used to remove contaminants from fluidal emissions or discharges. It is a special form of carbon, often derived from charcoal, and treated to make it capable of adsorbing and retaining certain chemical substances.

Activated carbon adsorption — A treatment technology based on the principle that certain organic constituents preferentially adsorb to organic carbon.

Acute exposure — A single large exposure or dose to a chemical, generally occurring over a short period (usually 24 to 96 h).

Acute toxicity — The development of symptoms of poisoning or the occurrence of adverse health effects after exposure to a single dose or multiple doses of a chemical within a short period of time.

Adsorption — The removal of contaminants from a fluid stream by concentration of the constituents onto a solid material. It is the physical process of attracting and holding molecules of other chemical substances on the surface of a solid, usually by the formation of chemical bonds. A substance is said to be adsorbed if the concentration in the boundary region of a solid (e.g., soil) particle is greater than in the interior of the contiguous phase.

Air sparging — The process of blowing air through a liquid for mixing purposes in order to strip volatile materials (i.e., VOCs) or to add oxygen. Usually refers to the highly controlled injection of air into a contaminant plume in the soil *saturated* zone.

Air stripping — A remediation technique that involves the physical removal of dissolved-phase contamination from a water stream.

Aliphatic compounds — Organic compounds in which the carbon atoms exist as either straight or branched chains; examples include pentane, hexane, and octane.

Antagonism — The interference or inhibition of the effects of one chemical substance by the action of other chemicals.

Aquifer — A geological formation, group of formations, or part of a formation which is capable of yielding significant/usable quantities of groundwater to wells and/or springs.

Arithmetic mean (also, Average) — A measure of central tendency for data from a normal distribution, defined for a set of n values by the sum of values divided by n:

$$X_m = \frac{\sum_{i=1}^{n} X_i}{n}$$

Aromatic compounds — Organic compounds that contain carbon molecular ring structures; examples include benzene, toluene, ethylbenzene, and xylenes. These compounds are reasonably soluble, volatile, and mobile in the subsurface environment, and are a very useful indicator of contaminant migration.

Asphalt batching — Also referred to as *asphalt incorporation*, is a method for treating hydrocarbon-contaminated soils. It is a relatively new remedial technique that involves the incorporation of petroleum-laden soils into hot asphalt mixes as a partial substitute for stone aggregate. This mixture can then be utilized for pavings.

Attenuation — Any decrease in the amount or concentration of a pollutant in an environmental matrix as it moves in time and space. It is the reduction or removal of contaminant constituents by a combination of physical, chemical, and/or biological factors acting upon the contaminated media.

Average concentration — A mathematical average of contaminant concentration(s) from more than one sample, typically represented by the arithmetic mean or the geometric mean for environmental samples.

Average daily dose (ADD) — The average dose calculated for the duration of exposure, and used to estimate risks for chronic noncarcinogenic effects of environmental contaminants. This is defined by:

$$\text{ADD (mg/kg-day)} = \frac{\text{contaminant concentration} \times \text{contact rate}}{\text{body weight}}$$

Background level — The normal ambient environmental concentration of a chemical constituent. It may include both naturally occurring concentrations and elevated levels resulting from non-site-related human activities.

Benchmark risk — A threshold level of risk, typically prescribed by regulations, above which corrective measures will almost certainly have to be implemented to mitigate the risks.

Bioaccumulation — The retention and concentration of a chemical by an organism. It is a build-up of a chemical in a living organism, which occurs when the organism takes in more of the chemical than it can rid itself of in the same length of time and stores the chemical in its tissue, etc.

Bioassay — Measuring the effect(s) of environmental exposures by intentional exposure of living organisms to a chemical.

Bioaugmentation — A process in which specially selected bacteria cultures that are predisposed to metabolize some target compound(s) are added to impacted media, along with the nutrient materials, to encourage degradation of the contaminants of concern.

Bioconcentration — The accumulation of a chemical substance in tissues of organisms (such as fish) to levels greater than levels in the surrounding media (such as water) for the organism's habitat; often used synonymously with bioaccumulation.

Bioconcentration factor (BCF) — A measure of the amount of selected chemical substances that accumulates in humans or in biota. It is the ratio of the concentration of a chemical substance in an organism at equilibrium to the concentration of the substance in the surrounding environmental medium.

Biodegradable — Capable of being metabolized by a biologic process or an organism.

Biodegradation — Decomposition of a substance into simpler substances by the action of microorganisms, usually in soil. It may or may not detoxify the material which is decomposed.

Biomagnification — The serial accumulation of a chemical by organisms in the food chain, with higher concentrations occurring at each successive trophic level.

Bioremediation — Also called *biorestoration*, is a viable and cost-effective remediation technique for treating a wide variety of contaminants (such as petroleum and aromatic hydrocarbons, chlorinated solvents, and pesticides). It relies on microorganisms (especially bacteria and fungi) to transform hazardous compounds found in environmental matrices into innocuous or less toxic metabolic products. *Natural in situ bioremediation* involves the attenuation of contaminants by indigenous (native) microorganisms without any manipulation.

Biostimulation — A process whereby the addition of selected amounts of nutrient materials stimulate or encourage the growth of the indigenous bacteria in soil, resulting in the degradation of some target contaminant(s).

Biota — All living organisms which are found within a prescribed volume or space.

Cancer — A disease characterized by malignant, uncontrolled invasive growth of body tissue cells. It refers to the development of a malignant tumor or abnormal formation of tissue.

Cancer potency factor — Health effect information factor commonly used to evaluate health hazard potentials for carcinogens. It is usually represented by the cancer slope factor.

Capillary zone — The unsaturated area between the ground surface and the water table.

Carcinogen — A chemical or substance capable of producing cancer in living organisms.

Carcinogen Assessment Group (CAG) — A group within the U.S. EPA responsible for the evaluation of carcinogen bioassay results and also estimates of the carcinogenic potency of various chemicals.

Carcinogenic — Capable of causing, and tending to produce or incite cancer in living organisms.

Carcinogenicity — The ability of a chemical to cause cancer in a living organism.

Chronic — Of long-term duration.

Chronic daily intake (CDI) — The receptor exposure, expressed in mg/kg-day, averaged over a long period of time.

Chronic exposure — The long-term, usually low-level exposure to chemicals, i.e., the repeated exposure or doses to a chemical over a long period of time. It may cause latent damage that does not appear until a later period in time.

Chronic toxicity — The occurrence of symptoms, diseases, or other adverse health effects that develop and persist over time, after exposure to a single dose or multiple doses of a chemical delivered over a relatively long period of time.

Cleanup — Actions taken to abate the situation involving the release or threat of release of contaminants that could potentially affect human health and/or the environment. This typically involves a process to remove or attenuate contamination levels, in order to restore the impacted media to a useable state.

Cleanup level — The contaminant concentration goal of a remedial action, i.e., the media contaminant level to be attained through a remedial action.

Closure — All activities involved in taking a hazardous waste facility out of service and securing it for the time required by applicable regulations and laws. Site closures follow the implementation of appropriate site restoration and monitoring programs.

Confidence interval (CI) — A statistical parameter used to specify a range and the probability that an uncertain quantity falls within this range.

Confidence limits — The upper and lower boundary values of a range of statistical probability numbers.

Confidence limits, 95 percent (95% CL) — The limits of the range of values within which a single evaluation/analysis will be included 95% of the time. For large samples (i.e., n >30),

$$95\% \ CL \ = \ X_m \ \pm \ \frac{1.96\sigma}{n^{0.5}}$$

where CL is the confidence level, and σ is the estimate of the standard deviation of the mean (X_m). For a limited number of samples (n $\leq$30), a confidence limit or confidence interval may be be estimated from

$$CL \ = \ X_m \ \pm \ \frac{ts}{n^{0.5}}$$

where t is the value of the student *t*-distribution (refer to standard statistical texts) for the desired confidence level and degrees of freedom, $(n - 1)$.

Consequence — The impacts resulting from a receptor response due to specified exposures, or loading or stress conditions.

Contaminant — Any physical, chemical, biological, or radiological material that can potentially have adverse impacts on environmental media, or that can adversely impact public health and the environment. It represents any undesirable substance/material that normally is not present in the environmental media of concern.

Contaminant migration — The movement of a contaminant from its source through other matrices/media such as air, water, or soil. A *contaminant migration pathway* is the path taken by the contaminants as they travel from the contaminated site through various environmental media.

Contaminant plume — A body of contaminated groundwater or vapor originating from a specific source and spreading out due to influences of factors such as local groundwater conditions or soil vapor flow patterns. It represents the volume of groundwater or vapor that contains the contaminants released from a pollution source.

Contaminant release — The ability of a contaminant to enter into other environmental media/matrices (e.g., air, water or soil) from its place/point of origin.

Corrective action — Action taken to correct a problematic situation. A typical example involves the remediation of chemical contamination in soil and groundwater.

Cost-effective alternative — The most cost-effective alternative is the lowest cost alternative that is technologically feasible and reliable, and which effectively mitigates and minimizes environmental damage. It generally provides adequate protection of public health, welfare, or the environment.

Data quality objectives (DQOs) — Qualitative and quantitative statements developed by analysts to specify the quality of data that, at a minimum, is needed and expected from a particular data collection activity (or site characterization activity). It is determined based on the end use of the data to be collected.

Decision analysis — A process of systematic evaluation of alternative solutions to a problem where the decision is made under uncertainty. The approach is comprised of a conceptual and systematic procedure for analyzing complex sets of alternative solutions in a rational manner in order to improve the overall performance of a decision-making process.

Decision framework — A management tool designed to facilitate rational decision making on environmental contamination problems.

Degradation — The physical, chemical, or biological breakdown of a complex compound into simpler compounds and byproducts.

de Minimus — A legal doctrine dealing with levels associated with insignificant vs. significant issues relating to human exposures to chemicals that present very low risk. It is the level below which one need not be concerned.

Dermal exposure — Exposure of an organism or receptor through skin absorption.

Detection limit — The minimum concentration or weight of analyte that can be detected by a single measurement with a known confidence level. *Instrument detection limit (IDL)* represents the lowest amount that can be distinguished from the normal "noise" of an analytical instrument, i.e., the smallest amount of a chemical detectable by an analytical instrument under ideal conditions. *Method detection limit (MDL)* represents the lowest amount that can be distinguished from the normal "noise" of an analytical method, i.e., the smallest amount of a chemical detectable by a prescribed or specified method of analysis.

Diffusion — The migration of molecules, atoms, or ions from one fluid to another in a direction tending to equalize concentrations.

Dissolved product — the water-soluble fuel components of contaminant releases.

Dose — The amount of a chemical taken in by potential receptors on exposure. It is a measure of the amount of the substance received by the receptor, whether human or animal, as a result of exposure, expressed as an amount of exposure (in mg) per unit body weight of the receptor (in kg).

Dose-response — The quantitative relationship between the dose of a chemical and an effect caused by exposure to such substance.

Dose-response curve — A graphical representation of the relationship between the degree of exposure to a chemical substance and the observed or predicted biological effects or response.

Dose-response evaluation — The process of quantitatively evaluating toxicity information and characterizing the relationship between the dose of a chemical administered or received and the incidence of adverse health effects in the exposed population.

Dump — A site used for the disposal of solid wastes without environmental controls or safeguards.

Ecosystem — The interacting system of a biological community and its abiotic (i.e., nonliving) environment.

Ecotoxicity assessment — The measurement of effects of environmental toxicants on indigenous populations of organisms.

Effect (local) — The response produced from a chemical contact with an exposed receptor that occurs at the site of first contact.

Effect (systemic) — The response produced from a chemical contact with an exposed receptor that requires absorption and distribution of the chemical and tends to affect the receptor at sites away from the entry point(s).

Effective porosity — The ratio of the volume of interconnected voids through which fluid can flow to the total volume of material.

Endangerment assessment — A site-specific risk assessment of the actual or potential danger to human health and welfare, and also the environment, from the release of hazardous chemicals into various environmental media.

Endpoint — A biological effect used as index of the impacts of a chemical on an organism.

Environmental fate — The ultimate and intermediary destinies of a chemical after release or escape into the environment, and following transport through various environmental compartments. It is the movement of a chemical through the environment by transport in air, water, and soil culminating in exposures to living organisms. It represents the disposition of a material in various environmental compartments (e.g., soil, sediment, water, air, biota) as result of transport, transformation, and degradation.

Event tree analysis — A procedure that uses deductive logic to evaluate series of events which lead to an upset or accident scenario.

Exposure — The situation of receiving a dose of a chemical substance (or physical agent), or coming in contact with a hazard. It represents the contact of an organism with a chemical or physical agent available at the exchange boundary (e.g., lungs, gut, skin) during a specified time period.

Exposure assessment — The qualitative or quantitative estimation, or the measurement, of the dose or amount of a chemical to which potential receptors have been exposed, or could potentially be exposed. It comprises the determination of the magnitude, frequency, duration, route, and extent of exposure (to the chemicals or hazards of potential concern).

Exposure conditions — Factors (such as location, time, etc.) that may have significant effects on an exposed population's response to a hazard situation.

Exposure duration — The length of time that a potential receptor is exposed to the contaminants of concern in a defined exposure scenario.

Exposure frequency — The number of times (per year or per event) that a potential receptor would be exposed to site contaminants in a defined exposure scenario.

Exposure parameters — Variables used in the calculation of intake (e.g., exposure duration, inhalation rate, average body weight).

Exposure pathway — The course a chemical or physical agent takes from a source to an exposed population or organism. It describes a unique mechanism by which an individual or population is exposed to chemicals or physical agents at or originating from a contaminated site.

Exposure point — A location of potential contact between an organism and a chemical or physical agent.

Exposure route — The avenue by which an organism contacts a chemical, such as inhalation, ingestion, or dermal contact.

Exposure scenario — A set of conditions or assumptions about sources, exposure pathways, concentrations of chemicals, and potential receptors that aid in the evaluation and quantification of exposure in a given situation.

Extrapolation — The estimation of unknown values by extending or projecting from known values.

Feasibility study (FS) — The analysis and selection of alternative remedial or corrective actions for hazardous waste or contaminated sites. The process identifies and evaluates remedial alternatives by utilizing a variety of appropriate environmental, engineering, and economic criteria.

Field sampling plan (FSP) — A documentation that defines in detail the sampling and data gathering activities to be used in the investigation of a potentially contaminated site.

Free product — Chemical constituents that floats on groundwater, or that remains unadulterated as a contaminant pool in the environment.

Fugitive dust — Atmospheric dust arising from disturbances of particulate matter exposed to the air. Fugitive dust emissions consist of the release of chemicals from contaminated surface soil into the air, attached to dust particles.

Geometric mean — A measure of the central tendency for data from a positively skewed distribution (lognormal), given by:

$$X_{gm} = [(X_1)(X_2)(X_3)...(X_n)]^{1/n}$$

or,

$$X_{gm} = antilog\left\{\frac{\sum_{i=1}^{n} \log X_i}{n}\right\}$$

Groundwater — Water beneath the ground surface. It represents underground waters, whether present in a well-defined aquifer, or present temporarily in the vadose (unsaturated soil) zone.

Hazard — The inherent adverse effect that a chemical or other object poses. It is that innate character which has the potential for creating adverse and/or undesirable consequences. It defines the chance that a particular substance will have an adverse effect on human health or the environment in a particular set of circumstances which creates an exposure to that substance.

Hazard assessment — The evaluation of system performance and associated consequences over a range of operating and/or failure conditions. It involves gathering and evaluating data on types of injury or consequences that may be produced by a hazardous situation or substance.

Hazard identification — The systematic identification of potential accidents, upset conditions, etc. It is the recognition that a hazard exists and the definition of its characteristics. The process involves determining whether exposure to an agent can cause an increase in the incidence of a particular adverse health effect in receptors of interest.

Hazard index (HI) — The sum of several hazard quotients for multiple substances and/or multiple exposure pathways.

Hazard quotient (HQ) — The ratio of a single substance exposure level for a specified time period to a reference dose of that substance derived from a similar exposure period.

Hazard Ranking System (HRS) — A scoring system used by the EPA to assess the relative threat associated with actual or potential releases of hazardous substances at contaminated sites. The HRS is the primary way of determining whether a site is to be included on the National Priorities List (NPL).

Hazardous substance — Any substance that can cause harm to human health or the environment whenever excessive exposure occurs.

Hazardous waste — Wastes that are ignitable, explosive, corrosive, reactive, toxic, radioactive, pathological, or have some other properties that produces substantial risk to life. It is that byproduct which has the potential of causing detrimental effects on human health and/or the environment if not managed efficiently.

Heavy metals — Members of a group of metallic elements which are recognized as toxic and generally bioaccumulative. The term arises from the relatively high atomic weights of these elements.

Hot spot — Term used to denote zones where contaminants are present at much higher concentrations than the immediate surrounding areas. It represents a relatively small area which is highly contaminated within a study area.

Human equivalent dose — A dose which, when administered to humans, produces effects comparable to that produced by a dose in experimental animals.

Human health risk — The likelihood (or probability) that a given exposure or series of exposures to a hazardous substance will cause adverse health impacts on individual receptors experiencing the exposures.

Hydraulic conductivity — A measure of the ability of earth materials to transmit fluid, that is dependent on the type of fluid passing through the material.

Hydrocarbon — Organic chemicals/compounds, such as benzene, that contain atoms of both hydrogen and carbon.

Hydrophilic — Having greater affinity for water, or water-loving. Hydrophilic compounds tend to become dissolved in water.

Hydrophobic — Tending not to combine with water, or having less affinity for water. Hydrophobic compounds tend to avoid being dissolved in water and are more attracted to nonpolar liquids (e.g., oils) or solids.

Incineration — A thermal treatment/degradation process by which contaminated materials are exposed to excessive heat in a variety of incinerator types. The incineration process typically involves the thermal destruction of contaminants by burning under controlled conditions. Depending on the intensity of the heat, the contaminants of concern are volatilized and/or destroyed during the incineration process.

Incompatible wastes — Wastes, which when mixed with other materials without controls, may create fire, explosion, or other severe hazards.

Individual excess lifetime cancer risk — An upper bound estimate of the increased cancer risk, expressed as a probability, that an individual receptor could expect from exposure over a lifetime. It is a statistical concept and is not necessarily dependent on the average residency time in an area.

Ingestion — An exposure type whereby chemical substances enter the body through the mouth, and into the gastrointestinal system.

Inhalation — The intake of a substance by receptors through the respiratory tract system.

Intake — The amount of material inhaled, ingested, or dermally absorbed during a specified time period. It is a measure of exposure, expressed in mg/kg-day.

Integrated Risk Information System (IRIS) — A EPA database containing verified reference doses (RfDs) and slope factors (SFs), and up-to-date health risk and EPA regulatory information for numerous chemicals. It serves as a source of toxicity information for human health and environmental risk assessment.

Interim action — An action that initiates remediation of a contaminated site, but may also constitute part of the final remedy.

Investigation-derived wastes (IDWs) — Wastes generated in the process of collecting samples during a remedial investigation or site characterization activity. Such wastes must be handled according to all relevant and appropriate regulatory requirements. The wastes may include soil, groundwater, used personal protective equipment, decontamination fluids, and disposable sampling equipment.

K_d *(soil/water partition coefficient)* — Provides a soil- or sediment-specific measure of the extent of chemical partitioning between soil or sediment and water, unadjusted for the dependence on organic carbon.

K_{oc} *(organic carbon adsorption coefficient)* — Provides a measure of the extent of chemical partitioning between organic carbon and water at equilibrium.

K_{ow} *(octanol/water partition coefficient)* — Provides a measure of the extent of chemical partitioning between water and octanol at equilibrium.

K_w *(water/air partition coefficient)* — Provides a measure of the distribution of a chemical between water and air at equilibrium.

Landfarming — The application of biodegradable organic wastes onto a land surface and their incorporation into the surface soil so that they degrade more readily.

Landfill — A controlled site for the disposal of wastes on land, generally operated in accordance with regulatory safety and environmental compliance requirements.

Latent period — The time between the initial induction of a health effect from first exposure to a chemical and the manifestation or detection of actual health effects.

LC_{50} *(mean lethal concentration)* — The lowest concentration of a chemical in air or water that will be fatal to 50% of test organisms living in that media.

LD_{50} *(mean lethal dose)* — The median lethal dose, i.e., the single dose (ingested or dermally absorbed) required to kill 50% of a test animal group.

Leachate — Aqueous, often-contaminated, liquid generated when water percolates or trickles through waste materials or contaminated sites and collects components of those wastes. Leaching usually occurs at landfills as a result of infiltration of rainwater or snowmelt, and may result in hazardous chemicals entering soils, surface water, or groundwater.

Lifetime average daily dose (LADD) — The exposure, expressed as mass of a substance contacted and absorbed per unit body weight per unit time, averaged over a lifetime. It is usually used to calculate carcinogenic risks; it takes into account the fact that, whereas carcinogenic risk values are determined with an assumption of lifetime exposure, actual exposures may be for a shorter period of time.

Lifetime exposure — The total amount of exposure to a substance that a human would be subjected to in a lifetime.

Lifetime risk — Risk which results from lifetime exposure to a chemical substance.

LOAEL (lowest-observed-adverse-effect level) — The chemical dose rate causing statistically or biologically significant increases in frequency or severity of adverse effects between the exposed and control groups. It is the lowest dose level, expressed in mg/kg body weight/day, at which adverse effects are noted in the exposed population.

$LOAEL_a$ — LOAEL values adjusted by dividing by one or more safety factors.

LOEL (lowest-observed-effect-level) — The lowest exposure or dose level of a substance at which effects are observed in the exposed population; the effects may or may not be serious.

Matrix (or medium) — The predominant material comprising the environmental sample being investigated (e.g., soils, water, air).

Maximum Daily Dose (MDD) — The maximum dose calculated for the duration of receptor exposure, and used to estimate risks for subchronic or acute noncarcinogenic effects of environmental contaminants.

MCL (maximum contaminant level) — A legally enforceable maximum chemical concentration standard that is allowable in drinking water, issued by the U.S. EPA under authority of the SDWA.

MCLG (maximum contaminant level goal) — A nonenforceable health goal for public drinking water systems issued by the U.S. EPA under authority of the SDWA. It is also referred to as the recommended maximum contaminant level (RMCL).

Microbe — A microscopic or ultramicroscopic organism (e.g., bacterium or virus).

Mitigation — The process of reducing or alleviating a problem situation.

Modeling — The use of mathematical equations to simulate and predict real events and processes.

Monitoring — The measurement of concentrations of chemicals in environmental media or in tissues of humans and other biological receptors/organisms.

Monte Carlo simulation — A technique in which outcomes of events or variables are determined by selecting random numbers subject to a defined probability law.

National Contingency Plan (NCP) — The National Oil and Hazardous Substances Pollution Contingency Plan, commonly known as the NCP, is a regulation that establishes roles, responsibilities, and authorities for responding to hazardous substance releases. The NCP established the HRS as the principal mechanism for placing sites on the NPL.

National Priorities List (NPL) — A list of waste sites for which the U.S. EPA has assessed the relative threat of a site contamination on air, surface and groundwater, soil, and the population potentially at risk; this site listing, which is found under CERCLA (Section 105), is updated three times a year.

Neurotoxicity — Hazard effects that are poisonous to the nerve cells.

NOAEL (no-observed-adverse-effect level) — The chemical intakes at which there are no statistically or biologically significant increases in frequency or severity of adverse effects between the exposed and control groups (which means that statistically significant effects are observed at this level, but they are not considered to be adverse). It is the highest level at which a chemical causes no observable adverse effect in the species being tested or the exposed population.

$NOAEL_a$ — NOAEL values adjusted by dividing by one or more safety factors.

NOEL (no-observed-effect level) — The dose rate of chemical at which there are no statistically or biologically significant increases in frequency or severity of any effects between the exposed and control groups. It is the highest level at which a chemical causes no observable changes in the species or exposed populations under investigation.

Nonparametric statistics — Statistical techniques whose application is independent of the actual distribution of the underlying population from which the data were collected.

Nonthreshold chemical — Also called *zero threshold chemical*, refers to a substance which is known, suspected, or assumed to potentially cause some adverse response at any dose above zero.

Off-site — Areas outside the boundaries or limits of a presumed contaminated site.

On-site — The boundaries or limits of a presumed contaminated site.

Organic carbon content of soils or sediments (%) — This reflects the amount of organic matter present, and generally correlates with the tendency of chemicals to accumulate in the soil or sediment. The accumulation of chemicals in soils or sediments is frequently the result of adsorption onto organic matter. In general, the higher the organic carbon content of the soil or sediment, the more a contaminant will be adsorbed to the soil particles, rather than be dissolved in the water or gases permeating the soil or sediment.

Pathway — Any specific route which environmental contaminants take in order to travel away from the source and to reach potential receptors or individuals.

PEL (permissible exposure limit) — A maximum (legally enforceable) allowable level for a chemical in workplace air.

Permeability — A measure of a material's ability to transmit fluid. It is the capacity of a porous medium to transmit a fluid subjected to an energy gradient (the hydraulic gradient, in the case of water).

pH — A measure of the acidity or alkalinity of a material or medium, with a value of 7 representing neutral; a low pH means a highly acidic medium, whereas a high pH indicates an alkaline medium.

Pica — The behavior in toddlers and children (usually under age 6 years) involving the intentional eating/mouthing of large quantities of dirt and other objects.

PM-10, PM_{10} — Particulate matter with physical/aerodynamic diameter <10 μm. It represents the respirable particulate emissions.

Population-at-risk (PAR) — A population subgroup that is more susceptible to hazard or chemical exposures. It represents that group which is more sensitive to a hazard or chemical than is the general population.

Population excess cancer burden — An upper bound estimate of the increase in cancer cases in a population as a result of exposure to a carcinogen.

Porosity — The ratio of the volume of void spaces in earth materials to the total volume of the material. The wider the range of grain sizes, the lower the porosity.

Potency — A measure of the relative toxicity of a chemical.

Potentially responsible party (PRP) — Those identified by the U.S. EPA as potentially liable under CERCLA for cleanup costs at specified waste sites.

ppb (parts per billion) — An amount of substance in a billion parts of another material; also expressed by μg/kg or μg/L.

ppm (parts per million) — An amount of substance in a million parts of another material; also expressed by mg/kg or mg/L.

ppt (parts per trillion) — An amount of substance in a trillion parts of another material; also expressed by ng/kg or ng/L.

Practical quantitation limit (PQL) — Also called *sample quantitation limit (SQL)*. It is the lowest level that can be reliably achieved within specified limits of precision and accuracy during routine laboratory operating conditions. It represents a detection limit that has been corrected for sample characteristics, sample preparation, and analytical adjustments such as dilution. Typically, the PQL or SQL will be about 5 to 10 times the chemical-specific detection limit.

Preliminary assessment (PA) — A survey and evaluation whereby sites are characterized with respect to their potential to release significant amounts of contaminants into the environment.

Preliminary site appraisal — Process used for quick assessment of a site's potential to adversely affect the environment and/or public health.

Probability — The likelihood of an event occurring.

Promoter — A chemical that, when administered after an initiator has been given, promotes the change of an initiated cell, culminating in a cancer.

Proxy concentration — Assigned contaminant concentration value for situations where sample data may not be available, or when it is impossible to quantify accurately.

Qualitative — Description of a situation without numerical specifications.

Quality assurance (QA) — A system of activities designed to assure that the quality control system is performing adequately. It consists of the management of investigation data to assure that they meet the data quality objectives. This commonly includes designing appropriate protocols, ensuring they are carried out, and independently testing data quality.

Quality assurance project plan (QAPP) — A plan that describes protocols necessary to achieve the data quality objectives defined for a remedial investigation or site characterization.

Quality control (QC) — A system of specific efforts designed to test and control the quality of data obtained in an investigation. It consists of the management of activities involved in the collection and analysis of data to assure they meet the data quality objectives. It is the system of activities required to provide information as to whether the quality assurance system is performing adequately. Activities include following the sampling protocols, and routinely checking calibration of laboratory equipment.

Quantitation limit (QL) — The lowest level at which a chemical can be accurately and reproducibly quantitated. It usually is equal to the instrument detection limit (IDL) multiplied by a factor of 3 to 5, but varies for different chemicals and different samples.

Quantitative — Description of a situation presented in exact numerical terms.

Receptor — Members of a potentially exposed population, such as persons or organisms that are potentially exposed to concentrations of a particular chemical compound.

Reference concentration (RfC) — A concentration of a chemical substance in an environmental medium to which exposure can occur over a prolonged period without an expected adverse effect. The medium in this case is usually air, with the concentration expressed in mg of chemical per m^3 of air.

Reference dose (RfD) — The maximum amount of a chemical that the human body can absorb without experiencing chronic health effects, expressed in mg of chemical per kg body weight per day. It is the estimate of lifetime daily exposure of a noncarcinogenic substance for the general human population (including sensitive receptors) which appears to be without an appreciable risk of deleterious effects, consistent with the threshold concept.

Remedial action — Those actions consistent with a permanent remedy in the event of a release of a hazardous substance into the environment, meant to prevent or minimize such releases so that they do not migrate to cause substantial danger to present or future public health or welfare or the environment.

Remedial action objective — Cleanup objectives that specify the level of cleanup, area of cleanup (or area of attainment), and the time required to achieve cleanup (i.e., the restoration time frame).

Remedial alternative — An action considered in the feasibility study, that is intended to reduce or eliminate significant risks to human health and/or the environment at a contaminated site.

Remedial investigation (RI) — The field investigation of hazardous waste sites to determine pathways, nature, and extent of contamination, as well as to identify preliminary alternative remedial actions. It addresses data collection and site characterization to identify and assess threats or potential threats to human health and the environment posed by a site.

Remediation — The process of cleaning up of a potentially contaminated site, in order to prevent or minimize the potential release and migration of hazardous substances from the impacted media that could cause adverse impacts to present or future public health and welfare, or the environment.

Removal action — An action that is implemented to address a direct threat to human health or the environment.

Representative sample — A sample that is assumed *not* to be significantly different than the population of samples available.

Residual risk — The risk of adverse consequences that remains after corrective actions have been implemented.

Response — The reaction of a body or organ to a chemical substance or other physical, chemical, or biological agent.

Restoration time frame — Time required to achieve requisite cleanup levels or site restoration goals.

Risk — The probability or likelihood of an adverse consequence from a hazardous situation or hazard, or the potential for the realization of undesirable adverse consequences from impending events. It is a measure of the probability and severity of an adverse effect to health, property, or the environment.

Risk acceptance — The willingness of an individual, group, or society to accept a specific level of risk in order to obtain some gain or benefit.

Risk appraisal — The assessment of whether existing or potential biologic receptors are presently, or may in the future, be at risk of adverse effects as a result of exposures to contaminants originating at a contaminated site.

Risk assessment — A methodology that combines exposure assessment with health and environmental effects data to estimate risks to human or environmental target organisms which results from exposure to pollutants.

Risk control — The process to manage risks associated with a hazard situation. It may involve the implementation, enforcement, and reevaluation of the effectiveness of corrective measures from time to time.

Risk decision — The process used for making complex public policy decisions relating to the control of risks associated with hazardous situations.

Risk determination — The evaluation of the environmental and health impacts of contaminant releases.

Risk estimate — A description of the probability that a potential receptor exposed to a specified dose of a chemical will develop an adverse response.

Risk estimation — The process of quantifying the probability and consequence values for a hazard situation. It is the process used to determine the extent and probability of adverse effects of the hazards identified, and to produce a measure of the level of health, property, or environmental risks being assessed.

Risk evaluation — The complex process of developing acceptable levels of risk to individuals or society. It is the stage at which values and judgments enter into the decision process.

Risk group — A real or hypothetical exposure group composed of general or specific population groups.

Risk management — The steps and processes taken to reduce, abate, or eliminate the risk that has been revealed by a risk assessment. It is an activity concerned with decisions about whether an assessed risk is sufficiently high to present a public health concern, and about the appropriate means for controlling the risks judged to be significant.

Risk perception — The magnitude of the risk as it is perceived by an individual or population. It consists of the measured risk and the preconceptions of the observer.

Risk reduction — The action of lowering the probability of occurrence and/or the value of a risk consequence, thereby reducing the magnitude of the risk.

Risk-specific dose (RSD) — An estimate of the daily dose of a carcinogen which, over a lifetime, will result in an incidence of cancer equal to a given risk level. It is the dose associated with a specified risk level.

Sample blank — Blanks are samples considered to be the same as the environmental samples of interest except with regard to one factor whose influence on the samples is being evaluated. Blanks are used to ensure that contaminant concentrations actually reflect site conditions, and are not artifacts of the sample handling processes. The blanks consist of laboratory-prepared sample bottles of distilled or deionized water that accompany the empty sample bottles to the field as well as the samples returning to the laboratory, and are not opened until both the blanks and the actual site samples are analyzed.

Sample duplicate — Two samples taken from the same source at the same time and analyzed under identical conditions.

Sampling and analysis plan (SAP) — Documentation that consists of a quality assurance project plan (QAPP) and a field sampling plan (FSP).

Saturated zone — An underground geologic formation in which the pore spaces or interstitial spaces in the formation are filled with water under a pressure equal to or greater than atmospheric pressure.

Sediment — Soil that is normally covered with water. It generally is considered to provide a direct exposure pathway to aquatic life.

Sensitive receptor — Individual in a population who is particularly susceptible to health impacts due to exposure to a chemical pollutant.

Sensitivity analysis — A method used to examine the operation of a system by measuring the deviation of its nominal behavior due to pertubations in the performance of its components from their nominal values.

Site assessment — Process used to identify toxic substances that may be present at a site and to present site-specific characteristics that influence the migration of contaminants.

Site categorization — A classification of sites to reflect the uniqueness of each site.

Site cleanup — The decontamination of a site, initiated as a result of the discovery of contamination at a site or property.

Site mitigation — The process of cleaning up a contaminated site in order to return it to an environmentally acceptable state.

Slope factor (SF) — A plausible upper bound estimate of the probability of a response per unit intake of a chemical over a lifetime. It is used to estimate an upper bound probability of an individual developing cancer as a result of a lifetime of exposure to a particular level of a carcinogen.

Soil gas — The vapor or gas found in the unsaturated soil zone.

Soil vapor extraction (SVE) — Also known as *in situ soil venting, subsurface venting, vacuum extraction,* or *in situ soil stripping*, is a technique that uses soil aeration to treat subsurface zones of VOC contamination in soils.

Solubility — A measure of the ability of a substance to dissolve in a fluid.

Stabilization — The conversion of a substance into a form that will not readily change its physical or chemical characteristics.

Standard — A general term used to describe legally established values above which regulatory action will be required.

Standard deviation — The most widely used measure to describe the dispersion of a data set, defined for a set of n values as follows:

$$s = \left[\frac{\sum_{i=1}^{n} \left(X_i - X_m \right)^2}{\left(n - 1 \right)} \right]^{0.5}$$

where X_m is the arithmetic mean for the data set of n values.

Subchronic — Relating to intermediate duration, usually used to describe studies or exposure levels spanning 5 to 90 days duration.

Subchronic daily intake (SDI) — The exposure, expressed in mg/kg-day, averaged over a portion of a lifetime.

Subchronic exposure — The short-term, high-level exposure to chemicals, i.e., the maximum exposure or doses to a chemical over a portion of a lifetime.

Superfund — A commonly used name for the Comprehensive Environmental Response, Compensation, and Liability Act (CERCLA), also referred to as the "Trust Fund."

Surfactant — A surface-active chemical agent, usually made up of phosphates, used in detergents to produce lathering.

Synergism — An interaction of two or more chemicals that results in an effect that is greater than the sum of their effects taken independently. It is the effects from a combination of two or more events, efforts, or substances that are greater than would be expected from adding the individual effects.

Systemic — Relating to whole body, rather than individual parts of exposed individual or receptor.

Threshold — The lowest dose or exposure of a chemical at which a specified measurable effect is observed and below which such effect is not observed. *Threshold dose* is the minimum exposure dose of a chemical that will evoke a stipulated toxicological response. *Toxicological threshold* refers to the concentration at which a compound exhibits toxic effects.

Threshold chemical — Also, *non-zero threshold chemical*, refers to a substance which is known or assumed to have no adverse effects below a certain dose.

Threshold limit — A chemical concentration above which adverse health and/or environmental effects may occur.

Tolerance limit — The level or concentration of a chemical residue in media of concern above which adverse health effects are possible, and above which corrective action should therefore be undertaken.

Toxic — Harmful, or deleterious with respect to the effects produced by exposure to a chemical substance.

Toxicant — Any synthetic or natural chemical with an ability to produce adverse health effects. It is a poisonous contaminant that may injure an exposed organism.

Toxicity — The harmful effects produced by a chemical substance. It is the quality or degree of being poisonous or harmful to human or ecological receptors. It represents the property of a substance to cause any adverse physiological effects (on living organisms).

Toxicity assessment — Evaluation of the toxicity of a chemical based on all available human and animal data. It is the characterization of the toxicological properties and effects of a chemical substance, with special emphasis on the establishment of dose-response characteristics.

Toxic substance — Any material or mixture that is capable of causing an unreasonable threat to human health or the environment.

Treatment — A change in the composition or concentration of a waste substance so as to make it less hazardous, or to make it acceptable at disposal and re-use facilities. It involves the application of technological process to a contaminant or waste in order to render it nonhazardous or less hazardous or more suitable for resource recovery.

Trip blank — A trip blank is transported just like actual samples, but does not contain the chemicals to be analyzed. The purpose of this blank is to evaluate the possibility that a chemical could seep into samples (to adulterate them) during transportation to the laboratory.

Uncertainty — The lack of confidence in the estimate of a variable's magnitude or probability of occurrence.

Uncertainty factor (UF) — Also called *safety factor*, refers to a factor that is used to provide a margin of error when extrapolating from experimental animals to estimate human health risks.

Underground storage tank (UST) — A tank fully or partially located below the ground surface that is designed to hold gasoline or other petroleum products, or indeed other chemical products.

Unit cancer risk (UCR) — The excess lifetime risk of cancer due to a continuous lifetime exposure/dose of one unit of carcinogenic chemical concentration (caused by one unit of exposure in the low exposure region).

Unit Risk (UR) — A measure of the carcinogenic potential of a substance, when a dose is received through the inhalation pathway, that is based on several assumptions. It is an upper bound estimate of the probability of contracting cancer as a result of constant exposure over the individual lifetime to an ambient concentration of 1 $\mu g/m^3$.

Upper bound estimate — The estimate not likely to be lower than the true (risk) value.

Upper confidence limit, 95% (95% UCL) — The upper limit on a normal distribution curve below which the observed mean of a data set will occur 95% of the time. This is also equivalent to stating that there is, at most, a 5% chance of the true mean being greater than the observed value.

Vadose zone — Also called the *unsaturated soil zone*, is the zone between the ground surface and the top of the groundwater table.

Volatile organic compound (VOC) — Any organic compound that has a great tendency to vaporize, and is susceptible to atmospheric photochemical reactions.

Volatility — A measure of the tendency of a compound to vaporize from the liquid state.

Water table — The top of the saturated zone where confined groundwater is under atmospheric pressure.

APPENDIX C
IMPORTANT FATE AND
TRANSPORT PROPERTIES OF
ENVIRONMENTAL CONTAMINANTS

C.1 CONTAMINANT FATE AND TRANSPORT
IN THE ENVIRONMENT

The existence of contaminated sites may result in the release of chemicals into air (via volatilization and fugitive dust emissions), surface water (from surface runoff/ overland flow and groundwater seepage), groundwater (through infiltration/leaching), soils (due to erosion — including fugitive dust generation/deposition and tracking), sediments (from surface runoff/overland flow), and biota (due to biological uptake and bioaccumulation). Contaminants released into the environment are controlled by a complex set of processes including transport, transformation, degradation and decay, intermedia transfer, and biological uptake. In addition, many toxic chemicals are persistent and undergo complex interactions in more than one environmental medium. Environmental fate analyses can be used to assess the movement of chemicals between environmental compartments.

The fate of chemical compounds released into the environment forms an important basis for evaluating the exposure of biological receptors to hazardous chemicals. Multimedia transport models are generally employed in the prediction of the long-term fate of such chemicals in the environment.

C.2 CONTAMINANT DISTRIBUTION BETWEEN
ENVIRONMENTAL COMPARTMENTS

Chemicals present in one environmental medium are affected by several complex processes and phenomena, facilitating transfers into other media. The potential for intermedia transfer of pollutants from the soil medium to other media is particularly significant; in fact, contaminated soil can be a major source for the contamination of groundwater, atmospheric air, subsurface soil gas, sediments, and surface water. For example, chemical constituents having a moderate to high degree of mobility can leach from soils into groundwater; volatile constituents may contribute to subsurface gas in the vadose zone and also possible releases into the atmosphere. Conversely, the potential for intermedia transport of constituents from other media into soils does exist; for example, chemical constituents may be transported to soils via atmospheric deposition, and also through releases of subsurface gas.

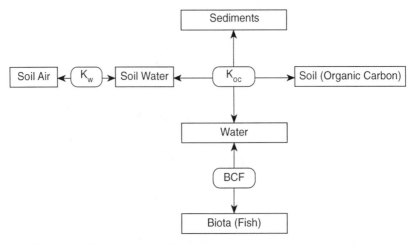

Figure C.1   Contaminant partitioning between environmental compartments.

The affinity that contaminants have for soil affects their mobility by retarding transport. For instance, hydrophobic or cationic contaminants that are migrating in solution are subject to retardation effects. The hydrophobicity of a contaminant can greatly affect its fate, which explains some of the different rates of contaminant migration occurring in the subsurface environment. Also, the phenomenon of adsorption is a major reason why the sediment zones of surface water bodies/systems may become highly contaminated with specific organic and inorganic chemicals. On the other hand, a number of natural processes work to lessen or attenuate contaminant concentrations in the environment; the mechanisms of natural attenuation include dispersion/dilution, ion exchange, precipitation, adsorption and absorption, filtration, gaseous exchange, and biodegradation.

In general, the distribution of organic chemicals among environmental compartments can be defined in terms of the simple equilibrium expressions shown in Figure C.1 (Swann and Eschenroeder, 1983), where the K_w, K_{oc}, and BCF symbols refer to partitioning coefficients that are elaborated in the next section. The general assumption is that all environmental compartments are well mixed to achieve equilibrium between them.

The partitioning of inorganic chemicals is somewhat different from organic constituents. Typically, metals generally exhibit relatively low mobilities in soils (Evans, 1989). Rather, the inorganics will tend to adsorb onto soils (which may become airborne or be transported by surface erosion) and sediments (that may be transported in water).

C.3 IMPORTANT FATE AND TRANSPORT PROPERTIES AND PARAMETERS

As pollutants are released into various environmental media, several factors contribute to their migration and transport. For instance, in the groundwater system, the solutes in the porous media will move with the mean velocity of the solvent by advective mechanism; in addition, other mechanisms governing the spread of contaminants include hydraulic dispersion and molecular diffusion (which is caused by the random Brownian motion of molecules in solution that occurs whether the solution in the porous media is stationary or has an average motion). Furthermore, the transport and concentration of the solute(s) are affected by reversible ion exchange with soil grains, the chemical degeneration with other constituents, fluid compression and expansion, and in the case of radioactive wastes by the radioactive decay.

Examination of a chemical's physical and chemical properties can often allow an estimation of its environmental partitioning. Qualitative analysis of the fate of a chemical can also be made by analogy with other chemicals whose fate properties are well documented; that is, if the chemical under investigation is structurally similar to a previously well-studied one, some parallels can be drawn to the environmental fate of the analogue. In addition, several site characteristics may influence the environmental fate of chemicals, including the amount of ambient moisture, humidity levels, temperatures and wind speed, the geologic, hydrologic, pedologic and watershed characteristics, topographic features of the site and vicinity, vegetative cover of site and surrounding area, and land use characteristics. Several important properties affecting environmental fate and/or intermedia transfers of environmental contaminants are briefly annotated below. Further details, including elaboration of a variety of estimation methods for evaluating the important parameters, are elaborated elsewhere in the literature (e.g., Lyman et al., 1990; Swann and Eschenroeder, 1983).

Physical State

Chemical compounds may occur in either solid or fluid state. Solid contaminants are generally less susceptible to release and migration than liquids. However, processes such as leaching, erosion, and/or runoff, and physical transport of contaminated materials can act as significant release mechanisms.

Water Solubility

The solubility of a chemical in water is the maximum amount of the chemical that will dissolve in pure water at a specified temperature, usually expressed in terms of weight per weight (e.g., ppb, ppm, mg/kg) or weight per volume (e.g., ppb, ppm, μg/L, mg/L). Several different approaches to the estimation of water solubility are described in the literature (e.g., Lyman et al., 1990).

Solubility is an important factor affecting a chemical constituent's release and subsequent migration and fate in the surface water and groundwater environments. In fact, among the various parameters affecting the fate and transport of organic chemicals in the environment, water solubility is one of the most important, especially with regards to hydrophilic compounds.

In general, highly soluble chemicals are easily and quickly distributed by the hydrologic system. Such chemicals tend to have relatively low adsorption coefficients for soils and sediments, and also relatively low bioconcentration factors in aquatic biota. Furthermore, they tend to be more readily biodegradable.

Diffusion and Dispersion

Diffusive processes create mass spreading due to molecular diffusion in response to concentration gradients. The higher the diffusivity, the more likely a chemical is to move in response to concentration gradients. Thus, diffusion coefficients are used to describe the movement of a molecule in a liquid or gas medium as a result of

differences in concentration; it can also be used to calculate the dispersive component
of chemical transport.

Dispersive processes create mass mixing due to system heterogeneities (e.g.,
velocity variations). Consequently, for example, as a pulse of contaminant plume
migrates through a soil system, the peaks in concentration are decreased by spreading.

Volatilization

Volatilization is the process by which a chemical compound evaporates into the
vapor phase from another environmental compartment. The volatilization of chemicals is an important mass-transfer phenomenon. Knowledge of volatilization rates is
necessary to determine the amount of chemical that enters the atmosphere and the
change of pollutant concentrations in the source media. The transfer process from the
source (e.g., water body, sediments, soil) to the atmosphere is dependent on the
physical and chemical properties of the chemical compound in question, the presence
of other pollutants, the physical properties of the source media, and the atmospheric
conditions. Volatility is indeed a very important parameter for hazard assessments.
Several estimation methods for evaluating this parameter are elaborated in the literature (e.g., Lyman et al., 1990).

Henry's Law Constant

Henry's law constant provides a measure of the extent of chemical partitioning
between air and water at equilibrium. It indicates the relative tendency of a constituent
to volatilize from aqueous solution to the atmosphere, based on the competition
between its vapor pressure and water solubility. This parameter is important in
determining the potential for intermedia transport into air. As an example of its
application, the concentration in soil gas is related to the concentration in an underlying aquifer by:

$$\frac{C_{sg}}{C_w} = \frac{H}{RT}$$

where:

C_{sg} = concentration of the chemical in soil gas (mg/m^3)
C_w = concentration of the chemical in groundwater (µg/L)
H = Henry's law constant (atm-m^3/mol)
R = gas constant (atm-m^3/mol K)
T = temperature (K)

Several other forms of expressing H also exist in the technical literature (e.g., Lyman
et al. 1990).

Contaminants with low Henry's law constant values will tend to favor the
aqueous phase and will therefore volatilize to the atmosphere more slowly than
constituents with high values. As a general guideline, for H values in the range of
10^{-7} to 10^{-5} atm-m^3/mol volatilization is low, for H between 10^{-5} and 10^{-3} atm-m^3/mol
volatilization is not rapid but possibly significant, and for H $> 10^{-3}$ atm-m^3/mol
volatilization is rapid. The variation in H between chemicals is indeed extensive.

Vapor Pressure

Vapor pressure is the pressure exerted by a chemical vapor in equilibrium with its solid or liquid form at any given temperature. It is a relative measure of the volatility of a chemical in its pure state, and is an important determinant of the rate of volatilization. It is used to calculate the rate of volatilization of a pure substance from a surface, or in estimating a Henry's law constant for chemicals with low water solubility.

In general, the higher the vapor pressure, the more likely a chemical exists in significant quantities in a gaseous state. Thus, constituents with high vapor pressure are more likely to migrate from soil and groundwater, to be transported in air. Numerous estimation procedures exist in the technical literature (e.g., Lyman et al. 1990) for the estimation of this parameter.

Boiling Point (BP)

The boiling point is the temperature at which the vapor pressure of a liquid is equal to the atmospheric pressure on the liquid. At this temperature, the substance transforms from a liquid into a vapor phase. Besides being an indicator for the physical state of a chemical, the BP also provides an indication of its volatility. Other physical properties, such as critical temperature and latent heat (or enthalpy) of vaporization, may be predicted by use of its normal BP as an input.

Water/Air Partition Coefficient (K_w)

The water/air partition coefficient (K_w) relates the distribution of a chemical between water and air. It consists of an expression that represents the reciprocal of Henry's law constant (H), given by:

$$K_w = \frac{C_{water}}{C_{air}} = \frac{1}{H}$$

where:

C_{air} = concentration of the chemical in air (mg/L)
C_{water} = concentration of the chemical in water (mg/L)

Octanol/Water Partition Coefficient (K_{ow})

The octanol/water partition coefficient (K_{ow}) is defined as the ratio of a chemical's concentration in the octanol phase (organic) to its concentration in the aqueous phase of a two-phase octanol/water system, represented by (Lyman et al. 1990):

$$K_{ow} = \frac{\text{concentration in octanol phase}}{\text{concentration in aqueous phase}}$$

This dimensionless parameter provides a measure of the extent of chemical partitioning between water and octanol at equilibrium. It has become a particularly important parameter in studies of the environmental fate of organic chemicals.

K_{ow} can be used to predict the magnitude of an organic constituent's tendency to partition between the aqueous and organic phases of a two-phase system, such as surface water and aquatic organisms. The higher the value of K_{ow}, the greater the tendency of an organic constituent to adsorb to soil or waste matrices containing appreciable organic carbon or to accumulate in biota. It has been found to be related to water solubility, soil/sediment adsorption coefficients, and bioaccumulation factors for aquatic life.

In general, chemicals with low K_{ow} (<10) values may be considered relatively hydrophilic, whereas those with high K_{ow} (>10^4) values are very hydrophobic. Thus, the greater the K_{ow}, the more likely a chemical is to partition to octanol than to remain in water. The hydrophilic chemicals tend to have high water solubilities, small soil or sediment adsorption coefficients, and small bioaccumulation factors for aquatic life. High K_{ow} values are generally indicative of a chemical's ability to accumulate in fatty tissues and therefore bioaccumulate in the foodchain. It is also a key variable in the estimation of skin permeability.

The Organic Carbon Adsorption Coefficient (K_{oc})

The sorption characteristics of a chemical may be normalized to obtain a sorption constant based on organic carbon which is essentially independent of any soil. The organic carbon adsorption coefficient (K_{oc}) provides a measure of the extent of partitioning of a chemical constituent between soil or sediment organic carbon and water at equilibrium.

K_{oc} is the ratio of the amount of constituent adsorbed per unit weight of organic carbon in the soil or sediment to the concentration of the constituent in aqueous solution at equilibrium. Also called the organic carbon partition coefficient, K_{oc} is a measure of the tendency for organics to be adsorbed by soil and sediment, and is expressed by:

$$K_{oc}\ [mL/g] = \frac{\mu g \text{ chemical absorbed per g soil organic carbon}}{\mu g \text{ chemical dissolved per mL of water}}$$

The extent to which an organic constituent partitions between the solid and solution phases of a saturated or unsaturated soil, or between runoff water and sediment, is determined by the physical and chemical properties of both the constituent and the soil (or sediment). The K_{oc} is chemical-specific and largely independent of the soil or sediment properties. The tendency of a constituent to be adsorbed to soil, however, is dependent on its properties and also on the organic carbon content of the soil or sediment.

When constituents have a high K_{oc}, they have a tendency to partition to the soil or sediment. This value is a measure of the hydrophobicity of a chemical. The more highly sorbed, the more hydrophobic (or the less hydrophilic) a substance becomes. Values of K_{oc} typically range from 1 to 10^7; the higher the K_{oc}, the more likely a chemical is to bind to soil or sediment than to remain in water. Several estimation methods for evaluating this parameter are elaborated in the technical literature (e.g., Lyman et al., 1990).

Soil-Water Partition Coefficient (K_d)

The mobility of contaminants in soil depends not only on properties related to the physical structure of the soil, but also on the extent to which the soil material will retain, or adsorb, the chemical constituents. The extent to which a constituent is adsorbed depends on the chemical properties of the constituent and of the soil.

Therefore, the sorptive capacity must be determined with reference to a particular constituent and soil pair. The soil-water partition coefficient (K_d) is generally used to quantify soil sorption. It provides a soil- or sediment-specific measure of the extent of chemical partitioning between soil or sediment and water, unadjusted for dependence on organic carbon.

The distribution of a chemical between water and adjoining soil or sediment may indeed be described by an equilibrium expression that relates the amount of chemical sorbed to soil or sediment to the amount in water at equilibrium. For most environmental concentrations, the soil/water distribution coefficient, K_d, can be approximated by:

$$K_d \, [mL/g] = \frac{\left(\text{concentration of adsorbed chemical in soil, } C_s\right)}{\left(\text{concentration of chemical in solution in water, } C_w\right)}$$

or

$$K_d \, [mL/g] = \frac{\left[\left(\mu g \text{ chemical per g soil}\right)\right]}{\left[\left(\mu g \text{ chemical per g water}\right)\right]}$$

On this basis, K_d describes the sorptive capacity of the soil and allows estimation of the concentration in one medium, given the concentration in the adjoining medium. For hydrophobic contaminants,

$$K_d = f_{oc} K_{oc}$$

where f_{oc} is the fraction of organic carbon in the soil.

K_d is the ratio of the adsorbed contaminant concentration to the dissolved concentration at equilibrium. The higher the value of K_d, the less mobile is the contaminant; this is because, for large values of K_d, most of the chemical remains stationary and attached to soil particles due to the high degree of sorption. Thus, the higher the K_d the more likely a chemical is to bind to soil or sediment than to remain in water.

Bioconcentration Factor (BCF)

The bioconcentration factor (BCF) is the ratio of the concentration of a chemical constituent in an organism or whole body (e.g., a fish) or specific tissue (e.g., fat) to the concentration in water at equilibrium, given by:

$$BCF = \frac{\left(\text{concentration in biota}\right)}{\left(\text{concentration in surrounding medium}\right)}$$

$$= \frac{\left[\left(\mu g \text{ chemical per g biota, e.g., fish}\right)\right]}{\left[\left(\mu g \text{ chemical per g medium, water}\right)\right]}$$

The partitioning of a chemical between water and biota (fish) also gives a measure of the hydrophobicity of the chemical. Ranges of BCFs for various constituents and organisms can be used to predict the potential for bioaccumulation, and therefore to determine whether sampling of the biota is a necessary part of a site characterization program.

In fact, the accumulation of chemicals in aquatic organisms is of increased concern as a significant source of environmental and health hazard. The BCF indicates the degree to which a chemical residue may accumulate in aquatic organisms, coincident with ambient concentrations of the chemical in water; it is a measure of the tendency of a chemical in water to accumulate in the tissue of an organism. The concentration of the chemical in the edible portion of the organism's tissue can be estimated by multiplying the concentration of the chemical in surface water by the fish BCF for that chemical. Indeed, the BCF is an estimate of the bioaccumulation potential for biota in general, not just for fish. Thus, the average concentration in fish or biota is given by

$$C_{\text{fish-biota}} \ (\mu g/kg) = C_{\text{water}} \ (\mu g/L) \times BCF$$

where C_{water} is the concentration in water. This parameter is indeed an important determinant for human intake via ingestion of aquatic foods.

Values of BCF typically range from 1 to over 10^6. Constituents exhibiting a BCF greater than 1.0 are potentially bioaccumulative. Generally, constituents exhibiting a BCF greater than 100 cause the greatest concern (USEPA, 1987). Several estimation methods for evaluating this parameter are elaborated in the technical literature (e.g., Lyman et al., 1990).

Degradation/Chemical Half-Lives

Degradation, whether biological, physical, or chemical, is often reported in the literature as a half-life, which is usually measured in days. It is usually expressed as the time it takes for one-half of a given quantity of a compound to be degraded.

Half-lives are used as measures of persistence, since they indicate how long a chemical will remain in various environmental media. Long half-lives (e.g., greater than a month or a year) are characteristic of persistent constituents. Media-specific half-lives provide a relative measure of the persistence of a chemical in a given medium, although actual values can vary greatly depending on site-specific conditions. For example, the absence of certain microorganisms at a site, or the number of microorganisms, can influence the rate of biodegradation and, therefore, the half-life for specific compounds. As such, half-life values should be used only as a general indication of a chemical's persistence in the environment (USEPA, 1987). In general, however, the greater the half-life, the more persistent a chemical is likely to be.

Biodegradation

Biodegradation is one of the most important environmental processes affecting the breakdown of organic compounds. It results from the enzyme-catalyzed transformation of organic constituents, primarily from microorganisms. The ultimate fate of a constituent introduced into several environmental systems (e.g., soil, water, etc.) may be any compound other than the parent compound that was originally released into the environment. For example, trichloroethylene (TCE) biodegrades to produce 1,2-dichloroethylene (1,2-DCE), vinyl chloride, and other compounds; from a toxicity viewpoint 1,2-DCE is less toxic and vinyl chloride more toxic than TCE.

Biological degradation may also initiate other chemical reactions such as oxygen depletion in microbial degradation processes, creating anaerobic conditions and the initiation of redox potential-related reactions. The biodegradation potential should therefore be carefully evaluated in the design of monitoring programs.

Photolysis

Photolysis can be an important dissipative mechanism for specific chemical constituents in the environment. Similar to biodegradation, photolysis (or photodegradation) may cause the ultimate fate of a constituent introduced into an environmental system (e.g., surface water, soil, etc.) to be different from the constituent originally released. Hence, the photodegradation potential should be carefully evaluated in designing sampling and analysis as well as monitoring programs.

Chemical Degradation (Hydrolysis and Oxidation/Reduction)

Similar to photodegradation and biodegradation, chemical degradation, primarily through hydrolysis (i.e., a chemical transformation process in which an organic molecule reacts with water, forming a new carbon-oxygen bond and cleaving the carbon bonding with the original molecule) and oxidation/reduction (redox) reactions, can also act to change chemical constituent species from what the parent compound used to be when it was first introduced to the environment. For instance, oxidation may occur as a result of chemical oxidants being formed during photochemical processes in natural waters. Similarly, reduction of constituents may take place in some surface water environments (primarily those with low oxygen levels). Hydrolysis of organics usually results in the introduction of a hydroxyl group (–OH) into a constituent structure. Hydrated metal ions (particularly those with a valence ≥ 3) tend to form ions in aqueous solution, thereby enhancing species solubility.

Retardation Factor

Retardation is the chemical-specific, dynamic process of adsorption to and desorption from aquifer materials. In the assessment of the environmental fate and transport properties of chemical contaminants, reversible equilibrium and controlled sorption may be simulated by the use of a retardation factor or coefficient.

A retardation factor, R, can be calculated for a contaminant as a function of the chemical's organic carbon partition coefficient (K_{oc}), and also the bulk density (β) and porosity (n) of the medium through which the contaminant is moving. Typically, retardation factors are calculated for linear sorption in accordance with the following formula:

$$R = 1 + (\beta K_d)n$$

where $K_d = K_{oc} \times f_{oc}$, and f_{oc} is the organic carbon fraction. Thus, the retardation coefficient is reduced to the following relationship:

$$R = [1 + \beta K_{oc} f_{oc} n]$$

In general, if a compound is strongly adsorbed then it also means this particular compound will be highly retarded. In the aquifer system, the retardation factor gives a measure of how fast a compound moves relative to groundwater (Nyer, 1993). For example, a retardation factor of two indicates that the specific compound is traveling at one-half the groundwater flow rate. This will usually become a very important parameter in the design of groundwater remediation systems.

Miscellaneous Equilibrium Constants Associated with Speciation

Equilibrium constants are important predictors of a compound's chemical state in solution. For example, a constituent which is dissociated (ionized) in solution will be more soluble and therefore more likely to be released into the environment, and more likely to migrate in a surface water body. Many inorganic constituents, such as heavy metals and mineral acids, can occur as different ionized species depending on pH. Organic acids, such as the phenolic compounds, exhibit similar behavior. It should also be noted that ionic metallic species present in the environment may have a tendency to bind to particulate matter, if present in a surface water body, and to settle out to the sediment over time and distance. In fact, heavy metals are removed in natural attenuation by ion exchange reactions, whereas trace organics are removed primarily by adsorption. Metallic species also generally exhibit bioaccumulative properties.

Typically, when metallic species are present in the environment, the sampling and analysis of both sediment and biota would be a necessary part of a site characterization or remedial investigation effort.

REFERENCES

Evans, L.J. 1989. Chemistry of Metal Retention by Soils. *Environ. Sci. Technol.,* Vol. 23, No. 9, 1047–56.

Lyman, W.J., W.F. Reehl and D.H. Rosenblatt. 1990. *Handbook of Chemical Property Estimation Methods: Environmental Behavior of Organic Compounds.* American Chemical Society, Washington, D.C.

Nyer, E.K. 1993. *Practical Techniques for Groundwater and Soil Remediation.* Lewis Publishers, Boca Raton, FL.

Swann, R.L. and A. Eschenroeder (Eds.). 1983. *Fate of Chemicals in the Environment.* ACS Symp. Ser. 225, American Chemical Society, Washington, D.C.

USEPA. 1987. RCRA Facility Investigation (RFI) Guidance, EPA/530/SW-87/001. U.S. Environmental Protection Agency, Washington, D.C.

APPENDIX D
HEALTH AND SAFETY REQUIREMENTS FOR THE INVESTIGATION OF CONTAMINATED SITES

D.1 THE HEALTH AND SAFETY PLAN

The purpose of a health and safety plan (HSP) is to identify, evaluate, and control health and safety hazards, and to provide for emergency response during site characterization activities at a potentially contaminated site. All personnel entering the case site will generally have to comply with the applicable HSP. Also, the scope and coverage of the HSP may be modified or revised to encompass any changes that may occur at the site, or in the working conditions following the development of the initial HSP. Box D.1 shows the typical outline/elements of a HSP. Typically, a description of pertinent site background information should be fully elaborated as part of the HSP.

In general, every project should start with a health and safety review (at which all site personnel sign a review form), a tailgate safety meeting (to be attended by all site activities personnel), and a safety compliance agreement (that should be signed by all persons entering the site, i.e., both site personnel and site visitors).

BOX D.1
Typical Outline of a Health and Safety Plan

- Introduction (consisting of site information and the identification of responsible personnel, such as the health and safety officer and an on-site safety manager)
- General safety requirements (including a discussion of safety requirements to be met, and the type of emergency equipment required)
- Employee protection program
- Decontamination procedures
- Health and safety training
- Emergency response (including emergency telephone numbers, personal injury actions, acute exposure to toxic materials responses, and directions to nearest hospital or medical facility)

D.2 RESPONSIBLE PERSONNEL

Contractors, subcontractors, and other investigative teams are required to implement the HSP during site characterization and remedial activities. The responsible personnel will include a Health and Safety Officer (HSO) and an On-site Health and Safety Manager (OHSM); the responsibilities of the HSO and the OHSM may be assumed by the same individual.

The HSO has the primary responsibility for ensuring that the policies and procedures of the HSP are implemented by the OHSM. The HSO ensures that all personnel designated to work at the site:

- Have been declared fit for the specific tasks by a physician or other qualified health professional
- Are able to wear air-purifying respirators
- Would have received adequate hazardous waste site operations and emergency response training

The HSO is also responsible for providing the appropriate safety monitoring equipment and other resources necessary to implement the HSP. Significant deviations/changes to the original HSP must be approved by the HSO. Typically, the HSO will have the authority to resolve outstanding health and safety issues that come up during site operations.

The OHSM supervises all site activities and is responsible for implementing the HSP. The OHSM is responsible for providing copies of the HSP to the site crew (including subcontractors) and for advising the site crew on all health and safety matters. Specific tasks generally required to be performed by the OHSM are shown in Box D.2.

The OHSM has stop-work authority if a dangerous or potentially dangerous and unsafe situation exists at the site. Consequently, the OHSM must be notified of any changes in site conditions during the course of the project that may have an impact on the safety of personnel, the environment, or property.

D.3 GENERAL SAFETY REQUIREMENTS

Box D.3 identifies the minimum safety requirements to adopt during the investigation of potentially contaminated site problems. In addition, the OHSM should take steps to protect personnel engaged in site activities from physical hazards such as falling objects or tripping over hoses, pipes, tools, or equipment, slipping on wet or uneven surfaces, insufficient or faulty protective equipment, insufficient or faulty tools and equipment, overhead or below-ground electrical hazards, heat stress and strain, insect and reptile bites, inhalation of dust, etc. The OHSM should also monitor and check the work habits of the site crew to ensure that they are safety-conscious.

D.3.1 Safety and Emergency Needs

Several safety and emergency supplies are generally required on-site at all times, including:

BOX D.2
Specific Task Requirements of the OHSM

- Inspect site prior to start of work, especially for areas of concern that may have been omitted in the HSP or other areas that require special attention
- Conduct daily safety briefings and site-specific training for on-site personnel prior to commencing work
- Modify or develop health and safety procedures, after consultation with the HSO, when site or working conditions change
- Maintain adequate safety supplies and equipment on site
- Maintain and supervise site control, decontamination, and contamination-reduction procedures
- Investigate all accidents and incidents that occur during site activities
- Discipline or dismiss personnel whose conduct do not meet the requirements of the HSP, or whose conduct may jeopardize the health and safety of the site crew
- Immediate notification of the emergency response authorities (e.g., the Fire and Police Departments), and the implementation of evacuation procedures during an emergency situation
- Coordination of emergency response equipment, and also coordination of transportation of affected personnel to the nearest emergency medical care (including subsequent personnel decontamination) in the event of an emergency

BOX D.3
Minimum Safety Requirements for Contaminated Site Investigation Activities

- At least one copy of the HSP should be available at the site at all times
- In general, field work will preferably be conducted during daylight hours only. The OHSM will have to grant special permission for any field activities beyond daylight hours
- There should be at least two persons in the field/site at all times of site activities
- Eating, drinking, and smoking should be restricted to a designated area, and all personnel should be required to wash their hands and faces before eating, drinking, or smoking
- Shaking off and blowing dust or other materials from potentially contaminated clothing or equipment should be prohibited
- The OHSM should take steps to protect personnel engaged in site activities from potential contacts with splash water generated from decontamination of samplers, augers, etc.

- First aid kits containing bandaging materials (e.g., band aids, adhesive tape, gauze pads and rolls, butterfly bandages, and splints), antibacterial ointments, oxygen, pain killers (e.g., aspirin, acetaminophen), etc.
- Plastic sheetings
- Shovels and tools
- Warning tapes
- Personal protection equipment, including respirators, protective suits (e.g., Tyvek suits), hard hats, goggles, boots, and gloves
- Potable water with electrolytic solution

In general, the use of appropriate types of protection equipment will ensure the safety of the field crew. For instance, monitoring well installation operations, proposed soil sampling procedures, and other excavation activities will typically disturb potentially impacted soils and allow associated dusts to become airborne. These dusts may migrate into the workers' breathing zone and pose health threats to the affected individuals by becoming deposited onto their lung tissues and mucous membranes. Criteria that meets stipulated threshold limit values should be met during the field operations, else adequate corrective measures should be implemented to avoid potential risks to workers and other potential receptors in the vicinity of the site. Also, drilling operations will normally generate constant high-decibel noise levels. Workers should be protected from high-frequency hearing deficiency (otherwise known as tinnitus) that is caused by chronic exposures to high-decibel noise and high-level impact noise. The use of ear plugs and other hearing protection apparatus should help minimize worker exposures to excessive and damaging noise from field activities.

D.4 EMPLOYEE PROTECTION PROGRAM

The level of personnel protection equipment (PPE) to be worn by field personnel are identified and enforced by the HSO and OHSM. The levels of protection may change as additional information is acquired. In general, if a higher level of protection above that specified in the HSP is needed, then approval by the HSO will be required. For instance, at a given facility or site, it may initially be determined that work crews and field sampling teams will require Level C PPE consisting of a full-face or half-face respirators fitted with approved HEPA filters, chemical resistant coveralls (such as Tyvek or Saranex), hard hats, gloves with chemical-resistant outer shells and chemical-resistant linings, chemical-resistant boots, ear plugs or other hearing protection apparatus, and safety glasses or goggles during drilling and sampling activities. However, changes in site conditions may create a situation in which a higher level (e.g., Level D), or a lower level (e.g., Level B or Level A) PPE may be required. Such changes should be closely monitored by the OHSM. All personnel utilizing higher level (e.g., Level C or D) PPE must have successfully passed a thorough physical examination and must receive the consent of a qualified physician prior to engaging in any on-site activities that utilize these types of equipment.

In all cases of contaminated site investigation programs, any additional precautions necessary to prevent exposures during drilling, sampling, and decontamination procedures should be implemented as site and weather conditions change. At all times, sufficient drinking water and safety equipment should be made available as necessary. Decontamination and emergency procedures are also an important part of the overall employee protection program.

D.4.1 Decontamination Procedures

All site personnel as well as equipment used for site-related activities should be subject to a thorough decontamination process. Separate decontamination stations should be established for personnel, equipment, and machinery. Personnel decontamination typically consist of portable showers and clean clothes provided to the employees. The OHSM should establish the equipment decontamination area and provide the decontamination station with a basin of soapy water, a rinse basin with plain water, thick plastic base sheeting, and a waste container with a disposable plastic bag.

The decontamination area should be clearly delineated, highly visible, and accessible to all personnel engaged in site investigation and remedial activities. Small equipment such as hand tools may be decontaminated at this station. A decontamination area should also be designated for large equipment; this area will consist of a thick plastic floor covering (of minimum thickness ≈ 10 mm) bermed to collect decontamination runoff (that will be disposed of appropriately) and wash and rinse solutions.

A decontamination crew should be established with the purpose of decontaminating site activities personnel and equipment. The decontamination of the field activities personnel should be the primary focus of the decontamination crew. After decontamination of the site personnel, all equipment should then be decontaminated. Following decontamination of the equipment, the decontamination crew would then decontaminate themselves.

D.4.2 Health and Safety Training

All personnel working at a project site should have participated in adequate health and safety training. This should be ascertained and certified by the OHSM before an individual is allowed to enter into and work at a potentially contaminated site. Typically, all personnel working at the site are required to have participated in a 40-h health and safety training course, in accordance with 29 CFR Part, 1910.120(e)(2). Proof of current certification in this training must be presented to the OHSM before an individual is allowed to enter or work on the site. The OHSM should also conduct on-site health and safety training covering items required by the HSP. Additional safety briefings should be provided by the OHSM if the scope of work changes in a manner that will potentially affect personal health and safety of the workers.

All personnel engaged in site activities are required to participate in the training by the OHSM. The OHSM should indeed make a copy of the HSP available to all personnel before any field investigations or remedial activities commence. All persons involved in these activities should sign a health and safety review sheet, certifying that they have read, understand, and agree to comply with the stipulations of the HSP.

D.4.3 Emergency Response

Several provisions for an emergency response plan should be carried out whenever there is a fire, explosion, or release of hazardous material which could threaten human health or the environment. The decision as to whether or not a fire, explosion, or hazardous material(s) release poses a real or potential hazard to human health or the environment is to be made by the OHSM.

If a situation requires outside assistance, the appropriate response parties should be contacted using mobile/cellular phones to be carried to the site, and/or phones located in nearby facilities that should have been identified at the start of the project.

Emergency phone numbers should be compiled and included in the HSP. Also, the directions to the nearest hospital or medical facility, including a map clearly showing the shortest route from the site to the hospital or medical facility, should be kept with the HSP at the site.

In all of the emergency response situations, the emergency transport and medical personnel should immediately be notified as to the type and degree of injury, as well as the extent and nature of contamination to the injured or affected individual(s).

Personal Injury

If injury occurring at the site is not immediately life-threatening, the individual should be decontaminated and then emergency first aid provided. If the injury is serious, the appropriate emergency response agencies should be notified, who will arrange transportation for the individual to the nearest medical facility. If required, life-saving procedures should be initiated. The OHSM will provide information on any acute exposures and materials safety data sheets (MSDS) to appropriate medical personnel.

Acute Exposure to Toxic Materials

The potential human exposure routes to toxic materials affecting site workers typically include inhalation, dermal contact, ingestion, and eye contact. Different procedures will generally be required to deal with worker exposures via the various routes.

For inhalation exposures, the following steps should be followed: move the victim to fresh air, call for help, decontaminate as best as possible, and notify the appropriate emergency medical services to transport the individual to the nearest medical facility.

For dermal contact, if life-saving procedures are required, then: first notify the appropriate emergency medical services for transportation of the individual to the nearest medical facility, then decontaminate as best as possible and initiate life-saving procedures if qualified. In a dermal contact situation where life-saving procedures are not required: decontaminate the individual by using copious amounts of soap and water, wash/rinse the affected area for at least 15 minutes, complete decontamination, and then provide appropriate medical attention.

For ingestion exposures: call for help, decontaminate as best as possible, and arrange for the appropriate emergency medical services to transport the individual to the nearest medical facility.

In the case of eye contact, decontaminate as best as possible and arrange for the appropriate emergency medical services to transport the individual to the nearest medical facility. If life-saving procedures are not required, then rinse eyes using copious amounts of water for at least 15 min, complete decontamination, and then provide appropriate medical attention.

The OHSM should provide information on any acute exposures and MSDS to the appropriate medical personnel. The MSDS for the contaminants that are anticipated to be encountered at the site would already have been compiled and appended to the HSP.

APPENDIX E
SELECTED ENVIRONMENTAL MODELS AND DATABASES POTENTIALLY RELEVANT TO THE MANAGEMENT OF CONTAMINATED SITE PROBLEMS

E.1 SELECTED ENVIRONMENTAL MODELS

Table E.1 consists of selected models that may be applied to some aspect of contaminated site management problems. The choice of one particular model over another will generally be problem-specific. This listing is by no means complete and exhaustive.

E.2 SELECTED DATABASES

Several databases relating to numerous chemical substances that exist within the scientific community may find extensive useful applications in the management of contaminated site problems. Two example databases of general interest to environmental contamination management problems — one for its international appeal, and the other for its wealth of risk assessment support information — are presented below. The presentation is meant to demonstrate the overall wealth of scientific information that already exist, and that should be consulted to provide the relevant support in the design of corrective action programs for contaminated site problems. Also, the U.S. Superfund program maintains an information directory through the EPA's Office of Emergency and Remedial Response in Washington, D.C.; this directory identifies and describes several sources of information that should prove useful in contaminated site management problems. The Superfund information directory also identifies, references, and provides guidance and documentation on several EPA and non-EPA databases and information libraries that may be useful for similar purposes.

E.2.1 Overview of the International Register of Potentially Toxic Chemicals

In, 1972, the United Nations Conference on the Human Environment, held in Stockholm, recommended the setting up of an international registry of data on chemicals likely to enter and damage the environment. Subsequently, in 1974, the Governing Council of the United Nations Environment Programme (UNEP) decided to establish both a chemicals register and a global network for the exchange of information the

Table E.1 Listing of Selected Environmental Models Potentially Applicable to the Evaluation of Environmental Contamination Problems

Model Name	Environmental Compartment	Model Description	Model Uses	Sources of Model Information and/or Developer
AIRTOX (Air Toxics risk management framework)	Air	AIRTOX is a decision analysis model for air toxics risk management. The framework consists of a structural model that relates emissions of air toxics to potential health effects, and a decision tree model that organizes scenarios evaluated by the structural model	AIRTOX evaluates the magnitude of health risks to a population, a specific source's contribution to the total health risk, and the cost effectiveness of current and future emission control measures	EPRI (Electric Power Research Institute), Palo Alto, CA
TOXIC	Air	TOXIC is a microcomputer program that calculates the incremental risk to the hypothetical maximum exposed individual from hazardous waste incineration. It calculates exposure to each pollutant individually, using a specified dispersion coefficient (which is the ratio of pollutant concentration in air in $\mu g/m^3$ to pollutant emission rate in gm/s)	TOXIC is used in hazardous waste facility risk analysis. It is a flexible and convenient tool for performing inhalation risk assessments for hazardous waste incinerators. It gives point estimates of inhalation risks	Rowe Research and Engineering Associates, Alexandria, VA
BOXMOD (Box Model)	Air	BOXMOD is an interactive steady-state, simple atmospheric area source box model for screening chemicals. It is applicable to regions containing many diffuse emission sources within its boundaries, such as in an urban area	BOXMOD calculates a single annual average concentration applicable to the entire region based on a uniform area emission rate. It is used for detailed screening assessments of contaminated areas	USEPA's Office of Air Quality Planning and Standards (OAQPS), Research Triangle Park, NC
GAMS (GEMS Atmospheric Modeling Subsystem)	Air	GAMS is an integrated atmospheric modeling system that can be used to estimate annual average concentrations, annual exposure, and lifetime and annual incidence of excess cancer cases	GAMS is used for refined exposure and risk analyses	USEPA's Office of Air Quality Planning and Standards (OAQPS), Research Triangle Park, NC

Model	Media	Description	Developer
SHORTZ and LONGZ (Atmospheric Dispersion Models)	Air	SHORTZ is a computer algorithm designed to use sequential short-term meteorological inputs to calculate chemical concentrations for averaging times. It is applicable in areas of both flat and complex terrain, and can accommodate receptors that are both above and below source elevations. LONGZ is the long-term average version of SHORTZ. It is an area source computer program designed for application to square, ground-level, area sources	SHORTZ calculates the short-term ground-level pollutant concentrations produced at large number of receptors, by emissions from various sources. LONGZ, used in near-field concentration estimates, calculates vertical dispersion based on the integral of the vertical term in the Gaussian equation between the downwind edge and the upwind edge of the source area	USEPA's Office of Air Quality Planning and Standards (OAQPS), Research Triangle Park, NC
ISCLT (Industrial Source Complex Long-Term Model)	Air	ISCLT is a long-term sector-averaged environmental model that uses statistical wind summaries to calculate annual ground-level concentrations or dispersion values	ISCLT is an air model that calculates annual ground-level concentrations or deposition values, and estimates risk and exposure level using these values. It is used for modeling long-term air exposures associated with point and areal sources of air emissions	USEPA's Office of Air Quality Planning and Standards (OAQPS), Research Triangle Park, NC
ISCST (Industrial Source Complex Short-Term Model)	Air	ISCST is a short-term model that uses finite line source approach to model area sources. Each square area source is modeled as a single line segment oriented normal to the wind direction. The model does not accurately account for source-receptor geometry	ISCST algorithm is used to model short-term air exposures associated with air emissions. It predicts zero concentration for a receptor located within an area source	USEPA's Office of Air Quality Planning and Standards (OAQPS), Research Triangle Park, NC
PTPLU (PoinT PLUme model)	Air	PTPLU is a Gaussian plume dispersion model that estimates the location of the maximum short-term concentration in the atmosphere from a single point source (as a function of stability and wind speed)	PTPLU is a point-source Gaussian plume dispersion algorithm, used for detailed screening analyses. It is used for estimating worst-case hourly concentrations for steady-state conditions	Environmental Sciences Research Laboratory, Office of Research and Development, USEPA, Research Triangle Park, NC

Table E.1 (continued)   Listing of Selected Environmental Models Potentially Applicable to the Evaluation of Environmental Contamination Problems

Model Name	Environmental Compartment	Model Description	Model Uses	Sources of Model Information and/or Developer
FDM (Fugitive Dust Model)	Air	FDM is a computerized analytical air quality model specifically designed for computing concentration and deposition impacts from fugitive dust sources. The sources may be point, line or areal; it contains no plume rise algorithm. The model is generally based on Gaussian plume formulation for computing concentrations, with improved gradient-transfer deposition algorithm. FDM accounts for deposition losses as well as pollutant dispersion	FDM models both short-term and long-term average particulate emissions from surface mining and similar sources. A primary use is for computation of concentrations and depositions rates resulting from emission sources such as hazardous waste sites where fugitive dust is a concern. Concentration and deposition are computed at all user-selected receptor locations	Support Center for Regulatory Air Models, Office of Air Quality Planning and Standards (OAQPS), USEPA, Research Triangle Park, NC
PAL (Point, Area, Line-Source model)	Air	PAL is a steady-state Gaussian plume model, recommended for source dimensions of tens to hundreds of meters. It uses Gaussian model equations for a finite line segment	PAL calculates concentrations at user-specified accuracy levels	USEPA's Office of Air Quality Planning and Standards (OAQPS), Research Triangle Park, NC
CREAMS (The Chemical, Runoff, and Erosion from Agricultural Management Systems model)	Surface water	CREAMS is a numerical, finite-difference method of solution, used to estimate sorption and degradation of chemicals, and erosion of single land segments. Inputs required include watershed characteristics and chemical properties	CREAMS is used to estimate sorption and degradation of chemicals, and erosion of single land segments	U.S. Department of Agriculture (USDA) U. S. Corps of Engineers
QUAL2E (Enhanced Stream Water Quality Model)	Surface water	QUAL2E is a steady-state numerical, finite-difference model for conventional pollutants in branching streams and well-mixed lakes. It includes conservative substances, temperature,	QUAL2E is used for waste load allocation and permitting	Center for Water Quality Modeling, Environmental Research Laboratory, Office

Name	Type	Description	Source
		coliform bacteria, biochemical oxygen demand (BOD), dissolved oxygen (DO), nitrogen, phosphorus, and algae	of Research and Development, USEPA, Athens, GA
EXAMS-II (Exposure Analysis Modeling System)	Surface water	EXAMS is a steady-state contaminant fate and transport numerical, finite-difference model, offering one-, two-, or three-dimensional compartmental solutions in surface water bodies. It is based on a series of equations which account for interactions between the canonical aquatic environment into which a chemical is released, the chemistry of a given chemical and the toxicant loading quantities. It includes process models of the physical, chemical, and biological phenomena governing the transport and fate of compounds	Environmental Research Laboratory, Office of Research and Development, USEPA, Athens, GA
REACHSCA (Reach Scan model)	Surface water	REACHSCA is a simple dilution model used to estimate steady-state chemical concentration in surface water bodies due to continuous loading from a single discharging facility	Office of Toxic Substances, USEPA, Washington, D.C.
WTRISK (Waterborne Toxic Risk Assessment model)	Surface water	WTRISK provides a framework that employs a risk assessment methodology in which mathematical models simulating chemical fate and transport processes can be linked to determine pollutant concentrations in all appropriate environmental media. These predicted concentrations are then used as input values for modeling nearby population exposures and potential health risks	EPRI (Electric Power Research Institute), Palo Alto, CA

EXAMS simulates the fate of organic chemicals in surface water bodies (i.e., rivers, lakes, reservoirs, estuaries). Model can estimate the time-varying and/or steady-state concentrations of the chemical in the water body in various phases (dissolved, sediment, sorbed, biosorbed). It has been designed to evaluate the consequences of longer-term, primarily time-averaged chemical loadings that ultimately result in trace-level contamination of aquatic systems. Suitable for modeling synthetic organic chemicals for freshwater, nontidal aquatic systems

REACHSCA is used to estimate steady-state chemical concentration in surface water bodies (mainly river reaches)

WTRISK provides a flexible framework for the risk assessment of toxics in surface water. It aids the estimation of the source terms or quantities of toxics emitted into the environment

Table E.1 (continued) Listing of Selected Environmental Models Potentially Applicable to the Evaluation of Environmental Contamination Problems

Model Name	Environmental Compartment	Model Description	Model Uses	Sources of Model Information and/or Developer
MINTEQA1 (Metals Exposure Analysis modeling system)	Surface water	MINTEQA1 is an interactive complex computer program consisting of a steady-state 3-dimensional numerical, finite-difference compartmental model	MINTEQA1 is designed for modeling the equilibrium states for metal loadings in freshwater, nontidal aquatic systems	Battelle Pacific Northwest Laboratories, Richland, WA Environmental Research Laboratory, USEPA, College Station Rd., Athens, GA
SARAH (Surface water backcalculation procedure)	Surface water	SARAH is a semianalytical steady-state, 1-dimensional surface water exposure assessment model. It models contaminated leachate plume feeding a downgradient surface waterbody (stream or river). It employs a Monte Carlo simulated generic environment. Bioaccumulation in fish, degradation, sorption, dilution, and volatilization are included	SARAH is used to model contaminated leachate plume feeding a downgradient stream or river	
MEXAMS (Metals Exposure Analysis Modeling System)	Surface water	MEXAMS is a steady-state, 3-dimensional compartmental model. It is a combination of two models (MINTEQ and EXAMS) designed for modeling metal loadings. The model links a complex speciation model with an aquatic transport/fate model, and should help discriminate between the fraction of metal that is dissolved and in bioavailable form, and the fraction that is complexed and rendered relatively nontoxic.	MEXAMS is designed for modeling metals loadings, and is suitable for freshwater, nontidal aquatic systems	Battelle Pacific Northwest Laboratories, Richland, WA Center for Water Quality Modeling, Environmental Research Laboratory, USEPA, College Station Rd., Athens, GA
PDM (Probabilistic Dilution Model)	Surface water	PDM models exceedances of specified concentration levels in streams. The model estimates are based on statistical distribution of daily volume flow, and on solution of mass balance dilution evaluation	PDM estimates the percentage of time that a given concentration level may be exceeded in receiving surface water bodies	Office of Toxic Substances, Exposure Evaluation Division, USEPA, Washington, D.C.

Model	Type	Description	Source
TOXIWASP	Surface water	TOXIWASP is a dynamic (time-varying) 3-dimensional model for simulating transport and fate of toxic chemicals in water bodies. Both organic chemicals and sediments are simulated by the model. The model attempts to account for the full array of chemical transformations and sediment-chemical exchange processes through a set of transport, chemical transformation, and mass loading submodels. The model accounts for advection dispersion, bed sedimentation and erosion, pore water diffusion, hydrolysis, photolysis, oxidation, biodegradation, and volatilization	TOXIWASP calculates the concentrations for every segment of modeled waterbody, including surface water, subsurface water, surface bed and subsurface bed. It is a chemical and stream quality model	Center for Water Quality Modeling, Environmental Research Laboratory, USEPA, College Station Rd., Athens, GA
SWIP (The Survey Waste Injection Program)	Groundwater	SWIP is a saturated zone 3-dimensional numerical model which simulates contaminant (chemical or thermal) movement in an aquifer, including well-pumping effects. It incorporates finite difference approximation with options for several matrix solution techniques	SWIP simulates chemical and/or thermal contaminant movement in an aquifer	USGS (U.S. Geological Survey)
SOLUTE (Solute transport model)	Groundwater	SOLUTE is a basic program package of analytical models for solute transport in groundwater. The package includes several subprograms, including UNITS (for conversion of units); ERFC (to calculate error functions and complimentary error functions); ONED1 (for solute transport in 1-D); WMPLUME and SLUG (for solute transport in 2-D); RADIAL and LTIRD (for 2-dimensional radial flow); and PLUME3D and SLUG3D (for 3-dimensional transport)	SOLUTE is used in solute transport modeling, to estimate receptor exposure concentration distributions	International Ground Water Modeling Center (IGWMC), Holcomb Research Institute, Butler University, Indianapolis, IN

Table E.1 (continued)   Listing of Selected Environmental Models Potentially Applicable to the Evaluation of Environmental Contamination Problems

Model Name	Environmental Compartment	Model Description	Model Uses	Sources of Model Information and/or Developer
AT123D (Analytical Transient 1-, 2-, 3-Dimensional simulation model)	Groundwater	AT123D is a saturated zone analytical transient 1-, 2-, 3-dimensional simulation model, that is used to simulate chemical movement and waste transport in the aquifer system. It predicts the spread of contaminant plume through groundwater, and can handle constant as well as time-varying chemical release to groundwater. Chemical processes include both adsorption and degradation. Output concentrations are time-varying	AT123D is an environmental model that predicts spread of a contaminant plume (chemical, thermal, or radioactive) through groundwater (saturated zone) and estimates the chemical concentration within groundwater at positions on a user-specified three-dimensional grid	Oak Ridge National Laboratory (ORNL), Environmental Sciences Division, Oak Ridge, TN
MYGRT (Migration of solutes in the subsurface environment)	Groundwater	MYGRT is a 1-, 2-dimensional fate model that provides a method for computing the fate of reacting or nonreacting inorganic chemicals released to the groundwater environment. The simulation is based on quasi-analytical solutions to the conservation of mass equations, including advection, dispersion, and retardation. Simulation of both continuous and finite-duration solute releases are possible. It gives results that are "ball-park" approximations to the dynamic and complex problem of solute migration predictions	MYGRT allows users to calculate concentration of solute at an elapsed time from the release time. Calculations can also be made to estimate time taken for a chemical to travel the distances of interest	EPRI (Electric Power Research Institute), Palo Alto, CA
SUTRA (Saturated-Unsaturated Transport model)	Groundwater	SUTRA is a 2-dimensional numerical, finite-element, and integrated finite-element solution technique. It is a	SUTRA may be applied to analyze groundwater contaminant transport problems and aquifer restoration	USGS, Water Resources Department

Model	Medium	Description	Application	Source
TRANS	Groundwater	solute transport simulation model, that may be used to model natural or human-induced chemical species transport, including processes of solute sorption, production, and decay. TRANS is a 2-dimensional numerical model that uses random walk solution technique. Concentration distribution in aquifer represents a vertically averaged value over the saturated thickness of the aquifer	designs. It predicts fluid movement and the transport of either energy or dissolved substances in a subsurface environment. TRANS is used to predict groundwater pollution problems	Illinois State Water Survey
SHALT (Solute, Heat, and Liquid Transport)	Groundwater	SHALT is a 2-dimensional numerical, finite element model for predicting liquid flow, heat transport, and solute transport in a regional groundwater flow system. Fractured media may be modeled by treating the fractured rock as a continuum. Model output consists of the pressure, concentration, and temperature distribution at each time step	SHALT is used to predict liquid flow, heat transport, and solute transport in a regional groundwater flow system	Inland Waters Directorate, Environment Canada; Atomic Energy of Canada, Ltd.
VHS (Vertical and Horizontal Spread model)	Groundwater	VHS is a steady-state analytical model used to simulate the dispersion of contaminants and to calculate the contaminant concentrations at a receptor point or well directly downgradient of a waste disposal area	VHS is a groundwater dilution model used for predicting steady-state contaminant concentrations at receptor locations	USEPA, Research Triangle Park, NC
MOC (Method-Of-Characteristics model for solute transport)	Groundwater	MOC is a finite-difference computer model that is applicable to 1- or 2-dimensional problems involving steady-state or transient flow. The model computes changes in concentration over time caused by the processes of convective transport, hydrodynamic dispersion, and mixing (or dilution) from fluid sources. Model assumes that	MOC is used for calculating changes in the concentration of dissolved (non-reactive) chemical species in flowing groundwater, by the model simulating solute transport in flowing groundwater. The purpose of the simulation is to compute the concentration of a dissolved chemical species in an aquifer at any specified place and time	USGS, The Holcomb Research Institute, IN

Table E.1 (continued)   Listing of Selected Environmental Models Potentially Applicable to the Evaluation of Environmental Contamination Problems

Model Name	Environmental Compartment	Model Description	Model Uses	Sources of Model Information and/or Developer
		solute is nonreactive and that gradients of fluid density, viscosity, and temperature do not affect the velocity distribution; aquifer may be heterogeneous and/or anisotropic. Model couples the groundwater flow equation with the solute transport equation. The model is based on a rectangular, block-centered, finite difference grid. The method of characteristics is used in the model to solve the solute transport equation. By coupling the flow equation with the solute-transport equation, the model can be applied to both steady-state and transient flow problems		
FEMWASTE1 (Finite-Element Model of Waste Transport)	Soil-groundwater (unsaturated and saturated zones)	FEMWASTE1 is a 2-dimensional transient model for predicting waste transport through saturated-unsaturated porous media under dynamic groundwater conditions. Solution is by finite-element-weighted residual method. It models interzone transfers, incorporates convection and dispersion, simulates degradation of nonconservative substances, accounts for adsorption, and is capable of modeling layered, heterogeneous soil zones. The model is applied to a single chemical species without considering	FEMWASTE1 is used to simulate the fate/transport of contaminants in saturated and unsaturated porous media	Oak Ridge National Laboratory (ORNL), Environmental Sciences Division, Oak Ridge, TN

Model	Media	Description	Source
RITZ (Regulatory and Investigative Treatment Zone model)	Soil vadose (unsaturated) zone	RITZ is useful in predicting contaminant transport of residual constituents in settings similar to a land treatment area with respect to contaminant transport. the effects of other chemicals that may be present in the porous media. RITZ is a 1-dimensional unsaturated zone analytical model for pollutant transport. It can account for soil solution, volatilization and atmospheric losses, and biological degradation of chemicals. It considers the effect of an oil phase on pollutant transport. The model was designed to predict fate of contaminants in a land treatment scenario and considers downward movement of chemicals. Output includes mass transport to groundwater that can become input to a groundwater dilution model	Office of Environmental Processes and Effects Research, Oak Ridge, TN
SESOIL (Seasonal Soil compartment model)	Soil vadose (unsaturated) zone	Estimates the rate of vertical chemical transport and transformation in the soil column. Also has capability to simulate contaminant transport in washload at soil surface and volatilization rates to the atmosphere. SESOIL is a seasonal soil compartment model that estimates the rate of vertical chemical transport and transformation in the soil column in terms of mass and concentration distributions among the soil, water, and air phases in the unsaturated soil zone. It is designed for long-term environmental fate simulations of pollutants in the vadose zone. It is a 1-dimensional unsaturated zone model for both organics and inorganics	Office of Toxic Substances, Exposure Evaluation Division, USEPA, Washington, D.C.
PRZM (Pesticide Root Zone Model)	Soil vadose (unsaturated) zone	PRZM provides pollutant velocity distribution and concentration data for organic substances. Degradation is also simulated. PRZM is an unsaturated zone dynamic numerical, finite-difference, 1-dimensional transport model which simulates the vertical movement of pesticides in unsaturated soil, within and below the plant root zone. Time-varying transport, including advection and dispersion, is represented by the	USEPA, Environmental Research Laboratory, Athens, GA

Table E.1 (continued) Listing of Selected Environmental Models Potentially Applicable to the Evaluation of Environmental Contamination Problems

Model Name	Environmental Compartment	Model Description	Model Uses	Sources of Model Information and/or Developer
		model. It has two major components — hydrology and chemical transport. It accommodates various release rates and schedules		
SWAG (Simulated Waste Access to Groundwater)	Soil vadose (unsaturated) zone	SWAG is a 3-compartment analytical computer model for organic pollutant transport that considers transformations in the soil-geological matrix	SWAG predicts organic pollutant transport into groundwater	Office of Environmental Processes and Effects Research, Oak Ridge, TN
HELP (Hydrologic Evaluation of Landfill Performance model)	Soil vadose (unsaturated) zone	HELP is a quasi two-dimensional deterministic numerical, finite-difference model that computes a daily water budget for a landfill represented as a series of horizontal layers. It models leaching from landfills to unsaturated soil beneath a landfill. It models both organics and inorganics, using rainfall and waste solubility to model leachate concentrations leaving the landfill	HELP is used for water balance computation, and for the estimation of chemical emissions and for leachate quality assessment	USEPA's National Computer Center, Research Triangle Park, NC U.S. Corps of Engineers
HSSM (Hydrocarbon Spill Screening Model)	Multimedia	HSSM is a screening tool for light nonaqueous-phase liquid (LNAPL) impacts to the water table. The model consists of separate modules for addressing LNAPL flow through the vadose zone, LNAPL spreading in the capillary fringe at the water table, and dissolved LNAPL groundwater transport to potential receptor exposure locations. These modules are based on simplified conceptualizations of the flow	HSSM offers a simplified "order-of-magnitude" analysis for emergency response, initial phases of site investigations, facilities siting and permitting, and underground storage tank programs. The HSSM model is intended for the simulation of subsurface releases of LNAPLs, and is used to estimate the impacts of this type of pollutant on water table aquifers	USEPA's Robert S. Kerr Environmental Research Laboratory, Ada, OK

Name	Type	Description		Source
MULTIMED (Multimedia Exposure Assessment Model)	Multimedia	and transport phenomena which were used so that the resulting model would be a practical, though approximate, tool for simulating the transport and transformation of contaminants released from a hazardous waste disposal facility into the multimedia environment. The MULTIMED model simulates release to air and soil, including the unsaturated (vadose) and saturated zones, and possible interception of the subsurface contaminant plume by a surface stream. It further simulates movement through the air, soil, groundwater, and surface water media to humans and other potentially affected receptors. Uncertainties in parameter values used in the model are quantified using Monte Carlo simulation techniques	MULTIMED is used to simulate the movement of contaminants leaching from a waste disposal facility. It is intended for general exposure and risk assessments of waste facilities and for analyses of the impacts of engineering and management controls	Environmental Research Laboratory, Office of Research and Development, USEPA, Athens, GA
SITES (Contaminated Sites Risk Management System)	Multimedia	SITES is a flexible interactive computerized decision support tool for organizing relevant information and for conducting risk management analyses of contaminated sites. It has the dimensionality to model multiple chemicals, pathways, population groups, health effects, and remedial actions. The model uses information from diverse sources, such as site investigations, transport and fate modeling, behavioral and exposure estimates, and toxicology. It explicitly addresses the many uncertainties and allows for quick sensitivity analyses.	SITES is a computer-based integrating framework used to help evaluate and compare site investigation and remedial action alternatives in terms of health and environmental effects and total economic costs/impacts. The decision-tree structure in SITES allows for explicit examination of key uncertainties and the efficient evaluation of numerous scenarios. The model's design and computer implementation facilitates extensive sensitivity analysis	EPRI (Electric Power Research Institute), Palo Alto, CA

Table E.1 (continued)   Listing of Selected Environmental Models Potentially Applicable to the Evaluation of Environmental Contamination Problems

Model Name	Environmental Compartment	Model Description	Model Uses	Sources of Model Information and/or Developer
GEMS (Graphical Exposure Modeling System)	Multimedia	Both deterministic and probabilistic analysis methods are possible. GEMS is an operating environment that houses a variety of programs for performing exposure assessment studies. It is an interactive computer system for environmental models, physicochemical property estimation, and statistical analysis. GEMS has graphic display capabilities	GEMS is an interactive management tool that allows quick and meaningful analysis of environmental problems. It allows a user to estimate chemical properties, assess fate of chemicals in receiving environments, model resulting chemical concentrations, determine the number of people potentially exposed, and estimate the resultant human exposure and risk	USEPA, Research Triangle Park, NC Office of Pesticide and Toxic Substances, USEPA, Washington, D.C. General Sciences Corporation (GSC), Laurel, MD
PCGEMS (Personal Computer version of the Graphical Exposure Modeling System)		PCGEMS is a complete information management tool designed to help in performing exposure assessment studies. It is able to work interactively with GEMS on the VAX clusters	PCGEMS can be used to estimate chemical properties and perform simulation studies of chemical release in air, soil, and groundwater systems. The environmental modeling programs allows for the simulation of the migration and transformation of chemicals through the air, water, soil, and groundwater subsystems	
ENPART (Environmental Partitioning Model)	Multimedia (air, water, and soil)	ENPART uses simple physical-chemical data to estimate equilibrium concentration ratios of a chemical between the environmental compartments of air, water, and soil. More of a screening tool	ENPART is a multimedia model that estimates equilibrium concentration ratios of a chemical between the environmental compartments of air, water, and soil. It also, provides a second level of concentration and mass partitioning called dynamic partitioning. Subsequently, potential exposures in each environmental compartment to the chemical are estimated	Office of Toxic Substances, Exposure Evaluation Division, USEPA, Washington, D.C.

Name	Type	Description	Source
POSSM (PCB On-Site Spill Model)	Multimedia	POSSM is an exposure assessment methodology. It consists of a chemical transport and fate model capable of considering all of the key processes controlling chemical losses from a spill site (including volatilization, leaching to groundwater, and chemical washoff from the land surface due to runoff/erosion). Contains several relevant subprograms, including PTDIS (air model), RIVLAK (surface water model), GROUND (ground water model), and EXPOSE (exposure intake model)	POSSM provides a quantitative framework for estimating general public exposure levels associated with spills from utility electrical equipment. On-site environmental concentrations can be estimated with POSSM; off-site concentrations can be estimated with one of three relatively simple transport and fate models for air (PTDIS), surface water (RIVLAK), and ground water (GROUND) that are also incorporated into the methodology. Given estimates of on-site and/or off-site concentrations and the characteristics and activity patterns of the receptors of concern, inhalation, ingestion, and dermal exposure levels can be calculated using the program, EXPOSE. Methodology was developed primarily for PCBs, but is applicable to a wide range of organic chemicals	EPRI (Electric Power Research Institute), Palo Alto, CA
INPOSSM (Interactive PCB On-Site Spill Model)	Multimedia	INPOSSM is an interactive PCB exposure assessment model. It includes a chemical transport and fate model capable of considering key processes controlling chemical losses from a PCB or organic chemical spill site, including volatilization, leaching to groundwater, and chemical washoff from the land surface due to runoff/erosion	INPOSSM uses a Monte Carlo simulation model in evaluating the variability of predicted chemical losses	EPRI (Electric Power Research Institute), Palo Alto, CA
MCPOSSM (Monte Carlo PCB On-Site Spill Model)	Multimedia	MCPOSSM is a chemical spill exposure assessment methodology providing a quantitative framework for estimating uncertainties of chemical levels associated with spills. The core of the methodology is the PCB On-Site Spill	MCPOSSM provides a distribution of concentrations over time and probabilities of exceeding specified levels (i.e., probability of a worst-case level)	EPRI (Electric Power Research Institute), Palo Alto, CA

Table E.1 (continued)   Listing of Selected Environmental Models Potentially Applicable to the Evaluation of Environmental Contamination Problems

Model Name	Environmental Compartment	Model Description	Model Uses	Sources of Model Information and/or Developer
		Model (POSSM), and a Monte Carlo chemical transport and fate model capable of considering key processes controlling chemical losses from a spill site, including volatilization, leaching to groundwater, and chemical washoff from the land surface due to runoff/erosion		
TOX-RISK (Toxic Risk)	Multimedia	TOX-RISK is a package that provides data entry and management, computation of maximum likelihood estimates of risks or dose with confidence bounds, and graphs of dose-response functions fitted to the data provided by the user. TOX-RISK has the ability to fit 10 different models to data which are readily available in animal bioassays — including the Multistage, One Stage, Two Stage, Three Stage, Four Stage, Five Stage, Six Stage, Weibull, Mantal-Bryan, and the LogNormal model	TOX-RISK is a menu-driven interactive software package for performing health-related risk assessments. These risk assessments produce quantitative estimates of risks from quantal dose-response data	EPRI (Electric Power Research Institute), Palo Alto, CA
TOX-SCREEN	Multimedia (air, surface water, soil)	TOX-SCREEN is a multimedia screening-level model which assesses the potential fate of toxic chemicals released to the air, surface water, or soil. It incorporates equations developed for media-specific models and allows users to describe compartments directly	TOX-SCREEN is a screening device used to identify chemicals that are unlikely to pose problems even under conservative assumptions	Office of Pesticides and Toxic Substances, USEPA, Washington, D.C.

Name	Media	Description	Description	Source
MICROBE-SCREEN	Multimedia (air, surface water, soil)	MICROBE-SCREEN is a screening level multimedia model for assessing potential fate of microorganisms released to air, surface water, or soil. The model deals specifically with passively dispersed microorganisms, such as bacteria	MICROBE-SCREEN is a screening device to estimate transport and densities of microorganisms in various environments	General Sciences Corp. (GSC), Laurel, MD
UTM-TOX (Unified Transport Model for Toxic Materials)	Multimedia (air, soil, water)	UTM-TOX is a multimedia model designed for predicting the dispersion of pollutants through air, soil, and water. It links together several media-specific models	UTM-TOX can be used to track a pollutant through the ecosystem, budget for the partitioning of contaminants, calculate concentration of pollutant in many compartments of the ecosystem, and reliably assess the impact of the pollutant	Oak Ridge National Laboratory, Oak Ridge, TN
SMCM (Spatial-Multimedia-Compartmental-Model)	Multimedia (air, water, soil, biota)	SMCM is based on a modeling approach that estimates the multimedia partitioning of organic pollutants in local environments. It consists of coupled partial and ordinary differential equations that are solved simultaneously by finite-difference method, using operator-splitting techniques. It is a hydrid transport and fate model that makes use of both uniform and nonuniform 1-dimensional compartments. It has the capability of simulating a variety of pollutant transport phenomena	SMCM is designed to predict the multimedia concentrations of organic chemicals in the environment. It is used for the analysis of the multimedia distribution of organic pollutants in the environment	The National Center for Intermedia Transport, UCLA, Los Angeles, CA
RISK*ASSISTANT	Multimedia	RISK*ASSISTANT is designed to assist the user in rapidly evaluating exposures and human health risks from chemicals in the environment at a particular site. User need only provide measurements or estimates of the concentrations of	RISK*ASSISTANT is used to evaluate exposures and human health risks from chemicals. It has the ability to tailor exposure and risk assessments to local conditions	USEPA, Research Triangle Park, NC Hampshire Research Institute, Alexandria, VA

Table E.1 (continued)   Listing of Selected Environmental Models Potentially Applicable to the Evaluation of Environmental Contamination Problems

Model Name	Environmental Compartment	Model Description	Model Uses	Sources of Model Information and/or Developer
		chemicals in the air, surface water, groundwater, soil, sediment, and/or biota. It provides an array of analytical tools, databases, and information-handling capabilities for risk assessment		California Department of Health Services (CDHS), Toxic Substances Control Program, Sacramento, CA New Jersey Department of Environmental Protection, Trenton, NJ
RISKPRO	Multimedia	RISKPRO is a complete software system designed to predict the environmental risks and effects of a wide range of human health-threatening situations. It consists of a multimedia/multipathway environmental pollution modeling system. It provides modeling tools to predict exposure from pollutants in the air, soil, and water	RISKPRO is used to evaluate receptor exposures and risks from environmental contaminants. It graphically represents its results through maps, bar charts, wind-rose diagrams, isopleth diagrams, pie charts, and distributional charts. Its mapping capabilities can also allow the user to create custom maps showing data and locations	General Sciences Corporation (GSC), Laurel, MD

| MSRM (Mixture and Systemic toxicant Risk Model) | Multimedia exposures | MSRM is an exposure assessment model that contains statistical methods and extrapolation models for using available toxicological and epidemiological data. Primarily, consists of noncancer risk assessment models and estimation categories, but also has cancer risk models included for completeness. It is applicable to single chemicals and mixtures | MSRM is used to estimate human health risks from exposure by various routes | Office of Health and Environmental Assessment, USEPA, Washington, D.C. |
| WET (Wastes-Environments-Technologies model) | Multimedia (air, surface water, groundwater) | WET is a RCRA risk/cost policy model that establishes a system to allow users to investigate how trade-offs of costs and risks can be made among wastes, environments, and technologies (W-E-Ts) in order to arrive at feasible regulatory alternatives. The system assesses waste streams in terms of likelihood and severity of human exposure to their hazardous constituents and models their behavior in three media — air, surface water, and groundwater. Groundwater exposure/risk score is tallied based on key flow and transport parameters | WET is used to assist policymakers in identifying cost-effective options that minimize risks to health and the environment | Office of Health and Environmental Assessment, USEPA, Washington, D.C. |

register would contain. The definition of the register's objectives was subsequently elaborated as follows:

- Make data on chemicals readily available to those who need it
- Identify and draw attention to the major gaps in the available information and encourage research to fill those gaps
- Help identify the potential hazards of using chemicals and improve people awareness of such hazards
- Assemble information on existing policies for control and regulation of hazardous chemicals at national, regional, and global levels

In, 1976, a central unit for the register, named the International Register of Potentially Toxic Chemicals (IRPTC), was created in Geneva, Switzerland with the main function of collecting, storing, and disseminating data on chemicals, and also to operate a global network for information exchange. IRPTC network partners, the designation assigned to participants outside the central unit, consist of National Correspondents appointed by governments, national and international institutions, national academies of science, industrial research centers, and specialized research institutions. Chemicals examined by the IRPTC have been chosen from national and international priority lists. The selection criteria used include the quantity of production and use, the toxicity to humans and ecosystems, persistence in the environment, and the rate of accumulation in living organisms.

IRPTC stores information that would aid in the assessment of the risks and hazards posed by a chemical substance to human health and environment. The major types of information collected include that relating to the behavior of chemicals and information on chemical regulation. Information on the behavior of chemicals is obtained from various sources such as national and international institutions, industries, universities, private databanks, libraries, academic institutions, scientific journals, and United Nations bodies such as the International Programme on Chemical Safety (IPCS). Regulatory information on chemicals is largely contributed by IRPTC National Correspondents. Specific criteria are used in the selection of information for entry into the databases. Whenever possible, IRPTC uses data sources cited in the secondary literature produced by national and international panels of experts to maximize reliability and quality. The data are then extracted from the primary literature. Validation is performed prior to data entry and storage on a computer at the United Nations International Computing Centre (ICC).

Following the successful implementation of the IRPTC databases, a number of countries created National Registers of Potentially Toxic Chemicals (NRPTCs) that are completely compatible with the IRPTC system. A more detailed information, and access to the IRPTC, may be obtained from the National Correspondent to the IRPTC, the USEPA, the National Technical Information Service (NTIS), the Agency for Toxic Substances and Disease Registry (ATSDR), or the National Academy of Sciences (NAS).

Types of Information in the IRPTC Databases
The complete IRPTC file structure consists of databases relating to the following subject matter and areas of interest:

- Legal
- Mammalian and Special Toxicity Studies

- Chemobiokinetics and Effects on Organisms in the Environment
- Environmental Fate Tests, and Environmental Fate and Pathways into the Environment
- Identifiers, Production, Processes and Waste

The IRPTC *Legal* database contains national and international recommendations and legal mechanisms related to chemical substances control in environmental media such as air, water, wastes, soils, sediments, biota, foods, drugs, consumer products, etc. This organization allows for rapid access to the regulatory mechanisms of several nations and to international recommendations for safe handling and use of chemicals.

The *Mammalian Toxicity* database provides information on the toxic behavior of chemical substances in humans; toxicity studies on laboratory animals are included as a means of predicting potential human effects. The Special Toxicity databases contain information on particular effects of chemicals on mammals, such as mutagenicity and carcinogenicity, as well as data on nonmammalian species when relevant for the description of a particular effect.

The *Chemobiokinetics and Effects on Organisms in the Environment* databases provide data that will permit the reliable assessment of the hazard of chemicals present in the environment to man. The absorption, distribution, metabolism, and excretion of drugs, chemicals, and endogenous substances are described in the Chemobiokinetics databases. The Effects on Organisms in the Environment databases contain toxicological information regarding chemicals in relation to ecosystems and to aquatic and terrestrial organisms at various nutritional levels.

The *Environmental Fate Tests,* and *Environmental Fate and Pathways into the Environment* databases assess the risk presented by chemicals to the environment.

The *Identifiers, Production, Processes and Waste* databases contain miscellaneous information about chemicals, including physical and chemical properties, hazard classification for chemical production and trade statistics of chemicals on worldwide or regional basis, information on production methods, information on uses and quantities of use for chemicals, data on persistence of chemicals in various environmental compartments or media, information on the intake of chemicals by humans in different geographical areas, sampling methods for various media and species, as well as analytical protocols for obtaining reliable data, recommended methods for the treatment and disposal of chemicals, etc.

The Role IRPTC in Contaminated Site Management

The IPRTC, with its carefully designed database structure, provides a sound model for national and regional data systems. More importantly, it brings consistency to information exchange procedures within the international community. The IPRTC is serving as an essential international tool for chemical hazards assessment, as well as a mechanism for information exchange on several chemicals. The wealth of scientific information contained in the IRPTC can serve as an invaluable database for contaminated site management problems.

E.2.2   Overview of the Integrated Risk Information System

The Integrated Risk Information System (IRIS), prepared and maintained by the Office of Health and Environmental Assessment of the U.S. Environmental Protection Agency (USEPA), is an electronic database containing health risk and regulatory information on several specific chemicals. It is an on-line database of chemical-specific risk information; it is also a primary source of EPA health hazard assessment and related information on several chemicals of environmental concern. IRIS was originally developed for EPA staff in response to a growing demand for consistent risk information on chemical substances for use in decision-making and regulatory activities.

The information in IRIS is accessible to those without extensive training in toxicology, but with some rudimentary knowledge of health and related sciences. To aid users in accessing and understanding the data in the IRIS chemical files, the following supportive documentation is provided:

- Alphabetical list of the chemical files in IRIS and list of chemicals by Chemical Abstracts Service (CAS) number.
- Background documents describing the rationales and methods used in arriving at the results shown in the chemical files.
- A user's guide that presents step-by-step procedures for using IRIS to retrieve chemical information.
- An example exercise in which the use of IRIS is demonstrated.
- Glossaries in which definitions are provided for the acronyms, abbreviations, and specialized risk assessment terms used in the chemical files and in the background documents.

The information in IRIS can be used to develop corrective action decision for potentially contaminated sites, through the application of risk assessment and risk management procedures. More detailed information, and access to IRIS may be obtained from the EPA (IRIS User Support, USEPA, Environmental Criteria and Assessment Office, Cincinnati, OH 45268), or from other independent/private database vendors (such as Chemical Information System [CIS] in Baltimore, Maryland, Dialog Information Services, Inc. [DIALOG] in Palo Alto, California, and National Library of Medicine [NLM] in Bethesda, Maryland).

Types of Information in IRIS

IRIS consists of a collection of computer files covering several individual chemicals. These chemical files contain descriptive and numerical information on several subjects, including:

- Oral and inhalation reference doses (RfDs) for chronic noncarcinogenic health effects
- Oral and inhalation slope factors and unit risks for chronic exposures to carcinogens
- Summaries of drinking water health advisories from EPA's Office of Drinking Water
- EPA regulatory action summaries
- Supplementary data on acute health hazards and physical/chemical properties of the chemicals

IRIS is a computerized library of current information that is updated periodically. An alphabetical and CAS number listing of chemicals in IRIS is included in this database. The information contained in Section I (Chronic Health Hazard Assessment for Noncarcinogenic Effects) and Section II (Carcinogenicity Assessment for Lifetime Exposure) of the chemical files represents a consensus opinion of EPA's Reference Dose Work Group or Carcinogen Risk Assessment Verification Endeavor Work Group, respectively. These two work groups include high-level scientists from EPA's program offices and Office of Research and Development. Individual EPA offices have conducted comprehensive scientific reviews of the literature available on the particular chemical, and have performed the first two steps of risk assessment: hazard evaluation and dose-response assessment. These assessments have been summarized in the IRIS format and reviewed and revised by the appropriate Work Group. As new information becomes available, these Work Groups reevaluate their work and revise IRIS files accordingly.

The Role of IRIS in Contaminated Site Management

IRIS is a tool which provides hazard identification and dose-response assessment information, but does not provide problem-specific information on individual instances of exposure. Combined with specific exposure information, the data in IRIS can be used to characterize the public health risks of a chemical of potential concern under specific scenarios, which can then facilitate the development of effectual corrective action decisions designed to protect public health.

EQUATIONS FOR ESTIMATING
POTENTIAL RECEPTOR EXPOSURES
TO SITE CONTAMINANTS

F.1 INTRODUCTION

An analysis of potential human exposures associated with contaminated site problems often involves a complexity of integrated evaluations and issues related to the variety of contaminant migration and exposure pathways (Figure F.1). The methods by which each type of exposure is evaluated are well documented in the literature (e.g., CAPCOA, 1990; CDHS, 1986; DTSC, 1994; USEPA, 1989a, 1989b). Receptor exposures for the different primary routes of contact are defined by the inhalation, ingestion, and dermal exposure relationships discussed in the following sections.

F.2 EQUATIONS FOR CALCULATING INHALATION EXPOSURES

Two major types of inhalation exposure pathways are generally considered in the investigation of potentially contaminated site problems. The primary pathway is inhalation of airborne particulates from fugitive dust, in which all individuals within approximately 80 km (≈50 mi) radius of a site are potentially impacted. A secondary exposure pathway relates to the inhalation of volatile compounds (i.e., airborne, vapor-phase chemicals). Potential inhalation intakes are estimated based on the length of exposure, the inhalation rate of the exposed individual, the concentration of contaminant in the inhaled air, and the amount retained in the lungs.

Receptor Inhalation Exposure to Particulates from Fugitive Dust

Box F.1 shows the relationship that is used to calculate potential receptor intakes as a result of the inhalation of wind-borne fugitive dust (CAPCOA, 1990; DTSC, 1994; USEPA, 1988, 1989a, 1989b). The contaminant concentration in air, C_a, is defined by the ground-level concentration (GLC) represented by the respirable (PM-10) particles, expressed in $\mu g/m^3$.

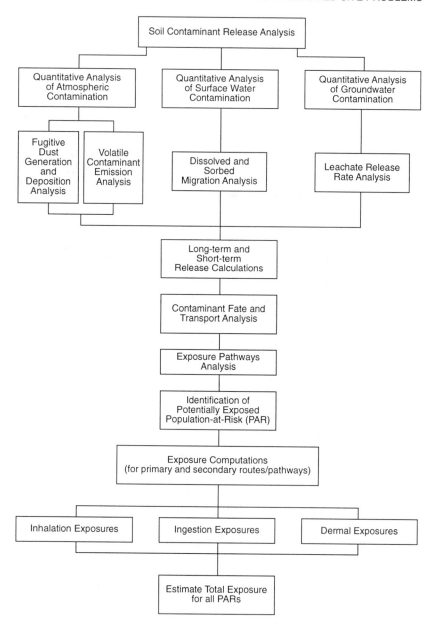

Figure F.1 A schematic of the analysis of potential exposures associated with contaminated site problems.

Receptor Inhalation Exposure to Volatile Compounds
  Box F.2 shows the relationship that is used to calculate potential receptor intakes as a result of the inhalation of airborne vapor-phase chemicals (CAPCOA, 1990; DTSC, 1994; USEPA, 1988, 1989a, 1989b). The vapor-phase contaminant concentration in air is assumed to be in equilibrium with the concentration in the release source.

BOX F.1
Equation for Calculating the Inhalation
of Contaminated Fugitive Dust

$$INH = \frac{\left(C_a \times IR \times RR \times ABS_S \times ET \times EF \times ED\right)}{(BW \times AT)}$$

where:

INH	=	inhalation intake (mg/kg-day)
C_a	=	chemical concentration of airborne particulates (mg/m^3)
IR	=	inhalation rate (m^3/hr)
RR	=	retention rate of inhaled air (%)
ABS_s	=	percent of chemical absorbed into the bloodstream (%)
ET	=	exposure time (hr/day)
EF	=	exposure frequency (days/year)
ED	=	exposure duration (years)
BW	=	body weight (kg)
AT	=	averaging time (period over which exposure is averaged — days)
	=	ED × 365 days/year, for noncarcinogenic effects
	=	70 years × 365 days/year, for carcinogenic effects

F.3 EQUATIONS FOR CALCULATING INGESTION EXPOSURES

Exposure through ingestion is a function of the concentration of the contaminant in the material ingested (e.g., soil, water, or food), the gastrointestinal absorption of the pollutant in solid or fluid matrix, and the amount ingested. In general, exposure to contaminants via the ingestion of contaminated fluids or solids may be estimated according to the following relationship (CAPCOA, 1990; DTSC, 1994; USEPA, 1988, 1989a, 1989b):

$$ING = \frac{\left(CONC \times IR \times CF \times FI \times ABS_S \times EF \times ED\right)}{(BW \times AT)}$$

where:

ING	=	ingestion intake, adjusted for absorption (mg/kg-day)
CONC	=	chemical concentration in media of concern (mg/kg or mg/L)
IR	=	ingestion rate (mg or L media material/day)
CF	=	conversion factor (10^{-6} kg/mg for solid media, or 1.00 for fluid media)
FI	=	fraction ingested from contaminated source (unitless)
ABS_s	=	bioavailability/gastrointestinal (GI) absorption factor (%)

BOX F.2
Equation for Calculating the Inhalation
of Vapor-Phase Contaminants

$$INH = \frac{\left(C_{av} \times IR \times RR \times ABS_s \times ET \times EF \times ED\right)}{(BW \times AT)}$$

where:

INH	=	inhalation intake (mg/kg-day)
C_{av}	=	chemical concentration in air (mg/m^3)
IR	=	inhalation rate (m^3/hr)
RR	=	retention rate of inhaled air (%)
ABS_s	=	percent of chemical absorbed into the bloodstream (%)
ET	=	exposure time (hr/day)
EF	=	exposure frequency (days/year)
ED	=	exposure duration (years)
BW	=	body weight (kg)
AT	=	averaging time (period over which exposure is averaged — days)
	=	ED $\times$ 365 days/year, for noncarcinogenic effects
	=	70 years $\times$ 365 days/year, for carcinogenic effects

EF	=	exposure frequency (days/years)
ED	=	exposure duration (years)
BW	=	body weight (kg)
AT	=	averaging time (period over which exposure is averaged — days)

The major types of ingestion exposure pathways that could affect contaminated site management decisions are presented below.

Receptor Exposure Through Ingestion of Drinking Water
Box F.3 contains the applicable relationship for estimating the exposure intake that occurs through the ingestion of drinking water.

Receptor Exposure Through Incidental Ingestion of Water
During Swimming Activities
Box F.4 shows the applicable relationship for estimating the exposure intake occurring as a result of the ingestion of contaminated surface water during recreational activities.

Receptor Exposure Through Ingestion of Food
Exposure from the ingestion of food can occur via the ingestion of plant products, fish, animal products, and mother's milk. The general relationship for the exposure intake through the ingestion of foods is shown in Box F.5.

BOX F.3
Equation for Calculating the Ingestion
of Contaminated Drinking Water

$$ING_{dw} = \frac{\left(C_w \times IR \times FI \times ABS_S \times EF \times ED\right)}{(BW \times AT)}$$

where:

ING_{dw}	=	ingestion intake, adjusted for absorption (mg/kg-day)
C_w	=	chemical concentration in water (mg/L)
IR	=	average water ingestion rate (L/day)
FI	=	fraction ingested from contaminated source (unitless)
ABS_s	=	bioavailability/gastrointestinal (GI) absorption factor (%)
EF	=	exposure frequency (days/years)
ED	=	exposure duration (years)
BW	=	body weight (kg)
AT	=	averaging time (period over which exposure is averaged — days)
	=	ED × 365 days/year, for noncarcinogenic effects
	=	70 years × 365 days/year, for carcinogenic effects

BOX F.4
Equation for Calculating the Ingestion of Contaminated
Surface Water During Recreational Activities

$$ING_r = \frac{\left(CW \times CR \times ABS_S \times ET \times EF \times ED\right)}{(BW \times AT)}$$

where:

ING_r	=	ingestion intake, adjusted for absorption (mg/kg-day)
CW	=	chemical concentration in water (mg/L)
CR	=	contact rate (L/hr)
ABS_s	=	bioavailability/gastrointestinal (GI) absorption factor (%)
ET	=	exposure time (hr/event)
EF	=	exposure frequency (events/year)
ED	=	exposure duration (years)
BW	=	body weight (kg)
AT	=	averaging time (period over which exposure is averaged — days)

BOX F.5
Equation for Calculating the Ingestion
of Contaminated Food Products

$$\text{ING}_f = \frac{\left(C_f \times IR \times CF \times FI \times ABS_s \times EF \times ED\right)}{(BW \times AT)}$$

where:

ING_f = ingestion intake, adjusted for absorption (mg/kg-day)
C_f = chemical concentration in food (mg/kg or mg/L)
IR = average food ingestion rate (mg or L/meal)
CF = conversion factor (10^{-6} kg/mg for solids and 1.00 for fluids)
FI = fraction ingested from contaminated source (unitless)
ABS_s = bioavailability/gastrointestinal (GI) absorption factor (%)
EF = exposure frequency (meals/year)
ED = exposure duration (years)
BW = body weight (kg)
AT = averaging time (period over which exposure is averaged — days)

Ingestion of plant products — Exposure through the ingestion of plant products, ING_p, is a function of the type of plant, a gastrointestinal absorption factor, and the fraction of plants ingested that are affected by pollutants. The calculation is done for each plant type according to the following relationship (CAPCOA, 1990; USEPA, 1989a):

$$\text{ING}_p = \frac{\left(CP_z \times PIR_z \times FI_z \times ABS_s \times EF \times ED\right)}{(BW \times AT)}$$

where:

ING_p = exposure intake from ingestion of plant products, adjusted for absorption (mg/kg-day)
CP_z = chemical concentration in plant type Z (mg/kg)
PIR_z = average consumption rate for plant type Z (kg/day)
FI_z = fraction of plant type Z ingested from contaminated source (unitless)
ABS_s = bioavailability/gastrointestinal (GI) absorption factor (%)
EF = exposure frequency (days/years)
ED = exposure duration (years)
BW = body weight (kg)
AT = averaging time (period over which exposure is averaged — days)

Bioaccumulation and ingestion of seafood — Exposure from the ingestion of contaminated fish (from contaminated surface water bodies) may be estimated using the following relationship (USEPA, 1987, 1988, 1989a):

$$ING_{sf} = \frac{\left(CW \times FIR \times CF \times BCF \times FI \times ABS_S \times EF \times ED\right)}{(BW \times AT)}$$

where:

ING_{sf}	=	total exposure, adjusted for absorption (mg/kg-day)
CW	=	chemical concentration in surface water (mg/L)
FIR	=	average fish ingestion rate (g/day)
CF	=	conversion factor (= 10^{-3} kg/g)
BCF	=	chemical-specific bioconcentration factor (L/kg)
FI	=	fraction ingested from contaminated source (unitless)
ABS_s	=	bioavailability/gastrointestinal (GI) absorption factor (%)
EF	=	exposure frequency (days/years)
ED	=	exposure duration (years)
BW	=	body weight (kg)
AT	=	averaging time (period over which exposure is averaged — days)

Ingestion of animal products — Exposure resulting from the ingestion of animal products, ING_a, is a function of the type of meat ingested (including animal milk products and eggs), a gastrointestinal absorption factor, and the fraction of animal products ingested that are affected by pollutants. The calculation is done for each animal product type according to the following relationship (CAPCOA, 1990; USEPA, 1989a):

$$ING_a = \frac{\left(CAP_Z \times APIR_Z \times FI_Z \times ABS_S \times EF \times ED\right)}{(BW \times AT)}$$

where:

ING_a	=	exposure intake through ingestion of plant products, adjusted for absorption (mg/kg-day)
CAP_z	=	chemical concentration in food type Z (mg/kg)
$APIR_z$	=	average consumption rate for food type Z (kg/day)
FI_z	=	fraction of product type Z ingested from contaminated source (unitless)
ABS_s	=	bioavailability/gastrointestinal (GI) absorption factor (%)
EF	=	exposure frequency (days/years)
ED	=	exposure duration (years)
BW	=	body weight (kg)
AT	=	averaging time (period over which exposure is averaged — days)

Ingestion of mother's milk — Exposure through the ingestion of a mother's milk, ING_m, is a function of the average chemical concentration in the mother's milk, the amount of mother's milk ingested, and a gastrointestinal absorption factor. This is estimated according to the following relationship (CAPCOA, 1990; USEPA, 1989a):

$$ING_m = \frac{(CMM \times IBM \times ABS_s \times EF \times ED)}{(BW \times AT)}$$

where:

ING_m = exposure intake through ingestion of mother's milk, adjusted for absorption (mg/kg-day)

CMM = chemical concentration in mother's milk — which is a function of the mother's exposure through all routes and the contaminant body half-life (mg/kg)

IBM = daily average ingestion rate for breast milk (kg/day)

ABS_s = bioavailability/gastrointestinal (GI) absorption factor (%)

EF = exposure frequency (days/years)

ED = exposure duration (years)

BW = body weight (kg)

AT = averaging time (period over which exposure is averaged — days).

Receptor Exposure Through Pica and Incidental Ingestion of Soil/Sediment

Exposures resulting from the incidental ingestion of contaminants sorbed onto soils is determined by multiplying the concentration of the contaminant in the medium of concern by the amount of soil ingested per day and the degree of absorption; the applicable relationship is shown in Box F.6 (CAPCOA, 1990; USEPA, 1988, 1989a, 1989b). In general, it normally is assumed that all ingested soil during receptor exposures comes from a contaminated source, so that FI becomes unity.

F.4   EQUATIONS FOR CALCULATING DERMAL EXPOSURES

Dermal intake is determined by the chemical concentration in the medium of concern, the body surface area in contact with the medium, the duration of the contact, the flux of the medium across the skin surface, and the absorbed fraction. The major types of dermal exposure pathways that could affect contaminated site management decisions are presented below.

Receptor Exposure Through Soils Contact/Dermal Absorption

The dermal exposures to chemicals in soils and sediments from a contaminated site may be estimated by applying the equation shown in Box F.7 (CAPCOA, 1990; DTSC, 1994; USEPA, 1988, 1989a,1989b).

BOX F.6
Equation for Calculating the Pica
and Incidental Ingestion of Soils

$$ING = \frac{\left(C_S \times SIR \times CF \times FI \times ABS_S \times EF \times ED\right)}{(BW \times AT)}$$

where:

ING	=	ingestion intake, adjusted for absorption (mg/kg-day)
C_s	=	chemical concentration in soil (mg/kg)
SIR	=	average soil ingestion rate (mg soil/day)
CF	=	conversion factor (10^{-6} kg/mg)
FI	=	fraction ingested from contaminated source (unitless)
ABS_s	=	bioavailability/gastrointestinal (GI) absorption factor (%)
EF	=	exposure frequency (days/years)
ED	=	exposure duration (years)
BW	=	body weight (kg)
AT	=	averaging time (period over which exposure is averaged — days)

BOX F.7
Equation for Calculating Dermal Exposures
Through Soil Contact

$$DEX = \frac{\left(C_S \times CF \times SA \times AF \times ABS_S \times SM \times EF \times ED\right)}{(BW \times AT)}$$

where:

DEX	=	absorbed dose (mg/kg-day)
C_s	=	chemical concentration in soil (mg/kg)
CF	=	conversion factor (10^{-6} kg/mg)
SA	=	skin surface area available for contact, i.e., surface area of exposed skin (cm²/event)
AF	=	soil to skin adherence factor, i.e., soil loading on skin (mg/cm²)
ABS_s =		skin absorption factor for chemicals in soil (%)
SM	=	factor for soil matrix effects (%)
EF	=	exposure frequency (events/year)
ED	=	exposure duration (years)
BW	=	body weight (kg)
AT	=	averaging time (period over which exposure is averaged — days)

BOX F.8
Equation for Calculating Dermal Exposures
Through Contacts with Contaminated Waters

$$DEX_W = \frac{(CW \times SA \times PC \times ET \times EF \times ED \times CF)}{(BW \times AT)}$$

where:

DEX_w	=	absorbed dose from dermal contact with chemicals in water (mg/kg-day)
CW	=	chemical concentration in water (mg/L)
SA	=	skin surface area available for contact, i.e., surface area of exposed skin (cm^2)
PC	=	chemical-specific dermal permeability constant (cm/hr)
ABS_s	=	skin absorption factor for chemicals in water (%)
CF	=	volumetric conversion factor for water (1 L/1000 cm^3)
ET	=	exposure time (hr/day)
EF	=	exposure frequency (days/year)
ED	=	exposure duration (years)
BW	=	body weight (kg)
AT	=	averaging time (period over which exposure is averaged — days)

Receptor Exposure Through Dermal Contact with Waters and Seeps

Dermal exposures to chemicals in water may occur during domestic use (such as bathing and washing), or through recreational activities (such as swimming or fishing). The dermal intakes of chemicals in ground or surface water and/or from seeps from a contaminated site may be estimated by the equation shown in Box F.8 (USEPA, 1988, 1989a, 1989b).

F.5 COMPUTATION OF INTAKE FACTORS FOR EXPOSURE ASSESSMENTS

Several exposure parameters are required in order to model the various exposure scenarios associated with contaminated site problems. Typically, default values may be obtained from the literature for some of the exposure parameters; Table F.1 shows some generic sets of values commonly used. More detailed information on the exposure parameters can be obtained from several sources (e.g., CAPCOA, 1990; OSA, 1992; USEPA, 1987, 1988, 1989a, 1989b, 1991). A spreadsheet for automatically calculating exposure factors as and when input parameters are changed to reflect site-specific problems may be developed to facilitate the computational efforts involved (Table F.2).

Table F.1 Examples of Case-Specific Parameters for Exposure Assessment

Parameter	Children up to 6 years	Children 6–12 years	Adult (>12 years)	Ref. Sources
Physical characteristics				
Average body weight	16 kg	29 kg	70 kg	(a,b,c)
Average total skin surface area	6980 cm²	10,470 cm²	18,150 cm²	(a,b,e,h)
Average lifetime	70 yrs	70 yrs	70 yrs	(a,b,c,e)
Average lifetime exposure period	5 yrs	6 yrs	58 yrs	(b,e)
Activity characteristics				
Inhalation rate	0.25 m³/hr	0.46 m³/hr	0.83 m³/hr	(b,e)
Retention rate of inhaled air	100%	100%	100%	(e)
Frequency of fugitive dust inhalation				
Off-site residents, schools, and passers-by	365 days/yr	365 days/yr	365 days/yr	(b,e)
Off-site workers	—	—	260 days/yr	(b,e)
Duration of fugitive dust inhalation (outside)				
Off-site residents, schools and passers-by	12 hr/day	12 hr/day	12 hr/day	(b,e)
Off-site workers	—	—	8 hr/day	(b,e)
Amount of soil ingested incidentally	200 mg/day	100 mg/day	50 mg/day	(a,b,c,e,h,i)
Frequency of soil contact				
Off-site residents, schools, and passers-by	330 days/yr	330 days/yr	330 days/yr	(b,e)
Off-site workers	—	—	260 days/yr	(b,e)
Duration of soil contact				
Off-site residents, schools, and passers-by	12 hr/day	8 hr/day	8 hr/day	(b,e)
Off-site workers	—	—	8 hr/day	(b,e)
Percentage of skin area contacted by soil	20%	20%	10%	(b,e,h)
Material characteristics				
Soil to skin adherence factor	0.75 mg/cm²	0.75 mg/cm²	0.75 mg/cm²	(a,b,e,f,g)
Soil matrix attenuation factor	15%	15%	15%	(d)

Note: The exposure factors represented here are for potential maximum exposures (for conservative estimates), and could be modified as appropriate to reflect the most reasonable exposure patterns anticipated. For instance, soil exposure will be reduced by snow cover and rainy days, thus reducing potential exposures for children playing in contaminated areas.

a USEPA, 1989: "Risk Assessment Guidance for Superfund, Vol. I. Human Health Evaluation Manual (Part A)," EPA/540/1-89/002.

b USEPA, 1989: "Exposure Factors Handbook," EPA/600/8-89/043.

c USEPA, 1988: "Superfund Exposure Assessment Manual," EPA/540/1-88/001.

d Hawley, J.K., 1985: "Assessment of Health Risk from Exposure to Contaminated Soil," Risk Anal. Vol. 5, No. 4, 289–302.

e Estimate based on site-specific conditions.

f Lepow, M.L., L. Bruckman, M. Gillette, S. Markowitz, R. Robino and J. Kapish, 1975: "Investigations into Sources of Lead in the Environment of Urban Children," Environ. Res. 10, 415–426.

g Lepow, M.L., M. Bruckman, L. Robino, S. Markowitz, M. Gillette and J. Kapish, 1974: "Role of Airborne Lead in Increased Body Burden of Lead in Hartford Children," Environ. Health Perspect. 6, 99–101.

h Sedman, R., 1989: "The Development of Applied Action Levels for Soil Contact: A Scenario for the Exposure of Humans to Soil in a Residential Setting," Environ. Health Perspect. 79, 291–313.

i Calabrese, E.J. et al. 1989: "How Much Soil Do Young Children Ingest: An Epidemiologic Study," Regul. Toxicol. Pharmacol. 10, 123–137.

Table F.2   Example Spreadsheet for Calculating Case-Specific Intake Factors in an Exposure Assessment

PATHWAY===> Group	Fugitive Dust Inhalation Pathway							
	IR	RR	ET	EF	ED	BW	AT	INH Factor
C(1-6)@NCarc	0.25	1	12	365	5	16	1825	1.88E-01
C(1-6)@Carc	0.25	1	12	365	5	16	25550	1.34E-02
C(6-12)@NCarc	0.46	1	12	365	6	29	2190	1.90E-01
C(6-12)@Carc	0.46	1	12	365	6	29	25550	1.63E-02
ResAdult@NCarc	0.83	1	12	365	58	70	21170	1.42E-01
ResAdult@Carc	0.83	1	12	365	58	70	25550	1.18E-01
JobAdult@NCarc	0.83	1	8	260	58	70	21170	6.76E-02
JobAdult@Carc	0.83	1	8	260	58	70	25550	5.60E-02

PATHWAY===> Group	Soil Ingestion Pathway							
	IR	CF	FI	EF	ED	BW	AT	ING Factor
C(1-6)@NCarc	200	1.00E-06	1	330	5	16	1825	1.13E-05
C(1-6)@Carc	200	1.00E-06	1	330	5	16	25550	8.07E-07
C(6-12)@NCarc	100	1.00E-06	1	330	6	29	2190	3.12E-06
C(6-12)@Carc	100	1.00E-06	1	330	6	29	25550	2.67E-07
ResAdult@NCarc	50	1.00E-06	1	330	58	70	21170	6.46E-07
ResAdult@Carc	50	1.00E-06	1	330	58	70	25550	5.35E-07
JobAdult@NCarc	50	1.00E-06	1	260	58	70	21170	5.09E-07
JobAdult@Carc	50	1.00E-06	1	260	58	70	25550	4.22E-07

PATHWAY===> Group	Soil Dermal Contact Pathway								
	SA	CF	AF	SM	EF	ED	BW	AT	DEX Factor
C(1-6)@NCarc	1396	1.00E-06	0.75	0.15	330	5	16	1825	8.87E-06
C(1-6)@Carc	1396	1.00E-06	0.75	0.15	330	5	16	25550	6.34E-07
C(6-12)@NCarc	2094	1.00E-06	0.75	0.15	330	6	29	2190	7.34E-06
C(6-12)@Carc	2094	1.00E-06	0.75	0.15	330	6	29	25550	6.30E-07
ResAdult@NCarc	1815	1.00E-06	0.75	0.15	330	58	70	21170	2.64E-06
ResAdult@Carc	1815	1.00E-06	0.75	0.15	330	58	70	25550	2.19E-06
JobAdult@NCarc	1815	1.00E-06	0.75	0.15	260	58	70	21170	2.08E-06
JobAdult@Carc	1815	1.00E-06	0.75	0.15	260	58	70	25550	1.72E-06

Notes:

Notations and units are as defined in the text.
INH Factor = inhalation factor for calculation of doses and intakes.
ING Factor = soil ingestion factor for calculation of doses and intakes.
DEX Factor = dermal exposure/skin absorption factor for calculation of doses and intakes.
C(1-6)@NCarc = noncarcinogenic effects for children aged 1–6 years.
C(1-6)@Carc = carcinogenic effects for children aged 1–6 years.
C(6-12)@NCarc = noncarcinogenic effects for children aged 6–12 years.
C(6-12)@Carc = carcinogenic effects for children aged 6–12 years.
ResAdult@NCarc = noncarcinogenic effects for resident adults.
ResAdult@Carc = carcinogenic effects for resident adults.
JobAdult@NCarc = noncarcinogenic effects for adult workers.
JobAdult@Carc = carcinogenic effects for adult workers.

REFERENCES

Calabrese, E.J. et al. 1989. "How Much Soil Do Young Children Ingest: An Epidemiologic Study," *Regul. Toxicol. Pharmacol.* 10, 123–137.

CAPCOA. 1990. Air Toxics "Hot Spots" Program. Risk Assessment Guidelines. California Air Pollution Control Officers Association, Sacramento.

CDHS. 1986. The California Site Mitigation Decision Tree Manual. California Department of Health Services, Toxic Substances Control Division, Sacramento.

DTSC. 1994. Preliminary Endangerment Assessment Guidance Manual (A guidance manual for evaluating hazardous substance release sites). California Environmental Protection Agency, Department of Toxic Substances Control, Sacramento.

Hawley, J.K. 1985. "Assessment of Health Risk from Exposure to Contaminated Soil," *Risk Anal.* Vol. 5, No. 4, 289–302.

Lepow, M.L., L. Bruckman, M. Gillette, S. Markowitz, R. Robino and J. Kapish, 1975. "Investigations into Sources of Lead in the Environment of Urban Children," *Environ. Res.* 10, 415–426.

Lepow, M.L., M. Bruckman, L. Robino, S. Markowitz, M. Gillette and J. Kapish, 1974. "Role of Airborne Lead in Increased Body Burden of Lead in Hartford Children," *Environ. Health Perspect.* 6, 99–101.

OSA. 1993. Supplemental Guidance for Human Health Multimedia Risk Assessments of Hazardous Waste Sites and Permitted Facilities. California Environmental Protection Agency, Sacramento.

Sedman, R. 1989. "The Development of Applied Action Levels for Soil Contact: A Scenario for the Exposure of Humans to Soil in a Residential Setting," *Environ. Health Perspect.* 79, 291–313.

USEPA. 1987. RCRA Facility Investigation (RFI) Guidance, EPA/530/SW-87/001, U.S. Environmental Protection Agency, Washington, D.C.

USEPA. 1988. Superfund Exposure Assessment Manual, Report No. EPA/540/1-88/001, OSWER Directive 9285.5-1, U.S. Environmental Protection Agency, Office of Remedial Response, Washington, D.C.

USEPA. 1989a. Exposure Factors Handbook, EPA/600/8-89/043. U.S. Environmental Protection Agency, Washington, D.C.

USEPA. 1989b. Risk Assessment Guidance for Superfund. Vol. I. Human Health Evaluation Manual (Part A). EPA/540/1-89/002. U.S. Environmental Protection Agency, Office of Emergency and Remedial Response, Washington, D.C.

USEPA. 1991. Risk Assessment Guidance for Superfund, Vol. I. Human Health Evaluation Manual. Supplemental Guidance. "Standard Default Exposure Factors" (Interim Final). Office of Emergency and Remedial Response. OSWER Directive: 9285.6-03, U.S. Environmental Protection Agency, Washington, D.C.

APPENDIX G
CARCINOGEN IDENTIFICATION
AND CLASSIFICATION

G.1 IDENTIFYING CARCINOGENS

An important issue in chemical carcinogenesis relates to initiators and promoters. A *promoter* is defined as an agent which results in an increase in cancer induction when it is administered some time after a receptor has been exposed to an *initiator*. A *co-carcinogen* differs from a promoter only in that it is administered at the same time as the initiator. Initiators, co-carcinogens, and promoters do not usually induce tumors when administered separately. *Complete carcinogens* act as both initiator and promoter (OSTP, 1985). Most regulatory agencies do not usually distinguish between initiators and promoters, because it is very difficult to confirm that a given chemical acts by promotion alone (OSHA, 1980; OSTP, 1985; USEPA, 1984). More generally, carcinogens may be categorized into the following identifiable groupings (IARC, 1982; USDHS, 1989):

- "Known human carcinogens", defined as those chemicals for which there is sufficient evidence of carcinogenicity from studies in humans to indicate a causal relationship between exposure to the agent and human cancer.
- "Reasonably anticipated to be carcinogens", referring to those chemical substances for which there is limited evidence for carcinogenicity in humans and/or sufficient evidence of carcinogenicity in experimental animals. Sufficient evidence in animals is demonstrated by positive carcinogenicity findings in multiple strains and species of animals, in multiple experiments, or to an unusual degree with regard to incidence, site or type of tumor, or age of onset.
- "Sufficient evidence" and "limited evidence" of carcinogenicity, used in the criteria for judging the adequacy of available data for identifying carcinogens, refer only to the amount and adequacy of the available evidence and not to the potency of carcinogenic effect on the mechanisms involved.

G.2 CARCINOGEN CLASSIFICATION SYSTEMS

A chemical's potential for human carcinogenicity is inferred from the available information relevant to the potential carcinogenicity of the chemical and from judgments as to the quality of the available studies. Two evaluation philosophies, one based on weight-of-evidence and the other on strength-of-evidence, have found common acceptance and usage. Systems that employ the weight-of-evidence evaluations consider and balance the negative indicators of carcinogenicity with those showing carcinogenic activity; schemes using the strength-of-evidence evaluations consider combined strengths of all positive animal tests (human epidemiology studies and genotoxicity) to rank a chemical without evaluating negative studies nor considering potency or mechanism (Huckle, 1991).

Weight-of-Evidence Classification

A weight-of-evidence approach is used by the USEPA to classify the likelihood that an agent in question is a human carcinogen. This is a classification system for characterizing the extent to which available data indicate that an agent is a human carcinogen (or for some other toxic effects such as developmental toxicity). A three-stage procedure is utilized as follows:

- Stage 1 — the evidence is characterized separately for human studies and for animal studies.
- Stage 2 — the human and animal evidence are integrated into a presumptive overall classification.
- Stage 3 — the provisional classification is modified (i.e., adjusted upwards or downwards), based on analysis of the supporting evidence.

The result is that chemicals are placed into one of five categories, in accordance with the USEPA Carcinogen Assessment Group (CAG) weight-of-evidence categories for potential carcinogens. Proposed guidelines for the classification of the weight-of-evidence for human carcinogenicity have been published by the EPA (USEPA, 1984); these guidelines are adaptations from those of the International Agency for Research on Cancer (IARC, 1984) and consist of the categorization of the weight-of-evidence into the following five groups (Groups A–E):

EPA Group	Reference Category
A	Human carcinogen (i.e., known human carcinogen)
B	Probable human carcinogen:
	B1 indicates limited human evidence
	B2 indicates sufficient evidence in animals and inadequate or no evidence in humans
C	Possible human carcinogen
D	Not classifiable as to human carcinogenicity
E	No evidence of carcinogenicity in humans (or, evidence of noncarcinogenicity for humans)

Group A: human carcinogen — For this group, there is sufficient evidence from epidemiologic studies to support a causal association between exposure to the agent and human cancer. The following three criteria must be satisfied before a causal association can be inferred between exposure and cancer in humans (Hallenbeck and Cunningham, 1988):

- No identified bias which could explain the association
- Possibility of confounding factors (i.e., variables other than chemical exposure level which can affect the incidence or degree of the parameter being measured) has been considered and ruled out as explaining the association
- Association is unlikely to be due to chance

This group is used only when there is sufficient evidence from epidemiologic studies to support a causal association between exposure to the agents and cancer.

Group B: probable human carcinogen — This group includes agents for which the weight-of-evidence of human carcinogenicity based on epidemiologic studies is "limited" and also includes agents for which the weight-of-evidence of carcinogenicity based on animal studies is "sufficient". The category consists of agents for which the evidence of human carcinogenicity from epidemiologic studies ranges from almost sufficient to inadequate. This group is divided into two subgroups, reflecting higher (Group B1) and lower (Group B2) degrees of evidence. Usually, category B1 is reserved for agents for which there is limited evidence of carcinogenicity to humans from epidemiologic studies; limited evidence of carcinogenicity indicates that a causal interpretation is credible but that alternative explanations such as chance, bias, or confounding could not be excluded. Inadequate evidence indicates that one of the following two conditions prevailed (Hallenbeck and Cunningham, 1988):

- There were few pertinent data, or
- The available studies, while showing evidence of association, did not exclude chance, bias, or confounding factors.

When there are inadequate data for humans, it is reasonable to regard agents for which there is sufficient evidence of carcinogenicity in animals as if they presented a carcinogenic risk to humans. Therefore, agents for which there is "sufficient" evidence from animal studies, and for which there is "inadequate" evidence from human (epidemiologic) studies, or "no data" from epidemiologic studies, would usually result in a classification as B2 (CDHS, 1986; Hallenbeck and Cunningham, 1988; USEPA, 1986).

Group C: possible human carcinogen — This group is used for agents with limited evidence of carcinogenicity in animals in the absence of human data. Limited evidence means that the data suggest a carcinogenic effect, but are limited for the following reasons (Hallenbeck and Cunningham, 1988):

- The studies involve a single species, strain, or experiment
- The experiments are restricted by inadequate dosage levels, inadequate duration of exposure to the agent, inadequate period of follow-up, poor survival, too few animals, or inadequate reporting
- An increase in the incidence of benign tumors only

Group C classification relies on a wide variety of evidence, including the following (Hallenbeck and Cunningham, 1988; USEPA, 1986)

- Definitive malignant tumor response in a single well-conducted experiment that does not meet conditions for "sufficient" evidence
- Tumor response of marginal statistical significance in studies having inadequate design or reporting

- Benign but not malignant tumors, with an agent showing no response in a variety of short-term tests for mutagenicity
- Responses of marginal statistical significance in a tissue known to have a high and variable background rate

Group D: not classifiable as to human carcinogenicity — This group is generally used for agents with inadequate animal evidence of carcinogenicity and also inadequate evidence from human (epidemiological) studies. Inadequate evidence means that because of major qualitative or quantitative limitations, the studies cannot be interpreted as showing either the presence or absence of a carcinogenic effect.

Group E: no evidence of carcinogenicity in humans — This group is used for agents for which there is evidence of noncarcinogenicity for humans, together with no evidence of carcinogenicity in at least two adequate animal tests in different species, or no evidence in both adequate animal and human (epidemiological) studies. The designation of an agent as being in this group is based on the available evidence and should not be interpreted as a definitive conclusion that the agent will not be a carcinogen under any circumstances.

Strength-of-Evidence Classification

Other varying carcinogen classification schemes exist within other regulatory and legislative agencies in Europe and elsewhere. Of particular interest, the IARC does its classification based on the strength-of-evidence philosophy. The corresponding IARC classification system, comparable or equivalent to the USEPA description presented above, is as follows:

IARC Group	Category
1	Human carcinogen (i.e., known human carcinogen)
2	Probable or possible human carcinogen:
	2A indicates limited human evidence (i.e., probable);
	2B indicates sufficient evidence in animals and inadequate or no evidence in humans (i.e., possible)
3	Not classifiable as to human carcinogenicity
4	No evidence of carcinogenicity in humans

Group 1: known human carcinogen — This group is generally used for agents with sufficient evidence from human (epidemiological) studies as to human carcinogenicity.

Group 2A: probable human carcinogen — This group is generally used for agents for which there is sufficient animal evidence, evidence of human carcinogenicity, or at least limited evidence from human (epidemiological) studies. These are probably carcinogenic to humans, with (usually) at least limited human evidence.

Group 2B: possible human carcinogen — This group is generally used for agents for which there is sufficient animal evidence and inadequate evidence from human (epidemiological) studies, or there is limited evidence from human (epidemiological) studies in the absence of sufficient animal evidence. These are probably carcinogenic to humans, but (usually) have no human evidence.

Group 3: not classifiable — This group is generally used for agents for which there is inadequate animal evidence and inadequate evidence from human (epidemiological) studies. There is sufficient evidence of carcinogenicity in experimental animals.

Group 4: noncarcinogenic to humans — This group is generally used for agents for which there is evidence for lack of carcinogenicity. They are probably not carcinogens.

G.3 EVALUATION OF CHEMICAL CARCINOGENICITY

The evaluation of a chemical's carcinogenicity involves the following two basic steps:

- Identification of potential carcinogens from among the contaminants of potential concern present at the problem site.
- Quantitative determination of the carcinogenic potency of the chemicals of potential concern; this is represented by the cancer slope factor.

Evidence of possible carcinogenicity in humans comes primarily from epidemiological studies and long-term animal exposure studies at high doses which have subsequently been extrapolated to humans. Results from these studies are supplemented with information from short-term tests, pharmacokinetic studies, comparative metabolism studies, molecular structure-activity relationships, and other relevant information sources.

REFERENCES

CDHS. 1986. The California Site Mitigation Decision Tree Manual. California Department of Health Services, Toxic Substances Control Division, Sacramento.

Hallenbeck, W.H. and K.M. Cunningham. 1988. *Quantitative Risk Assessment for Environmental and Occupational Health.* Lewis Publishers, Chelsea, MI.

Huckle, K. R. 1991. *Risk Assessment — Regulatory Need or Nightmare.* Shell Publications, Shell Center, London.

IARC. 1982. IARC Monographs on the Evaluation of the Carcinogenic Risk of Chemicals to Humans. Chemicals, Industrial Processes and Industries Associated with Cancer in Humans. Suppl. 4 (292 pp.) International Agency for Research on Cancer, Lyon, France.

IARC. 1984. IARC Monographs on the Evaluation of the Carcinogenic Risk of Chemicals to Humans, Vol. 33, World Health Organization, Lyon, France.

OSHA. 1980. Identification, Classification, and Regulation of Potential Occupational Carcinogens. *Federal Register*, 45, 5002–5296.

OSTP. 1985. Chemical Carcinogens: A Review of the Science and Its Associated Principles. *Federal Register,* 50, 10372–442.

USDHS. 1989. Public Health Service. Fifth Annual Report on Carcinogens. Summary. U.S. Department of Health and Human Services, Washington, D.C.

USEPA. 1984. Proposed Guidelines for Carcinogen, Mutagenicity, and Developmental Toxicant Risk Assessment. *Federal Register*, 49, 46294–46331.

USEPA. 1986. Guidelines for Carcinogen Risk Assessment. *Federal Register,* 51(185), 33992–34003, September 24, 1986.

APPENDIX H
DERIVATION OF TOXICITY PARAMETERS USED IN HUMAN HEALTH RISK ASSESSMENT

H.1 INTRODUCTION TO DOSE-RESPONSE ASSESSMENT

Dose-response assessment is the process of quantitatively evaluating toxicity information and characterizing the relationship between the dose of the contaminant administered or received (i.e., exposure to an agent) and the incidence of adverse health effects in the exposed populations. The process consists of estimating the potency of the specific compounds by use of dose-response relationships. Data are derived from animal studies or, less frequently, from studies in exposed human populations.

The risks of a substance cannot be ascertained with any degree of confidence unless dose-response relations are quantified, even if the substance is known to be toxic. Dose-response relationships are generally used to determine what dose of a particular chemical causes specific levels of toxic effects to potential receptors. There may be many different dose-response relationships for a substance if it produces different toxic effects under different conditions of exposure.

Dose-response curves are functional relationships between the amounts of a chemical substance and its morbidity/lethality. For some chemicals, a very small dose causes no observable effects, whereas a higher dose will result in some toxicity, and still higher doses cause even greater toxicity — up to the point of fatality; such chemicals are called threshold chemicals (curve B in Figure H.1). For other chemicals, such as most carcinogens, the threshold concept may not be applicable, in which case no minimum level is required to induce adverse and overt toxicity effects (curve A in Figure H.1). In general, the response of a toxicant depends on the mechanism of its action; in the simplest case, the response, R, is directly proportional to its concentration, [C], so that

$$R = k[C]$$

where k is a rate constant. This would be the case for a pollutant that metabolizes rapidly, but even so, the response and the value of the rate constant would tend to differ for different risk groups of individuals and for unique exposures. If the toxicant accumulates in the body, the response is defined by

$$R = k[C]t^n$$

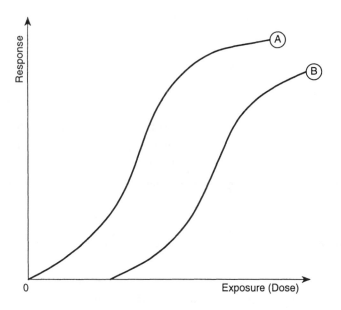

Figure H.1 A schematic representation of exposure-response relationships. (Illustration of
dose-response for A: nonthreshold chemicals, and B: threshold chemicals.)

where t is the time and n is a constant. For cumulative exposures, the response would
generally increase with time. Thus, the cumulative effect may show as linear until a
threshold is reached, after which secondary effects begin to affect and enhance the
responses. The cumulative effect may be related to what is referred to as the 'body
burden' (BB). The body burden is determined by the relative rates of absorption
(ABS), storage (ST), elimination (ELM), and biotransformation (BT), according to the
following relationship (Meyer, 1983):

$$BB = ABS + ST - ELM - BT$$

Each of the factors involved in the quantification of the body burden is dependent on
a number of biological and physicochemical factors. In fact, the response of an
individual to a given dose cannot be truly quantitatively predicted since it depends on
many extraneous factors, such as general health and diet of individual receptors or the
population-at-risk (PAR).

Three major classes of mathematical extrapolation models are often used for
relating dose and response in the subexperimental dose range:

    1. Tolerance Distribution Models, including Probit, Logit, and Weibull
    2. Mechanistic Models, including One-hit, Multi-hit, and Multi-stage
    3. Time-to-Occurrence Models, including Lognormal, and Weibull

Indeed, other independent models — such as linear, quadratic, and linear-*cum*-qua-
dratic, may also be employed for this purpose. The details of these are beyond the
scope of this discussion, but are elaborated elsewhere in the literature (e.g., CDHS,

1986). From the quantitative dose-response relationship, toxicity values can be derived and used to estimate the incidence of adverse effects occurring in humans at different exposure levels.

H.2 DERIVATION OF REFERENCE DOSES (RfDs) OR ACCEPTABLE DAILY INTAKES (ADIs)

The RfD is a "benchmark" dose operationally derived from the NOAEL by consistent application of general "order-of-magnitude" uncertainty factors (UFs) (also called "safety factors") that reflect various types of data sets used to estimate RfDs. In addition, a modifying factor (MF) is sometimes used that is based on professional judgment of the entire data base of the chemical. More generally stated, RfDs (and ADIs) are calculated by dividing a NOEL (i.e., the highest level at which a chemical causes no observable changes in the species under investigation), a NOAEL (i.e., the highest level at which a chemical causes no observable adverse effect in the species being tested), or a LOAEL (i.e., that dose rate of chemical at which there are statistically or biologically significant increases in frequency or severity of adverse effects between the exposed and appropriate control groups) which is derived from human or animal toxicity studies by one or more uncertainty and modifying factors.

To estimate the risk of acute exposures, levels of acceptable short-term exposure may be developed, representing for instance, the maximum one-day exposure levels that are anticipated not to result in adverse effects in most individuals. Where available, acute ADIs can be based on EPA's one-day drinking water health advisories. Where worker exposures are involved, OSHA's permissible exposure limits (PELs), or threshold limit values (TLVs) where no PEL has been established, may serve as ARARs for acute exposures. Also, in rather rare cases where only TLV data may be all that are available, acceptable intake levels may be derived by correcting for continuous exposure and further dividing by a safety factor (of 100) to account for highly sensitive segments of impacted populations.

When no toxicological information exists for a chemical, concepts of structure-activity relationships may have to be employed to derive acceptable intake levels by influence and analogy to closely related or similar compounds. In such cases, some reasonable degree of conservatism is suggested in any judgment call to be made.

H.2.1 Approach for Estimating RfDs (or ADIs)

RfDs are typically calculated using a single exposure level and uncertainty factors that account for specific deficiencies in the toxicological database. Both the exposure level and uncertainty factors are selected and evaluated in the context of the available chemical-specific literature. After all toxicological, epidemiologic, and supporting data have been reviewed and evaluated, a key study is selected that reflects optimal data on the critical effect. Dose-response data points for all reported effects are examined as a component of this review. USEPA (1989b) discusses specific issues of particular significance in this endeavor — including the types of response levels (ranked in order of increasing severity of toxic effects as NOEL, NOAEL, LOAEL, and FEL [the Frank effect level, defined as overt or gross adverse effects]) considered in deriving RfDs for systemic toxicants.

The RfD (or ADI) can be determined by using the following relationship:

$$\text{Human dose (e.g., ADI or RfD)} = \frac{\text{Experimental dose (e.g., NOAEL)}}{(UF \times MF)}$$

or, specifically:

$$RfD = \frac{NOAEL}{(UF \times MF)}$$

The uncertainty factor (also, safety factor) used in calculating the RfD reflects scientific judgment regarding the various types of data used to estimate RfD values. It is used to offset the uncertainties associated with extrapolation of data, etc. Generally, the UF consists of multiples of 10, each factor representing a specific area of uncertainty inherent in the available data. For example, a factor of 10 may be introduced to account for the possible differences in responsiveness between humans and animals in prolonged exposure studies. A second factor of 10 may be used to account for variation in susceptibility among individuals in the human population. The resultant UF of 100 has been judged to be appropriate for many chemicals. For other chemicals with databases that are less complete (for example, those for which only the results of subchronic studies are available), an additional factor of 10 (leading to a UF of 1000) might be judged to be more appropriate. For certain other chemicals based on well-characterized responses in sensitive humans (as in regard to the effect of fluoride on human teeth), an UF as small as 1 might be selected (Dourson and Stara, 1983).

The following general guidelines are useful in the process of selecting uncertainty and modifying factors for the derivation of RfDs (Dourson and Stara, 1983; USEPA, 1986b, 1989a, 1989b).

Guidelines for Standard Uncertainty Factors (UFs)

- Use a 10-fold factor when extrapolating from valid experimental results in studies using prolonged exposure to average healthy humans. This factor is intended to account for the variation in sensitivity among the members of the human population due to heterogeneity in human populations, and is referenced as "10H". Thus, if NOAEL is based on human data, a safety factor of 10 is usually applied to the NOAEL dose to account for variations in sensitivities between individual humans.
- Use an additional 10-fold factor when extrapolating from valid results of long-term studies on experimental animals when results of studies of human exposure are not available or are inadequate. This factor is intended to account for the uncertainty involved in extrapolating from animal data to humans and is referenced as "10A". Thus, if NOAEL is based on animal data, the NOAEL dose is divided by an additional safety factor of 10 to account for differences between animals and humans.
- Use an additional 10-fold factor when extrapolating from less than chronic results on experimental animals when there are no useful long-term human data. This factor is intended to account for the uncertainty involved in

extrapolating from less than chronic (i.e., subchronic or acute) NOAELs to chronic NOAELs and is referenced as "10S".
- Use an additional 10-fold factor when deriving an RfD from a LOAEL, instead of a NOAEL. This factor is intended to account for the uncertainty involved in extrapolating from LOAELs to NOAELs, and is referenced as "10L".
- Use an additional up to 10-fold factor when extrapolating from valid results in experimental animals when the data are "incomplete". This factor is intended to account for the inability of any single animal study to adequately address all possible adverse outcomes in humans, and is referenced as "10D".

Selecting a Modifying Factor (MF)

- Use professional judgment to determine the MF, which is an additional uncertainty factor that is greater than zero and less than or equal to 10. The magnitude of the MF depends upon the professional assessment of the scientific uncertainties of the study and a database not explicitly treated above; e.g., the completeness of the overall database and the number of species tested. The default value for the MF is 1.

In general, the choice of the UF and MF values reflect the uncertainty associated with the estimation of an RfD from different human or animal toxicity databases. For instance, if sufficient data from chronic duration exposure studies are available on the threshold region of a chemical's critical toxic effect in a known sensitive human population, then the UF used to estimate the RfD may be set at unity (1). That is, these data are judged to be sufficiently predictive of a population's subthreshold dose, so that additional UFs are not needed (USEPA, 1989b).

H.2.2 Determination of the RfD for a Hypothetical Example

Using the NOAEL — Consider the case of a study made on 250 animals (e.g., rats) that is of subchronic duration, yielding a NOAEL dosage of 5 mg/kg/day. Then,

$$UF = 10H \times 10A \times 10S = 1000$$

In addition, there is a subjective adjustment (MF) based on the high number of animals (250) per dose group:

$$MF = 0.75$$

These factors then give UF ¥ MF = 750, so that

$$RfD = \frac{NOAEL}{(UF \times MF)} = \frac{5}{750} = 0.007 \, (mg/kg/day)$$

Using the LOAEL — If the NOAEL is not available, and if 25 mg/kg/day had been the lowest dose tested that showed adverse effects, then

$$UF = 10H \times 10A \times 10S \times 10L = 10,000$$

Using again the subjective adjustment of MF = 0.75, one obtains

$$\text{RfD} = \frac{\text{LOAEL}}{(\text{UF} \times \text{MF})} = \frac{25}{7500} = 0.003 \; (\text{mg/kg/day})$$

H.2.3 Interconversions of RfD Values

RfD values for inhalation exposure are usually reported both as a concentration in air (mg/m^3) and as a corresponding inhaled dose (mg/kg-day). When determining the toxicity value for inhalation pathways, the inhalation reference concentration (RfC [mg/m^3]) should be used when available. The RfC can also be converted to equivalent RfD_i values (in units of dose [mg/kg-day]) by multiplying the RfC by a ventilation rate of 20 m^3/day (for adults) and dividing it by an average adult body weight of 70 kg. That is

$$\text{RfD}_i \; [\text{mg/kg - day}] = \frac{\text{RfC} \left[\text{mg/m}^3\right] \times 20 \; \text{m}^3/\text{day}}{70 \; \text{kg}} = 0.286 \; \text{RfC}$$

or, $\text{RfC} \left[\text{mg/m}^3 \text{ in air}\right] = 3.5 \; \text{RfD}_i$

Similarly, RfD values associated with oral exposures, and reported in mg/kg-day, can be converted to a corresponding concentration in drinking water (DWEL) as follows:

$$\text{DWEL} \left[\text{mg/L in water}\right] = \frac{\text{oral RfD} \; (\text{mg/kg - day}) \times \text{body weight (kg)}}{\text{ingestion rate} \; (\text{L/day})}$$

$$= \frac{\text{RfD}_o \; (\text{mg/kg - day}) \times 70 \; (\text{kg})}{2 \; (\text{L/day})} = 35 \; \text{RfD}_o$$

assuming a 2-L/day water consumption by a 70-kg adult.

H.3 DETERMINATION OF SLOPE FACTORS (SFs) AND UNIT CANCER RISKS (UCRs)

The cancer slope factor (also, cancer potency factor, or potency slope) is a measure of the carcinogenic toxicity of a chemical generally required for completing a health risk assessment. Exposure to any level of a carcinogen is usually considered to have a finite risk of inducing cancer associated with it; that is, carcinogenic exposure is generally not considered to have a no-effect threshold. The slope factor is the cancer risk (proportion affected) per unit of dose, and is usually expressed in milligrams of substance per kilogram body weight per day (mg/kg-day).

Scientific investigators have developed numerous models to extrapolate and estimate low-dose carcinogenic risks to humans from high-dose carcinogenic effects usually observed in experimental animal studies. Such models yield an estimate of the upper limit in lifetime risk per unit of dose (or the unit cancer risk, UCR, or unit risk,

UR). The USEPA generally uses the so-called linearized multistage model to generate UCRs. This model, known to make several conservative assumptions, results in highly conservative risk estimates, yielding overestimates of actual UCR for carcinogens; in fact, the actual risks may be substantially lower than those predicted by the upper bounds of this model.

Structural similarity factors, etc., can be used to estimate a cancer potency unit for a chemical not having one, but that is suspected to be carcinogenic. This is achieved, for instance, by estimating the geometric mean of a number of similar compounds whose UCRs are known, and using this as a surrogate value for the chemical with unknown UCR.

H.3.1   Derivations and Conversion of Cancer Potency Slope to Unit Risk Values

The unit risk estimates the upper bound probability of a "typical" or "average" person contracting cancer when continuously exposed to 1 μg/m^3 of the chemical over an average 70-year lifetime. Potency estimates are also given in terms of the potency slope factor (SF), which is the probability of contracting cancer due to exposure to a given lifetime dose, in units of mg/kg-day.

The SF can be converted to UCR (also, unit risk, UR, or unit risk factor, URF), by adopting several assumptions. The most critical factor is that the endpoint of concern must be a systematic tumor, so that potential target organs experience the same blood concentration of the active carcinogen regardless of the method of administration. This implies an assumption of equivalent absorption by the various routes of administration. The basis for these conversions is the assumption that at low doses, the dose-response curve is linear, so that:

$$P(d) = SF \times \{dose\}$$

where:

$$
\begin{aligned}
P(d) &= \text{response (probability) as a function of dose} \\
SF &= \text{potency slope factor (mg/kg-day)}^{-1} \\
\{dose\} &= \text{amount of chemical intake (mg/kg-day)}
\end{aligned}
$$

Inhalation potency factor — Risk associated with unit chemical concentration in air is estimated as follows:

$$\text{risk per } \mu g/m^3 \text{ (air)}$$

$$= \text{slope factor}\left(\text{risk per mg/kg/day}\right) \times \frac{1}{\text{body weight (kg)}} \times \text{inhalation rate}\left(m^3/day\right) \times 10^{-3}\left(mg/\mu g\right)$$

Thus, the inhalation potency can be converted to an inhalation UCR by applying the following conversion factor:

$$\{(kg\text{-}day)/mg\} \times \{1/70 \text{ kg}\} \times \{20 \text{ m}^3/day\} \times \{1 \text{ mg}/1000 \text{ } \mu g\} = 2.86 \times 10^{-4}$$

Hence, the lifetime excess cancer risk from inhaling 1 $\mu g/m^3$ concentration for a full lifetime is

$$UCR_i \, (\mu g/m^3)^{-1} = (2.86 \times 10^{-4}) \times SF_i$$

Conversely, the SF_i can be derived from the UCR_i as follows:

$$SF_i = (3.5 \times 10^3) \times UCR_i$$

The assumptions used involve a 70-kg body weight, and an average inhalation rate of 20 m^3/day.

Oral potency factor — Risk associated with unit chemical concentration in water is estimated as follows:

$$\text{risk per } \mu g/L \text{ (water)}$$

$$= \text{slope factor} \left(\text{risk per mg/kg/day}\right) \times \frac{1}{\text{body weight (kg)}} \times \text{ingestion rate} \left(L/day\right) \times 10^{-3} \left(mg/\mu g\right)$$

Thus, the ingestion potency can be converted to an oral UCR value by applying the following conversion factor:

$$\{(\text{kg-day})/mg\} \times \{1/70kg\} \times \{2L/day\} \times \{1mg/1000\mu g\} = 2.86 \times 10^{-5}$$

Hence, the lifetime excess cancer risk from ingesting 1 $\mu g/L$ concentration for a full lifetime is

$$UCR_o \, (\mu g/L)^{-1} = (2.86 \times 10^{-5}) \times SF_o$$

Or alternatively, the potency, SF_o, can be derived from the unit risk as follows:

$$SF_o = (3.5 \times 10^4) \times UCR_o$$

The assumptions used involve a 70-kg body weight, and an average water ingestion rate of 2 L/day.

Risk-specific concentrations in air — Risk-specific concentrations of chemicals in air is estimated from the unit risk in air as follows:

$$\text{Air concentration, } \mu g/m^3 = \frac{\text{specified risk level} \times \text{body weight}}{SF_i \times \text{inhalation rate} \times 10^{-3}}$$

$$= \frac{\text{specified risk level (R)}}{UCR_i} = \frac{1 \times 10^{-6}}{UCR_i \, (\mu g/m^3)^{-1}}$$

or,

$$= \frac{1 \times 10^{-6} \times 70 \text{ kg}}{SF_i \, (mg/kg/day)^{-1} \times 20 \text{ m}^3/day \times 10^{-3}} = \frac{3.5 \times 10^{-3}}{SF_i}$$

The assumptions used involve a specified risk level of 10^{-6}, a 70-kg body weight, and an average inhalation rate of 20 m^3/day.

Risk-specific concentrations in water — Risk-specific concentrations of chemicals in drinking water can be estimated from the oral slope factor; the water concentration corrected for an increased upper bound lifetime risk of R is given by:

$$\mu g/L \text{ in water} = \frac{(\text{specified risk level} \times \text{ body weight})}{SF_o \times \text{ingestation rate} \times 10^{-3}}$$

$$= \frac{\text{specified risk level}}{UCR_o} = \frac{1 \times 10^{-6}}{UCR_o \left(\mu g/L\right)^{-1}}$$

or,
$$= \frac{1 \times 10^{-6} \times 70 \text{ kg}}{SF_o \left(mg/kg/day\right)^{-1} \times 2 \text{ L/day} \times 10^{-3}} = \frac{3.5 \times 10^{-2}}{SF_o}$$

The assumptions used involve a specified risk level of 10^{-6}, a 70-kg body weight, and an average water ingestion rate of 2 L/day.

REFERENCES

Casarett, L.J. and J. Doull. 1975. *Toxicology: The Basic Science of Poisons.* Macmillan, New York.

CDHS. 1986. The California Site Mitigation Decision Tree Manual. California Department of Health Services, Toxic Substances Control Division, Sacramento.

Dourson, M.L. and J.F. Stara. 1983. Regulatory History and Experimental Support of Uncertainty (Safety) Factors. Regulatory Toxicology and Pharmacology, 3:224-238.

Klaassen, C.D., M.O. Amdur and J. Doull. (Eds.). 1986. *Casarett and Doull's Toxicology: The Basic Science of Poisons.* 3rd ed. Macmillan, New York.

Meyer, C.R. 1983. Liver Dysfunction in Residents Exposed to Leachate from a Toxic Waste Dump. *Environ. Health Perspect.* 48, 9–13.

USEPA. 1986a. Guidelines for Carcinogen Risk Assessment. *Federal Register,* 51(185), 33992–34003, CFR 2984, Sept. 24, 1986.

USEPA. 1986b. Guidelines for the Health Risk Assessment of Chemical Mixtures. *Federal Register,* 51(185), 34014–34025, CFR 2984, Sept. 24, 1986.

USEPA. 1989a. Exposure Factors Handbook, EPA/600/8-89/043. U.S. Environmental Protection Agency, Washington, D.C.

USEPA. 1989b. Interim Methods for Development of Inhalation Reference Doses. EPA/600/8-88/066F. Office of Health and Environmental Assessment, U.S. Environmental Protection Agency, Washington, D.C.

EQUATIONS FOR CALCULATING CARCINOGENIC RISKS AND NONCARCINOGENIC HAZARD INDICES

I.1   CARCINOGENIC RISKS

Cancer risk is a function of the lifetime average daily dose and the chemical-specific potency slope. The carcinogenic effects of selected chemicals of concern are calculated by multiplying the estimated chemical exposure level by the route-specific/chemical-specific cancer slope factor, according to the following relationship (CAPCOA, 1990; CDHS, 1986; USEPA, 1989):

$$CR = CDI \times SF$$

where:

CR = probability of an individual developing cancer (unitless)

CDI = chronic daily intake averaged over a lifetime, say 70 years (mg/kg-day)

SF = slope factor (1/[mg/kg-day])

This represents the linear low-dose cancer risk model, and is valid only at low risk levels (i.e., risks $<10^{-2}$).

For sites where chemical intakes are high, resulting in potential carcinogenic risks above the 10^{-2} level, the one-hit model is used, in accordance with the following relationship:

$$CR = [1 - \exp(-CDI \times SF)]$$

where the terms are the same as defined above for the low-dose model.

The aggregate cancer risk equation for multiple substances is obtained by summing the risks calculated for the individual chemicals using the above relationship(s). Thus, for multiple compounds:

$$\text{Total Risk} = \sum_{i=1}^{n} (CDI_i \times SF_i) \text{ for the linear low-dose model at low risk levels}$$

$$\text{Total Risk} = \sum_{i=1}^{n} (1 - \exp(-CDI_i \times SF_i)) \text{ for the one-hit model at high risk levels}$$

where:

$$CDI_i = \text{chronic daily intake for the } i^{th} \text{ contaminant}$$
$$SF_i = \text{slope factor for the } i^{th} \text{ contaminant}$$
$$n = \text{total number of carcinogens}$$

For multiple compounds and multiple pathways, assuming risk probabilities determined for each carcinogen are additive over all exposure pathways, the overall total cancer risk for all exposure pathways and all contaminants considered in the risk evaluation will be:

$$\text{Overall Total Risk} = \sum_{j=1}^{p} \sum_{i=1}^{n} (CDI_{ij} \times SF_{ij}) \text{ for the linear low-dose model at low risk levels}$$

$$\text{Overall Total Risk} = \sum_{j=1}^{p} \sum_{i=1}^{n} \{1 - \exp(-CDI_{ij} \times SF_{ij})\} \text{ for the one-hit model at high risk levels}$$

where:

$$CDI_{ij} = \text{chronic daily intake for the } i^{th} \text{ contaminant and } j^{th} \text{ pathway}$$
$$SF_{ij} = \text{slope factor for the } i^{th} \text{ contaminant and } j^{th} \text{ pathway}$$
$$n = \text{total number of carcinogens}$$
$$p = \text{total number of pathways or exposure routes}$$

The CDIs are estimated from the equations previously discussed in Appendix F for exposure calculations. The SF values are obtainable from various sources or databases, including the Integrated Risk Information System (IRIS) and the Health Effects Assessment Summary Tables (HEAST) maintained by the USEPA and other regulatory agencies, or can be derived by using the suggested approach discussed in Appendix H. The SFs have been established on a substance-specific basis for oral and inhalation exposures, taking into account evidence from both human epidemiologic and laboratory toxicologic studies.

I.2   NONCARCINOGENIC HAZARDS

The noncarcinogenic effects of selected chemicals of concern are calculated as the ratio of the estimated chemical exposure level and the route-specific/chemical-specific reference dose, according to the following relationship (CAPCOA, 1990; CDHS, 1986; USEPA, 1989):

$$\text{Hazard Quotient, HQ} = \frac{E}{RfD}$$

where:

$$E = \text{chemical exposure level or intake (mg/kg-day)}$$
$$RfD = \text{reference dose (mg/kg-day)}$$

The sum total of the hazard quotients for all chemicals of concern gives the hazard index for a given exposure pathway. The applicable relationship is

$$\text{Total Hazard Index, HI} = \sum_{i=1}^{n} \frac{E_i}{RfD_i}$$

where:

E_i = exposure level (or intake) for the ith contaminant
RfD_i = acceptable intake level (or reference dose) for i^{th} contaminant
n = total number of chemicals presenting noncarcinogenic effects

For multiple compounds and multiple pathways, the overall total noncancer risk for all exposure pathways and all contaminants considered in the risk evaluation will be

$$\text{Overall Total Hazard Index} = \sum_{j=1}^{p} \sum_{i=1}^{n} \frac{E_{ij}}{RfD_{ij}}$$

where:

E_{ij} = exposure level (or intake) for the ith contaminant and j^{th} pathway
RfD_{ij} = acceptable intake level (or reference dose) for i^{th} contaminant and j^{th} pathway/exposure route
n = total number of chemicals presenting noncarcinogenic effects

The E values are estimated from the equations previously discussed in Appendix F for exposure calculations. The RfD values are obtainable from databases such as IRIS and HEAST maintained by the USEPA and other regulatory agencies, or can be derived by using the suggested approach discussed in Appendix H. RfDs have been established by the USEPA as thresholds of exposure to toxic substances below which there should be no adverse health impact. These thresholds have been established on a substance-specific basis for oral and inhalation exposures, taking into account evidence from both human epidemiologic and laboratory toxicologic studies.

In a comprehensive evaluation, it becomes necessary to introduce the idea of physiologic endpoints in the calculation process, in which case chemicals affecting the same target organs (i.e., chemicals determined to have the same physiologic endpoint) are grouped together in the calculation of the total hazard index. That is, if HI >1.0, it often becomes necessary to segregate chemicals by organ-specific toxicity and recalculate the values, since strict additivity without consideration for target-organ toxicities could overestimate potential hazards.

REFERENCES

CAPCOA. 1990. Air Toxics "Hot Spots" Program. Risk Assessment Guidelines. California Air Pollution Control Officers Association, Sacramento.

CDHS. 1986. The California Site Mitigation Decision Tree Manual. California Department of Health Services, Toxic Substances Control Division, Sacramento.

USEPA. 1989. Risk Assessment Guidance for Superfund. Vol. I. Human Health Evaluation Manual (Part A). EPA/540/1-89/002. Office of Emergency and Remedial Response, U.S. Environmental Protection Agency, Washington, D.C.

APPENDIX J
EQUATIONS FOR DEVELOPING
RISK-BASED SITE RESTORATION GOALS

J.1 INTRODUCTION

The site cleanup level is a site-specific criterion that a remedial action would have to satisfy in order to keep exposures of potential receptors at or below an "acceptable level". The acceptable contaminant levels correspond to contaminant concentrations in specific environmental media that, when exceeded, present significant risks of adverse impact to potential receptors. This target level generally tends to drive the cleanup process for a contaminated site, and represents the maximum acceptable contaminant level for site restoration decisions.

J.2 DEVELOPMENT OF RISK-BASED SOIL CLEANUP CRITERIA

Algebraic manipulations of the hazard index or carcinogenic risk equations (Appendix I) and the exposure estimation equations (Appendix F) can be used to arrive at appropriate risk-based soil cleanup criteria necessary for a site remediation program.

J.2.1 Carcinogenic Chemicals

Box J.1 contains the relevant equation to use in the development of risk-based site restoration criteria for carcinogenic constituents present in soils at a contaminated site. This has been derived by back-calculating from the risk equations for inhalation of soil particulates, ingestion of soils, and dermal contact with soils. For chemicals with carcinogenic effects, a target cancer risk of 1×10^{-6} is usually used in the back-modeling efforts.

J.2.2 Noncarcinogenic Chemicals

Box J.2 contains the relevant equation to use in the development of risk-based site restoration criteria for noncarcinogenic constituents present in soils at a contaminated site. This has been derived by back-calculating from the hazard equations for inhalation of soil particulates, ingestion of soils, and dermal contact with soils. For chemicals with noncarcinogenic effects, a reference hazard index of 1.0 is usually used in the back-modeling efforts.

J.3 DEVELOPMENT OF RISK-BASED WATER CLEANUP CRITERIA

Algebraic manipulations of the hazard index or carcinogenic risk equations (Appendix I) and the exposure estimation equations (Appendix F) can be used to arrive at appropriate risk-based water cleanup criteria necessary for a site remediation program.

J.3.1 Carcinogenic Chemicals

Boxes J.3A and J.3B contain the relevant equations to use in the development of risk-based site restoration criteria for carcinogenic constituents present in water at a contaminated site. These have been derived by back-calculating from the risk equations for inhalation of contaminants in water, ingestion of water, and dermal contact with water. For chemicals with carcinogenic effects, a target cancer risk of 1×10^{-6} is usually used in the back-modeling efforts.

J.3.2 Noncarcinogenic Chemicals

Boxes J.4A and J.4B contain the relevant equations to use in the development of risk-based site restoration criteria for noncarcinogenic constituents present in water at a contaminated site. These have been derived by back-calculating from the hazard equations for inhalation of contaminants in water, ingestion of water, and dermal contact with water. For chemicals with noncarcinogenic effects, a reference hazard index of 1.0 is usually used in the back-modeling efforts.

BOX J.1
Equation for Calculating Risk-Based
Soil Cleanup Criteria
for Carcinogenic Chemical Constituents

$RBAG_c$

$$= \frac{TCR}{\left(\dfrac{EE \times ED \times CF}{BW \times AT \times 365}\right) \times \left\{\left[SF_i \times IR \times RR \times ABS_a \times AEF \times CF_a\right] + \left[\left(SF_o \times SIR \times FI \times ABS_{si}\right) + \left(SF_o \times SA \times AF \times ABS_{sd} \times SM\right)\right]\right\}}$$

$$= \frac{TCR \times (BW \times AT \times 365)}{\left(EF \times ED \times 10^{-6}\right) \times \left\{\left[SF_i \times IR \times RR \times ABS_a \times AEF \times CF_a\right] + SF_o\left[\left(SIR \times FI \times ABS_{si}\right) + \left(SA \times AF \times ABS_{sd} \times SM\right)\right]\right\}}$$

where:

$RBAG_c$	=	acceptable risk-based action goal of contaminant in soil (mg/kg)
TCR	=	target cancer risk, usually set at 10^{-6} (unitless)
SF_i	=	inhalation slope factor $((mg/kg\text{-}day)^{-1})$
SF_o	=	oral slope factor $((mg/kg\text{-}day)^{-1})$
IR	=	inhalation rate (m³/day)
RR	=	retention rate of inhaled air (%)
ABS_a	=	percent chemical absorbed into bloodstream (%)
AEF	=	air emissions factor, i.e., PM_{10} particulate emissions or volatilization (kg/m³)
CF_a	=	conversion factor for air emission term (10^6)
SIR	=	soil ingestion rate (mg/day)
CF	=	conversion factor (10^{-6} kg/mg)
FI	=	fraction ingested from contaminated source (unitless)
ABS_{si}	=	bioavailability absorption factor for ingestion exposure (%)
ABS_{sd}	=	bioavailability absorption factor for dermal exposures (%)
SA	=	skin surface area available for contact, i.e., surface area of exposed skin (cm²/event)
AF	=	soil to skin adherence factor, i.e., soil loading on skin (mg/cm²)
SM	=	factor for soil matrix effects (%)
EF	=	exposure frequency (days/years)
ED	=	exposure duration (years)
BW	=	body weight (kg)
AT	=	averaging time (i.e., period over which exposure is averaged) (years)

BOX J.2
Equation for Calculating Risk-Based
Soil Cleanup Criteria for the Noncarcinogenic Effects
of Chemical Constituents

$RBAG_{nc}$

$$= \frac{\text{Target Hazard Quotient}}{\left(\dfrac{EF \times ED \times 10^{-6}}{BW \times AT \times 365}\right) \times \left\{\left[\dfrac{IR \times RR \times ABS_a}{RfD_i} \times AEF \times CF_a\right] + \left[\dfrac{SIR \times FI \times ABS_{si}}{RfD_o}\right] + \left[\dfrac{SA \times AF \times ABS_{sd} \times SM}{RfD_o}\right]\right\}}$$

$$= \frac{THQ \times (BW \times AT \times 365)}{\left(EF \times ED \times 10^{-6}\right) \times \left\{\left[\dfrac{IR \times RR \times ABS_a}{RfD_i} \times AEF \times CF_a\right] + \dfrac{1}{RfD_o}\left[\left(SIR \times FI \times ABS_{si}\right) + \left(SA \times AF \times ABS_{sd} \times SM\right)\right]\right\}}$$

where:

$RBAG_{nc}$	=	acceptable risk-based action goal of contaminant in soil (mg/kg)
THQ	=	target hazard quotient (usually equal to 1)
RfD_i	=	inhalation reference dose (mg/kg-day)
RfD_o	=	oral reference dose (mg/kg-day)
IR	=	inhalation rate (m^3/day)
RR	=	retention rate of inhaled air (%)
ABS_a	=	percent chemical absorbed into bloodstream (%)
AEF	=	air emission factor, i.e., PM_{10} particulate emissions or volatilization (kg/m^3)
CF_a	=	conversion factor for air emission term (10^6)
SIR	=	soil ingestion rate (mg/day)
CF	=	conversion factor (10^{-6} kg/mg)
FI	=	fraction ingested from contaminated source (unitless)
ABS_{si}	=	bioavailability absorption factor for ingestion exposure (%)
ABS_{sd}	=	bioavailability absorption factor for dermal exposures (%)
SA	=	skin surface area available for contact, i.e., surface area of exposed skin (cm^2/event)
AF	=	soil to skin adherence factor, i.e., soil loading on skin (mg/cm^2)
SM	=	factor for soil matrix effects (%)
EF	=	exposure frequency (days/years)
ED	=	exposure duration (years)
BW	=	body weight (kg)
AT	=	averaging time (i.e., period over which exposure is averaged) (years)

BOX J.3A
Equation for Calculating Risk-Based
Water Cleanup Criteria for Nonvolatile
Carcinogenic Chemical Constituents

$RBAG_c$

$$= \frac{TCR}{\left(\dfrac{EF \times ED}{BW \times AT \times 365}\right) \times \left\{\left[\left(SF_o \times WIR \times FI \times ABS_{si}\right) + \left(SF_o \times SA \times K_p \times ET \times ABS_{sd} \times CF\right)\right]\right\}}$$

$$= \frac{TCR \times \left(BW \times AT \times 365\right)}{\left(EF \times ED\right) \times SF_o \left[\left(WIR \times FI \times ABS_{si}\right) + \left(SA \times K_p \times ET \times ABS_{sd} \times CF\right)\right]}$$

where:

$RBAG_c$ = acceptable risk-based action goal of contaminant in water (mg/L)

TCR = target cancer risk, usually set at 10^{-6} (unitless)

SF_o = oral slope factor ($[mg/kg\text{-}day]^{-1}$)

WIR = water ingestion rate (L/day)

CF = conversion factor (1 L/1000 cm^3 = 10^{-3} L/cm^3)

FI = fraction ingested from contaminated source (unitless)

ABS_{si} = bioavailability absorption factor for ingestion exposure (%)

ABS_{sd} = bioavailability absorption factor for dermal exposures (%)

SA = skin surface area available for contact, i.e., surface area of exposed skin (cm^2/event)

K_p = chemical-specific dermal permeability coefficient from water (cm^2/hr)

ET = exposure time during water contacts (e.g., during showering/bathing activity) (hr/day)

EF = exposure frequency (days/years)

ED = exposure duration (years)

BW = body weight (kg)

AT = averaging time (i.e., period over which exposure is averaged) (years)

BOX J.3B
Equation for Calculating Risk-Based
Water Cleanup Criteria for Volatile
Carcinogenic Chemical Constituents

$RBAG_c$

$$= \frac{TCR}{\left(\dfrac{EF \times ED}{BW \times AT \times 365}\right) \times \left\{\left[SF_i \times IR_w \times RR \times ABS_a \times CF_a\right] + \left[(SF_o \times WIR \times FI \times ABS_{si}) + \left(SF_o \times SA \times K_p \times ET \times ABS_{sd} \times CF\right)\right]\right\}}$$

$$= \frac{TCR \times (BW \times AT \times 365)}{(EF \times ED) \times \left\{\left[SF_i \times IR_w \times RR \times ABS_a \times CF_a\right] + SF_o\left[(WIR \times FI \times ABS_{si}) + \left(SA \times K_p \times ET \times ABS_{sd} \times CF\right)\right]\right\}}$$

where:

$RBAG_c$	=	acceptable risk-based action goal of contaminant in water (mg/L)
TCR	=	target cancer risk, usually set at 10^{-6} (unitless)
SF_i	=	inhalation slope factor ($[mg/kg\text{-}day]^{-1}$)
SF_o	=	oral slope factor ($[mg/kg\text{-}day]^{-1}$)
IR_w	=	intake from the inhalation of volatiles (sometimes equivalent to the amount of ingested water) (m^3/day)
RR	=	retention rate of inhaled air (%)
ABS_a	=	percent chemical absorbed into bloodstream (%)
CF_a	=	conversion factor for volatiles inhalation term ($1000\ L/1\ m^3 = 10^3\ L/m^3$)
WIR	=	water ingestion rate (L/day)
CF	=	conversion factor ($1\ L/1000\ cm^3 = 10^{-3}\ L/cm^3$)
FI	=	fraction ingested from contaminated source (unitless)
ABS_{si}	=	bioavailability absorption factor for ingestion exposure (%)
ABS_{sd}	=	bioavailability absorption factor for dermal exposures (%)
SA	=	skin surface area available for contact, i.e., surface area of exposed skin ($cm^2/event$)
K_p	=	chemical-specific dermal permeability coefficient from water (cm^2/hr)
ET	=	exposure time during water contacts (e.g., during showering/bathing activity) (hr/day)
EF	=	exposure frequency (days/years)
ED	=	exposure duration (years)
BW	=	body weight (kg)
AT	=	averaging time (i.e., period over which exposure is averaged) (years)

BOX J.4A
Equation for Calculating Risk-Based
Water Cleanup Criteria for Noncarcinogenic Effects
of Nonvolatile Chemical Constituents

$RBAG_{nc}$

$$= \frac{\text{Target Hazard Quotient}}{\left(\dfrac{EF \times ED}{BW \times AT \times 365}\right) \times \left\{\left[\left(\dfrac{WIR}{RfD_o} \times FI \times ABS_{si}\right)\right] + \left[\dfrac{SA \times K_p \times ET \times ABS_{sd} \times CF}{RfD_o}\right]\right\}}$$

$$= \frac{THQ \times (BW \times AT \times 365)}{(EF \times ED) \times \left\{\dfrac{1}{RfD_o}\left[(WIR \times FI \times ABS_{si}) + (SA \times K_p \times ET \times ABS_{sd} \times CF)\right]\right\}}$$

where:

$RBAG_{nc}$	=	acceptable risk-based action goal of contaminant in soil (mg/L)
THQ	=	target hazard quotient (usually equal to 1)
RfD_o	=	oral reference dose (mg/kg-day)
WIR	=	water intake rate (L/day)
CF	=	conversion factor (1 L/1000 cm^3 = 10^{-3} L/cm^3)
FI	=	fraction ingested from contaminated source (unitless)
ABS_{si}	=	bioavailability absorption factor for ingestion exposure (%)
ABS_{sd}	=	bioavailability absorption factor for dermal exposures (%)
SA	=	skin surface area available for contact, i.e., surface area of exposed skin (cm^2/event)
K_p	=	chemical-specific dermal permeability coefficient from water (cm^2/hr)
ET	=	exposure time during water contacts (e.g., during showering/bathing activity) (hr/day)
EF	=	exposure frequency (days/years)
ED	=	exposure duration (years)
BW	=	body weight (kg)
AT	=	averaging time (i.e., period over which exposure is averaged) (years)

BOX J.4B
Equation for Calculating Risk-Based
Water Cleanup Criteria for Noncarcinogenic Effects
of Volatile Chemical Constituents

$RBAG_{nc}$

$$= \frac{THQ}{\left(\dfrac{EF \times ED}{BW \times AT \times 365}\right) \times \left\{\left[\dfrac{IR_W \times RR \times ABS_a \times CF_a}{RfD_i}\right] + \left[\left(\dfrac{WIR}{RfD_o} \times FI \times ABS_{si}\right)\right] + \left[\dfrac{SA \times K_p \times ET \times ABS_{sd} \times CF}{RfD_o}\right]\right\}}$$

$$= \frac{THQ \times (BW \times AT \times 365)}{(EF \times ED) \times \left\{\left[\dfrac{IR_W \times RR \times ABS_a \times CF_a}{RfD_i}\right] + \dfrac{1}{RfD_o}\left[(WIR \times FI \times ABS_{si}) + (SA \times K_p \times ET \times ABS_{sd} \times CF)\right]\right\}}$$

where:

$RBAG_{nc}$	=	acceptable risk-based action goal of contaminant in soil (mg/L)
THQ	=	target hazard quotient (usually equal to 1)
RfD_i	=	inhalation reference dose (mg/kg-day)
RfD_o	=	oral reference dose (mg/kg-day)
IR_w	=	inhalation intake rate (m^3/day)
RR	=	retention rate of inhaled air (%)
ABS_a	=	percent chemical absorbed into bloodstream (%)
CF_a	=	conversion factor for volatiles inhalation term (1000 L/1 m^3 = 10^3 L/m^3)
WIR	=	water intake rate (L/day)
CF	=	conversion factor (1 L/1000 cm^3 = 10^{-3} L/cm^3)
FI	=	fraction ingested from contaminated source (unitless)
ABS_{si}	=	bioavailability absorption factor for ingestion exposure (%)
ABS_{sd}	=	bioavailability absorption factor for dermal exposures (%)
SA	=	skin surface area available for contact, i.e., surface area of exposed skin (cm^2/event)
K_p	=	chemical-specific dermal permeability coefficient from water (cm^2/hr)
ET	=	exposure time during water contacts (e.g., during showering/bathing activity) (hr/day)
EF	=	exposure frequency (days/years)
ED	=	exposure duration (years)
BW	=	body weight (kg)
AT	=	averaging time (i.e., period over which exposure is averaged) (years)

APPENDIX K
SELECTED UNITS AND MEASUREMENTS

Mass/Weight Units

g	gram(s)
ton (metric)	tonne = 1×10^6 g
kg	kilogram(s) = 10^3 g
mg	milligram(s) = 10^{-3} g
µg	microgram(s) = 10^{-6} g
ng	nanogram(s) = 10^{-9} g
pg	picogram(s) = 10^{-12} g

Approximate Mass Conversions

1 g	=	0.035 oz
1 ton	=	2,205 lb
1 kg	=	2.25 lb
1 mg	=	10^{-3} g
1 µg	=	10^{-6} g
1 ng	=	10^{-9} g
1 pg	=	10^{-12} g

Volumetric Units

cc or cm^3	cubic centimeter(s) = 10^{-3} L
mL	milliliter(s) = 10^{-3} L
L	liter(s) = 10^3 cm^3
m^3	cubic meter(s) = 10^3 L

Approximate Volume Conversions

1 cc	=	1 mL
1 mL	=	10^{-3} L
1 L	=	0.95 liquid quart
1 m^3	=	35 cubic feet

Environmental Concentration Units

ppm	parts per million
ppb	parts per billion
ppt	parts per trillion

Concentration Equivalents
1 ppm = mg/kg or mg/L
1 ppb = μg/kg or μg/L
1 ppt = ng/kg or ng/L

Concentrations in Soils or Other Solid Media:
mg/kg   mg chemical per kg weight of sampled medium

Concentrations in Water or Other Liquid Media:
mg/L   mg chemical per L of total liquid volume

Concentrations in Air Media:
mg/m^3 mg chemical per m^3 of total fluid volume

Units of Chemical Intake and Dose
mg/kg-day = milligrams of chemical exposure per unit body weight of exposed
receptor per day

Commonly Used Expressions
"Order of Magnitude"
Reference to each "order of magnitude" means the base parameter may vary by a factor of 10. It is often used in reference to the calculation of environmental quantities or risk probabilities.

Exponentials denoted by 10^λ
Superscripts refer to the number of times "10" is multiplied by itself. For example, $10^2 = 10 \times 10 = 100$; $10^3 = 10 \times 10 \times 10 = 1000$; $10^6 = 10 \times 10 \times 10 \times 10 \times 10 \times 10 = 1,000,000$.

Exponentials denoted by $10^{-\lambda}$
Negative superscript is equivalent to the reciprocal of the positive term, i.e., $10^{-\lambda}$ equals $1/10^\lambda$. For example, $10^{-2} = 1/10^2 = 1/(10 \times 10) = 0.01$; $10^{-3} = 1/10^3 = 1/(10 \times 10 \times 10) = 0.001$; $10^{-6} = 1/10^6 = 1/(10 \times 10 \times 10 \times 10 \times 10 \times 10) = 0.000001$.

Exponentials denoted by X.YZ E+λ
Number after the "E" indicates the power to which 10 is raised, and then multiplied by the preceding term (i.e., the number of times "10^λ" is multiplied by preceding term, or $X.YZ \times 10^\lambda$). For example, $1.00E-01 = 1.00 \times 10^{-1} = 0.1$; $1.23E + 04 = 1.23 \times 10^{+4} = 12,300$; $4.44E + 05 = 4.44 \times 10^5 = 444,000$.

INDEX

A

Absorbed dose, defined, 293
Absorption, defined, 293
Absorption adjustments, risk assessment,
    138
Absorption factor, defined, 293
Acceptable benchmark, risk assessment,
    142
"Acceptable contaminant level", cleanup
    criteria, 190
Acceptable risk, defined, 293
Action levels for carcinogenic
    contaminants, cleanup criteria,
    180
Activated carbon, defined, 293
Activated carbon adsorption
    defined, 293
    restoration methods/technologies,
        199
    technology, restoration methods/
        technologies, 199
Activated carbons systems, restoration
    methods/technologies, 199
Acute exposure, defined, 293
Acute toxicity, defined, 293
Adjustments for absorption, risk
    assessment, 138
Adsorption, defined, 294
Air contaminant
    emissions, risk assessment,
        determination of, 160
    emissions from contaminated sites,
        risk assessment, 160
    releases, risk assessment of, 159
Air emissions
    categories, 160
    from contaminated sites, risk
        assessment, 160
    sources, classification of, 160

Air sparging
    in combination with SVE, restoration
        methods/technologies, 210
    defined, 294
    restoration methods/technologies, 209
        design of, 210
Air stripping
    defined, 294
    restoration methods/technologies,
        208–209, 218
AIR3D, corrective action assessment,
    94
AIRTOX, 328
    corrective action assessment, 94
Aliphatic compounds, defined, 294
Alternate concentration limit
    applications, cleanup criteria,
        189
Ambient air concentrations, estimate of,
    165
Analysis
    design, site characterization, 38
    plan, site characterization, 39
    purpose of, site characterization, 40
Analytical program requirements, site
    characterization, 45
Analytical protocols, site
    characterization, 45
ANOVA, corrective action assessment,
    application of, 108
Antagonism, defined, 294
Anthropogenic levels, corrective action
    assessment, 90
APEA, 161
    component of, 161
    design of, 161
Applicable, 205, 212
Applications
    contaminated site problems, risk
        assessment, 117–174